TECHNIQUES IN PROTEIN CHEMISTRY

TECHNIQUES
IN
PROTEIN CHEMISTRY

by

J. LEGGETT BAILEY, Ph. D., F.R.I.C., F.L.S.
Twyford Laboratories,
London (U.K.)

Second revised and expanded edition

ELSEVIER PUBLISHING COMPANY
AMSTERDAM–LONDON–NEW YORK
1967

ELSEVIER PUBLISHING COMPANY

335 JAN VAN GALENSTRAAT, P.O. BOX 211, AMSTERDAM

AMERICAN ELSEVIER PUBLISHING COMPANY, INC.

52 VANDERBILT AVENUE, NEW YORK, N.Y. 10017

ELSEVIER PUBLISHING COMPANY LIMITED

RIPPLESIDE COMMERCIAL ESTATE, RIPPLEROAD, BARKING, ESSEX

1ST EDITION: 1962

FIRST REPRINT: 1964

2ND REVISED AND EXPANDED EDITION: 1967

FIRST REPRINT: 1968

LIBRARY OF CONGRESS CATALOG CARD NUMBER 66-16719

WITH 120 ILLUSTRATIONS AND 54 TABLES

PRINTED IN THE NETHERLANDS

To Professor A. C. Chibnall, Sc.D., F.R.S.

PREFACE

In writing this book I have been guided by the needs of biochemists who have had limited practical experience of the chemistry of proteins and who may wish to try their hand at a tentative characterization and structural examination of the protein or peptide part of their particular experimental system.

Some technical aspects of the subject have become rather complicated by numerous modifications in procedure and at times, it is difficult even for the specialist worker to decide on the respective merits of all the variants. My aim therefore has been to compile a practical handbook in which established analytical methods are discussed in detail, recent promising techniques are touched on briefly, and in which theoretical considerations are introduced only where they have a direct bearing on practice.

A recurring difficulty that has to be faced by all who work with proteins is the lack of a simple criterion of purity that might be applied to purified preparations. Crystallinity has long been abandoned as a criterion of homogeneity and quantitative solubility tests have been applied in very few cases. The minimum requirement generally acceptable is evidence of homogeneity from data provided by column chromatography, electrophoresis under different conditions of pH, and sedimentation in the ultracentrifuge. Two chapters in this book are devoted to column chromatography and electrophoresis of proteins. A discussion of the ultracentrifuge is not included since the information is available in special operational manuals. In the interests of brevity accounts of the more complicated physical equipment, *e.g.* counter-current distribution apparatus, are also omitted.

The final criterion for a single molecular species is the demonstration of a unique amino acid sequence and hence the major part of this book is concerned with methods for sequence determination, selective cleavage of peptide chains, fractionation of mixtures of peptides, amino acid analysis and other techniques required in structural work on proteins. The general pattern of presentation is such that each chapter deals with a particular topic and is more or less self-contained. There may be some advantage in this over the alternative plan of a stage by stage examination of the methods used in the elucidation of the complete structure of a particular protein.

Few biochemists will wish to proceed with a complete structural investigation of the protein molecules that interest them — at least not using the methods presently available. Great advances have been made in fields such as enzymology, intermediary metabolism and photosynthesis where little is known of the chemical nature of the protein enzymes involved. However, sooner or later all biochemical events must be explained at the molecular level in terms of organic and physical chemistry and a knowledge of the complete structure of proteins, nucleic acids and other large molecules will be required.

The brilliant studies of Sanger and his collaborators on insulin will ever remain a landmark in the development of modern protein chemistry. It may be that the methods employed by Sanger will not be applied again *in toto* for the structural investigation of other proteins, but it is to be hoped that the basically simple approach manifest throughout the insulin work will be maintained in future fashioning of methods.

It is generally recognized among teachers that biochemical concepts are difficult to communicate to young people at the pre-university level. I would suggest to any science masters among my readers that some of the simpler experimental methods outlined in this book might be adapted to fire the imagination of senior pupils. With guidance some original research contributions might even result.

London, July 1962 J.L.B.

ACKNOWLEDGEMENTS

During the preparation of this book I have been conscious of my indebtedness to the many distinguished research workers who have allowed me to quote from their original papers, and to the editors and publishing houses who have allowed me to reproduce copyright material so freely. Their willing co-operation is gratefully acknowledged, and in particular I wish to thank the following: – the Academic Press Inc., New York, for Tables 11 and 35; the American Chemical Society for Figs. 2, 3, 34, 35, 37, 38, 40, 66, 79, 99, 100 and 101 and for Tables 1, 12, 13, 37, 44 and 45; the American Society of Biological Chemists for Figs. 25, 33, 44, 45, 46, 47, 50, 52, 53, 54, 55, 59, 60, 61, 76, 96 and 97, and for Tables 14, 15, 19, 33, 34, 50; *Archives of Biochemistry and Biophysics* for Figs. 93 and 98; *Arkiv för Kemi* for Figs. 71, 74 and 113; the *Biochemical Journal* for Figs. 1, 14, 21, 22, 30, 42, 64, 75, 77, 81, 94, 95, 106, 107 and 114, and for Tables 21, 22, 23, 24, 26, 27, 28, 38, 39, 40, 42, 43; the *Bulletin de la Société de Chimie Biologique* for part of Table 27; the *Canadian Journal of Biochemistry and Physiology* for Figs. 6, 7, 8, 9 and 10; *Chemistry and Industry* for Fig. 103; J. & A. Churchill Ltd., London, for Table 40; the *Collection of Czechoslovak Chemical Communications* for Fig. 109; the Elsevier Publishing Company for Figs. 13, 15, 16, 17, 18, 22, 27, 63, 72, 73 and 82, and for Tables 6, 7, 8, 9, 30, 31, 32, 36, 48; Walter de Gruyter & Co., Berlin, for Figs. 48, 56, 104 [taken from *Z. physiol. Chemie*, 290 (1952) 1], and for part of Table 3; Interscience Publishers Inc., New York, for Figs. 116 and 117; the Long Island Biological Association for Fig. 32; the National Academy of Sciences, Washington, for Fig. 57; *Nature* for Figs. 23, 24, 26 and 70, and for Table 25; the New York Academy of Sciences for Fig. 65; the Reinhold Publishing Corporation, New York, for Table 11; Springer-Verlag, Heidelberg, for part of Fig. 20 and Dr. I. Smith for Fig. 19. If I have used material without due acknowledgement, this has been quite unintentional and I hope I shall not be taken too much to task for the omission.

My especial thanks are due to Dr. F. Sanger, F.R.S. for his expert criticism of sections of the book, Dr. W. H. Stein and Dr. H.

Fraenkel-Conrat for information in advance of publication, and the many friends who have offered valuable suggestions of which I have been glad to take advantage.

I should like to express my sincere gratitude to Professor H. D. Kay, C.B.E., F.R.S. for his encouragement and support, and to my colleagues at Twyford Laboratories who have offered me every assistance. Mr. S. M. Ridley brought much skill to the task of preparing the drawings and photographic material; the School of Photography, Ealing Technical College, kindly supplied the print for Fig. 11.

Finally I wish to thank my wife for help in correcting the proofs and Miss M. B. Jones for her painstaking work in preparing the manuscript for the press.

J.L.B.

PREFACE TO THE SECOND EDITION

There has been no diminution of activity among protein and peptide chemists in the years that have elapsed since the first edition of this book. On the contrary, advances made in separational methods, for example, have already become established in many laboratories and are applied routinely. In the interests of preserving the original character of the book – something handy to have in the laboratory – inevitably some old material has been sacrificed to make way for the new. Even so, considerable expansion of the text has resulted from the revision. A comprehensive treatise was not the aim, so a careful selection process has been applied, I hope wisely, to the vast array of material which constitutes the 'stock-in-trade' of protein and peptide chemists all over the world.

More detailed treatment has been given to such topics as gel filtration in various media and to electrophoresis in polyacrylamide gel. Rapid developments have taken place in automated amino acid analysis and the precision now obtainable requires of the biochemist a re-assessment of his skill in the manipulation of protein hydrolysates. The remarkable achievement of the total synthesis of the polypeptide chains of insulin is a major triumph in a field which has always been very much regarded as the province of the specialist. However, automated peptide synthesis is under development and biochemists will no doubt wish to understand the new principles and develop some practical facility for themselves. It was therefore thought desirable to include some information on well-tried procedures for the chemical synthesis of polypeptides.

Acknowledgements

My thanks are due to the following for permission to reproduce figures and tables: Academic Press Inc., New York, for Figs. 78, 83, 84, 85, 86, 88 and 89; the American Chemical Society for Figs. 42A and 42B and for Table 46; the American Society of Biological Chemists for Figs. 4, 39, 111 and for Table 18; the *Biochemical*

Journal for Figs. 90, 91, 92 and 108; the *Canadian Journal of Physiology* for Fig. 112; the Elsevier Publishing Company for Figs. 28A, 28B, 36, 43, 87, 115 and for Table 2; *Experientia* for Figs. 67, 68 and 69; *Helvetica Chimica Acta* for Fig. 12; Koninklijke Nederlandse Chemische Vereniging for Table 53; the National Academy of Sciences, Washington for Fig. 58; and the Pergamon Press, Oxford for Fig. 119.

Helpful criticisms and suggestions have been given by many friends and colleagues. For the task of preparing the manuscript I am indebted again to Miss M. B. Jones who also helped with the proof reading.

London, June 1966 J.L.B.

CONTENTS

CHAPTER 1

PAPER CHROMATOGRAPHY OF AMINO ACIDS AND PEPTIDES

GENERAL CONSIDERATIONS

Since the classical work of Consden, Gordon and Martin[1], and later that of Dent[2], an overwhelming number of systems have been devised for the separation of amino acids by paper chromatography. Some workers have concentrated on the separation of certain groups of amino acids, others have sought the maximum resolution of a large number of components. Many have achieved their end in an overnight run whereas a whole week has been needed for more ambitious separations. Kowkabany and Cassidy[3] have examined 22 grades of filter paper and 5 different solvent systems and 53 other types of paper with two different solvent systems. Following on the earlier study of the effects of inorganic salts on paper chromatograms by Consden and Gordon[4], Westall[5] and others, a very detailed study of the influence of buffers over a wide pH range on the migration of amino acids was made by McFarren[6]. Between the years 1944–1954 over 1200 major contributions to paper chromatography were published in addition to many more short notes on the subject. The literature today is not only bewildering to any newcomer to the field but also presents a problem to the reviewer who is required to be brief.

It is worth re-stating the more generally accepted basic facts about paper chromatography before dealing with particular systems. Three phases are involved in the sealed chromatographic unit; a moving liquid phase, a stationary phase at the surface of the cellulose fibres and a vapour phase which may or may not be in equilibrium with the liquid phase. The rate of migration of the moving liquid phase is a function in the first place of the physical properties of the solvent. Important factors are the viscosity, surface tension and density, all of which are dependent on temperature. Secondly, the rate of migration is a function of the pore size of the paper; a medium possessing very fine channels or capillaries will exert a greater suction force on the liquid than one in which wide channels

References p. 38

predominate. Paper usually differs in structure along the two dimensions of its surface and solvent will usually rise faster up a strip cut in the machine direction. Slightly damp paper tends to curl with with the axis of curl parallel to this direction. Heterogeneity in the paper can lead to different flow rates along either dimension resulting in lack of reproducibility. The migration of the boundary between dry and moist paper in descending chromatography is more nearly uniform than is the case in ascending chromatography. Even for downward flow, however, solvent near the solvent source is behaving differently to that at the advancing front. At the commencement of a chromatogram, during the initial surge, solvent is flowing over as well as through the saturated area. At the solvent front liquid creeps forward through the finest capillaries.

The rates of migration of solutes are governed by specific partition between the moving liquid phase and the stationary gel-like phase which is usually considered to be located in the amorphous regions of the cellulose. Good paper is at least 96% α-cellulose (insoluble in strong NaOH) and about 40% is in the amorphous state. Most of the water is held by the amorphous regions; chemically bound water exists in native cellulose to the extent of 5.9%. Consden, Gordon and Martin[1] have likened the stationary phase to a strong solution of soluble polysaccharide. Rates of migration of selected amino acids come close to the theory for simple liquid–liquid distribution but in view of the strong affinity of the basic amino acids for cellulose specific adsorption or ion-exchange effects appear to operate. It is usual to express migration rates according to the definition

$$R_F = \frac{\text{distance travelled by solute zone}}{\text{distance travelled by solvent front}}$$

In systems where the solvent is allowed to flow off the paper the R_F value is compared to that of a convenient reference substance *e.g.* leucine. As a physical constant depending on the partitioning effect the R_F value is greatly affected by temperature variations. Non-uniform operating conditions can easily arise from interactions between the vapour and liquid phases resulting from temperature variations during a run.

In a very fine exposition of the subject Hanes and co-workers[7–12] discuss most of the factors governing the operating chromatogram and show how apparent weaknesses in the technique, after proper study, may be exploited to most useful ends. The 1961 series of papers deserve a wide reading, as the conclusions apply equally well to the separation of families of compounds other than amino acids.

TWO-DIMENSIONAL SEPARATIONS

The classical procedure of Consden, Gordon and Martin[1] has been widely used in laboratories all over the world. Sheets of Whatman No. 1 paper are developed downwards with phenol-ammonia in one dimension for 24–72 h and after drying developed with collidine in the other dimension for 24–48 h. This phenol–ammonia collidine combination proved to be a most useful system but many workers found the published R_F values of limited use owing to variations in isomer composition of the collidine. A comprehensive diagram

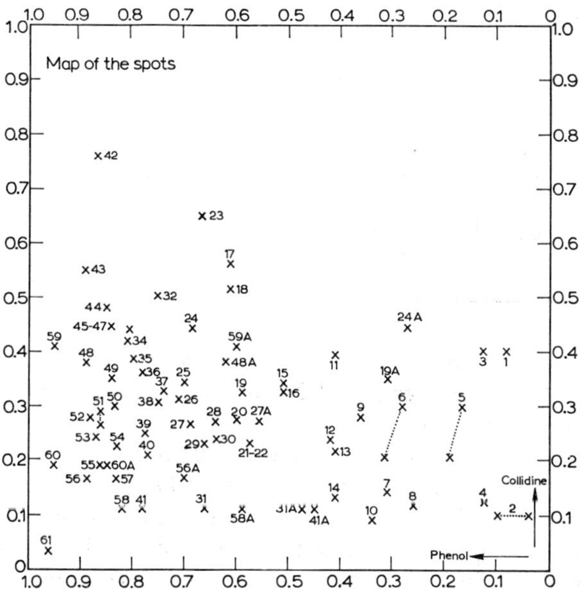

Fig. 1. Average positions of spots taken up on phenol–collidine two dimensional chromatograms. Phenol first solvent. Dent[2].

(For further explanatory text see next page)

α-Alanine (20)
β-Alanine (29)
Allothreonine (15)
α-Amino-*n*-butyric acid (26)
α-Aminoisobutyric acid (37)
γ-Aminobutyric acid (40)
ε-Aminohexanoic acid (55)
α-Amino-ε-hydroxy-caproic acid (38)
α-Amino octanoic acid (43)
δ-Aminopentanoic acid (57)
α-Aminophenylacetic acid (33)
Arginine (56)
Asparagine (13)
Aspartic acid (5)
Carnosine (54)
Citrulline (30)
Cystathionine (7)
Cysteic acid (1)
Diiodotyrosine (17)
Djenkolic acid (14)
Ethanolamine (48)
Ethanolamine–phosphoric acid (10)
Glucosamine (24)
Glutamic acid (6)
Glutamine (21)
Glutathione (4)
Glycine (12)
Histamine (59)
Histidine (27)
Homocysteic acid (3)
Hydroxylysine (31)
Hydroxyproline (28)
Lanthionine (8)
Leucine (45)
Isoleucine (46)
Lysine (58)
Methionine (34)
Methionine sulphone (25)
Methionine sulphoxide (39)
α-Methyl-α-amino-*n*-butyric acid (49)
Methyl histidine (51)
Monoiodotyrosine (23)
Norleucine (47)
Norvaline (35)
Ornithine (41)
Phenylalanine (44)
Proline (52)
Serine (9)
Serine-phosphoric acid (2)
Taurine (11)
β,β,β¹,β¹-Tetra-methylcystine (19)
Threonine (16)
Thyroxine (42)
Tryptophan (32)
Tyrosine (18)
Valine (36)

---

constructed by Dent showing the distribution of amino acids and other substances in phenol and 2,4-lutidine + 2,4,6-collidine (1:1 by vol) is shown in Fig. 1.

Phenol can be purified by steam distillation; details of a purification procedure have been described by Draper and Pollard[13]. The usual mixture is 900 g detached crystals to 100 g deionized water. To prepare a saturated solution this stock solution is shaken with excess water and the lower layer run off for use. The aqueous layer can be used in the bottom of a tank for the purpose of equilibration. Just before use 1% of concentrated ammonium hydroxide solution is added together with a few crystals of potassium cyanide to prevent formation of oxidation products catalyzed by traces of copper.

The following is a purification procedure for pyridine bases devised by A.J.P. Martin (see Ref.[14]). Bromine (5–10 ml) is added gradually with cooling and shaking to a litre of the base. The mixture is filtered through glass wool and excess bromine removed by the addition of a few crystals of sodium thiosulphate. The filtered liquid is allowed to stand over solid NaOH, filtered again and distilled.

The high toxicity and evil smell of collidine have tended to make this solvent less popular and *n*-butanol–acetic acid–water (4:1:5) is often used instead (Partridge[14]). A much used system in which collidine is replaced by butanol–acetic acid is the method of Levy and Chung[15].

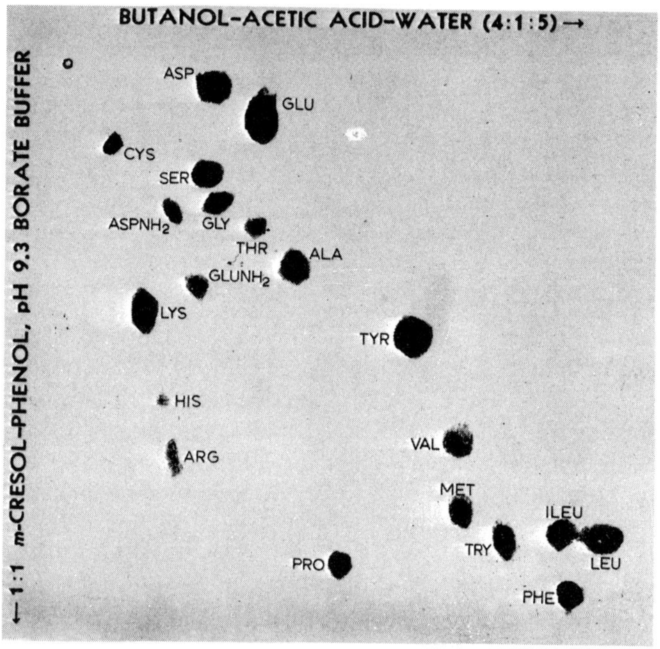

Fig. 2. Two-dimensional chromatogram by method of Levy and Chung[15]. ASP, aspartic acid; ALA, alanine; ASP.NH_2, asparagine; ARG, arginine; CI, citrulline; CYS.AC, cysteic acid; CYS, cystine; GLY, glycine; GLU, glutamic acid; GLU.NH_2, glutamine; HIS, histidine; HY.PRO, hydroxyproline; ILEU, isoleucine; LEU, leucine; LYS, lysine; MET, methionine; OR, ornithine; PHE, phenylalanine; PRO, proline; SER, serine; THR, threonine; TRY, tryptophan; TYR, tyrosine; VAL, valine.

Method of Levy and Chung

Following McFarren's idea of using aqueous buffers[6] these authors employed *m*-cresol–phenol (1:1), pH 9.3 borate buffer, and in view of the high pH an acid-washed paper, Whatman No. 52, was used. The butanol–acetic acid–water (4:1:5) irrigation is carried out first on a sheet of dimensions 22.5 × 18 in. in the long direction by the descending method. After about 16 h the paper is dried at 40°, trimmed top and bottom, the area occupied by the amino acids covered with a glass sheet and the rest of the paper sprayed with pH 9.3 borate buffer (200 ml of 0.1 *M* boric acid + 113.5 ml of 0.1 *N* NaOH). The dried paper is then run for 16 h in the other dimension with cresol–phenol (25 g of phenol, 25 g of *m*-cresol and 7 ml of pH 9.3 buffer). Fig. 2 shows a typical chromatogram. If a No. 1 paper is to be used the recommended mobile phase has the composition 30 g of *m*-cresol, 15 g of phenol and 7.5 ml of pH 8.3 borate buffer (300 ml of 0.1 *M* boric acid + 60 ml of 0.1 *N* NaOH). The use of buffered phenol usually leads to a clear chromatogram with the minimum of discoloration.

Method of Rockland and Underwood

A map constructed by Rockland and Underwood[16] illustrating a technique suitable for rapid routine screening of a large number of samples is shown in Fig. 3. Separations are carried out by ascending chromatography on 5 × 5 in. squares of Schleicher and Schüll 589 Blue Ribbon filter paper (Whatman equivalent No. 541) during about 12 h at 19°. The first solvent is *tert*.-butyl alcohol–water–formic acid prepared from 10 ml formic acid, 295 ml of deionized water and 695 ml of *tert*.-butyl alcohol. This solvent is allowed to migrate to within $\frac{1}{4}$ in. of the upper edge of the paper. The second solvent is undersaturated phenol prepared from 775 g phenol and 215 ml deionized water. Immediately before use 94 ml of the phenol solution is mixed with 1 ml concentrated ammonium hydroxide.

Method of Kamen and collaborators

Many proteins from plants and bacteria contain unusual amino acids and the method of Dus *et al.*[17] brings in a wider range of nin-

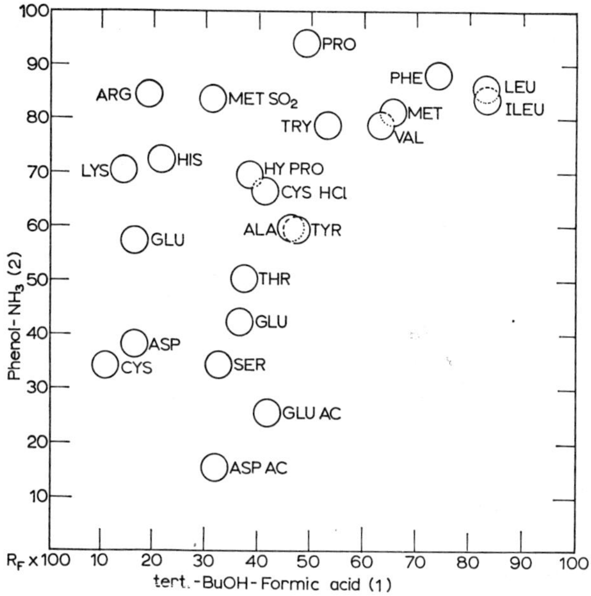

Fig. 3. Two-dimensional chromatogram by method of Rockland and Underwood[16].

hydrin-positive substances for identification (Fig. 4). The first solvent is *n*-butanol–methyl ethyl ketone–water–concentrated ammonium hydroxide (25:15:7:3) for a run of 45 h; the second solvent is *n*-butanol–acetic acid–water (4:1:5; 1 day old) for 22 h. The method was used in a study of a haem protein of the photoanaerobe, *Chromatium*.

Amino acid solutions for spotting

For spotting purposes it is useful to assay a given sample of a peptide or protein hydrolysate first by the quantitative ninhydrin method of Moore and Stein (see p. 79). A suitable load for application to paper is one micromole of total amino nitrogen in a volume of 5 to 10 μl. A standard mixture of amino acids is conveniently stored in 10% isopropanol in a concentration of 500 μg of α-amino nitrogen per ml. Alternatively 0.02 *M* solutions of amino acids in 0.1 *N* hydrochloric acid may be prepared and an aliquot neutralized with

References p. 38

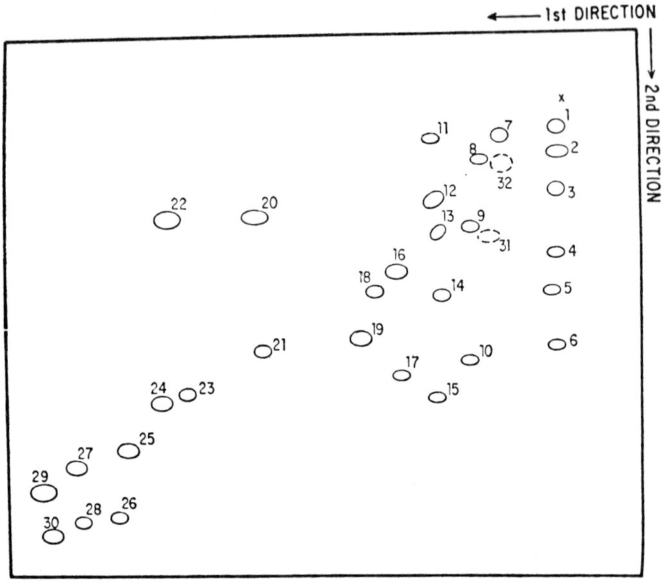

Fig. 4. Two-dimensional chromatogram by method of Dus, Bartsch and Kamen[17]. 1, cysteic acid; 2, homocysteic acid; 3, aspartic acid; 4, glutamic acid; 5, α-aminoadipic acid; 6, α-amino-pimelic acid; 7, lysine; 8, arginine; 9, glycine; 10, γ-aminobutyric acid; 11, histidine; 12, serine; 13, methionine sulphone; 14, β-alanine; 15, δ-aminovaleric acid; 16, alanine; 17, β-aminobutyric acid; 18, proline; 19, α-aminobutyric acid; 20, allothreonine; 21, tyrosine; 22, threonine; 23, valine; 24, methionine; 25, ethionine; 26, norleucine; 27, tryptophan; 28, iso + alloisoleucine; 29, phenylanine; 30, leucine; 31, hydroxyproline; 32, hydroxylysine.

1 *N* sodium hydroxide to bromothymol blue before application to paper. Micropipettes suitable for the application of solutions to paper are shown in Fig. 5; A is a commercial type and B is a simple

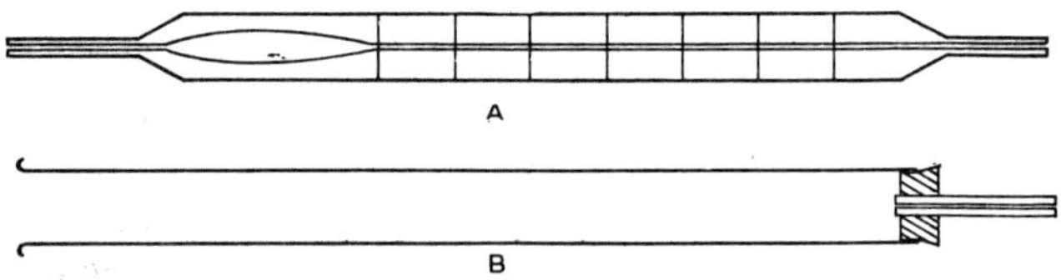

Fig. 5. Micropipettes for spotting purposes. A, commercial type; B, simple capillary type.

pipette made from a drawn out capillary tube. The latter type is calibrated by filling the pipette several times with 6 *N* HCl and emptying on to a small piece of filter paper. The filter paper is then dropped into some water and titrated with standard alkali.

ONE-DIMENSIONAL SEPARATIONS

In recent years considerable attention has been given to the problem of maximum resolution of amino acids by one-dimensional chromatography. The main drawbacks of two-dimensional chromatography are the following:

1. Only one sample can be separated for each run.
2. The amount of sample is limited as overloading leads to distortion of zones.
3. Heterogeneity in the paper and other defects give rise to irregular shaped diffuse spots.
4. The method is not ideal for quantitative work.

Among the very many systems proposed for one-dimensional chromatography the following are capable of near-perfect separations of the common amino acids.

1. Buffered phenols and alcohols (McFarren[6]).
2. Use of five different solvent systems involving 5 to 6 days' irrigation (Redfield and Guzman Barron[18]).
3. Use of three different solvent systems and irrigation periods of 5 days (Fowden[19,20])
4. Two solvent systems to resolve 16 amino acids sufficiently well for quantitative analysis (Roland and Gross[21]).
5. Repeated irrigation with the same solvent and intermediate drying of the paper. Separation of 13 amino acids and two pairs (Keil[22], Dubrovskaia *et al.*[23]).
6. Separation of 13–15 from a mixture of 18 amino acids over a period of 40 h using miscible buffered solvents, followed by rapid resolution of two pairs by simple reversed chromatography (Hanes and co-workers[7–12]).

Two of these systems have been developed for quantitative work

References p. 38

and will be described in detail. A useful diagram illustrating the patterns of separation of a number of published systems is shown in Fig. 6.

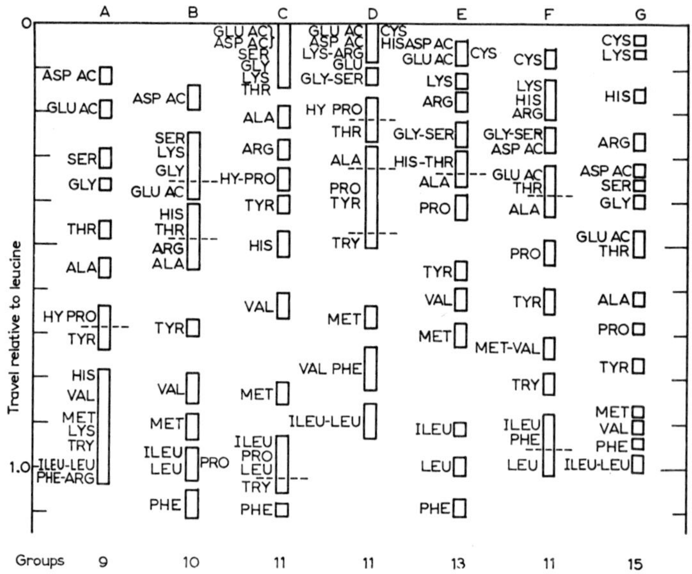

Fig. 6. Diagrammatic representation of one-dimensional separations (compiled by Wade *et al.*[8]) (See Fig. 2 for abbreviations.)
A. Phenol–aqueous phosphate buffer at pH 12[6]
B. Phenol–hydrochloric acid–water[24]
C. Phenol–aq. borate buffer at pH 8.4[6]
D. *n*-Butanol–ethanol–water[25,26]
E. *sec.*-Butanol–ammonia–water[21]
F. *n*-Butanol–acetic acid–water[15]
G. *n*-Butanol–acetic acid–water[22,23]

Quantitative method of Roland and Gross[21]

The ten amino acids, lysine, arginine, alanine, proline, tyrosine, valine, methionine, isoleucine, leucine and phenylalanine are resolved in the system 2-butanol–3% ammonium hydroxide (3:1) which was one of the many solvent systems studied by Block and Bolling[27] (see Fig. 6). Strips of Whatman No. 1 paper (8 × 25 in.) fitted with absorbent pads at the lower edge are irrigated by the descending method during a period of 48 h. The solvent is prepared by mixing 120 ml of 2-butanol with 40 ml of 3% ammonium hydroxide.

The second solvent is phenol which resolves the six amino acids cysteic acid, aspartic acid, glutamic acid, serine, glycine and threonine. Histidine which is not resolved as a discrete spot by either solvent may be determined using the Pauly reagent (see p. 23).

The quantitative determinations were carried out by Roland and Gross using the area density procedure of Block[28] after development of the chromatogram in a dip solution (0.2% ninhydrin in acetone containing 1% acetic acid). The phenol chromatograms were washed with ether after a preliminary drying. A comparison of the results of analysis of an actual acid hydrolysate of β-lactoglobulin by this method with values obtained by Stein and Moore[29] using a starch column is shown in Table 1.

TABLE 1

COMPARATIVE AMINO ACID VALUES OF β-LACTO-GLOBULIN[21]

Constituent	*Observed values (grams of amino acid per 100 grams protein)*	*Reported values (grams of amino acid per 100 grams protein)*
Arginine	2.8	2.91
Histidine	1.7	1.63
Lysine	13.0	12.58
Tyrosine	3.5	3.64
Tryptophan	2.0[a]	1.94
Phenylalanine	3.6	3.78
Cystine-cysteine	...	3.40
Methionine	3.0	3.22
Serine	3.9	3.96
Threonine	4.9	4.92
Leucine	15.7	15.50
Isoleucine	5.6	5.86
Valine	5.6	5.62
Glutamic acid	19.8	19.08
Aspartic acid	11.5	11.52
Glycine	1.6	1.39
Alanine	6.9	7.09
Proline	5.1	5.14

[a] Tryptophan was liberated by alkaline hydrolysis with saturated barium hydroxide in a small sealed tube under nitrogen atmosphere for 17 h at 107°.

Quantitative method of Hanes and co-workers

The work of the Toronto school has been published in two sections; the first communication[30] appeared in 1955 and the subsequent six papers[7–12] in 1961. The amino acid separations are achieved with miscible, buffered solvents in a 40 h run and should prove useful in structural work on proteins where analyses have to be performed on peptides of varying chain length. The method might be commended particularly to workers studying proteins in isolated areas where apparatus associated with ion-exchange chromatography is not readily available.

The preliminary screening of a number of miscible alcohols and buffers indicated high resolution in *n*-propanol–ethanol–aqueous pyrophosphate systems. Thus the system 50 vol. *n*-propanol–25 vol. ethanol–25 vol. aqueous pyrophosphate (P–E–PP) was capable of separating all the amino acids usually found in acid hydrolysates of proteins except the pairs aspartic + cysteic acids and serine + glycine. Sodium chloride was also added as it was found that the salt modified the distribution of water in the different phases of the system, accelerated the basic amino acids and removed the necessity for prior desalting of samples. In the earlier work a rather laborious washing stage was needed for preparation of the paper but as a result of improvements in production this stage is largely unnecessary.

Care must be taken in the preparation of the solvent and the following procedure is recommended[9]: 0.05 *M* pyrophosphate buffer, pH 8.9 ± 0.03, is prepared from 89.3 g of tetrasodium pyrophosphate decahydrate, 26.55 g of sodium chloride and 3800 ml of water; 22.0 ml of *N* hydrochloric acid are then added and the solution volume brought to 4 l with water. The P–E–PP system is made up from 200 ml of buffer solution, 300 ml of water, 4.8 g of sodium chloride, 500 ml of 95% ethanol and 1000 ml of *n*-propanol. The mixture is allowed to stand at 34° overnight and the supernatant liquid is drawn off for use leaving a deposit of salts. Whatman No. 3 paper is used for the separations and a special grade, Whatman No. 3 HR, $29\frac{1}{2} \times 11$ in. is now available. Before use the paper is uniformly impregnated with the pyrophosphate buffer solution. A feeder wick of Whatman No. 1 paper is stitched across the top of the paper and an absorption pad of No. 3 paper is attached to the lower edge.

Solvent flow is adjusted to give a 22–24 in. travel of leucine in 40 h at 34°. Preconditioning of the paper is necessary and for tanks of dimensions $13 \times 9\frac{1}{4} \times 28\frac{1}{4}$ in. a mixture of 9 ml of water and 4 ml of *n*-propanol should be allowed to vaporize before the commencement of the run. The exact composition of the preconditioning mixture depends on accurate knowledge of the water content of the paper and the reader should consult the original literature for a discussion of this aspect.

A small section of the P–E–PP chromatogram is cut out for separation of glycine and serine by *ascending* chromatography using the auxiliary phenol–aq. pyrophosphate solvent. The latter is prepared by adding 100 ml of the pyrophosphate buffer solution and 30 ml of water to 400 g of phenol.

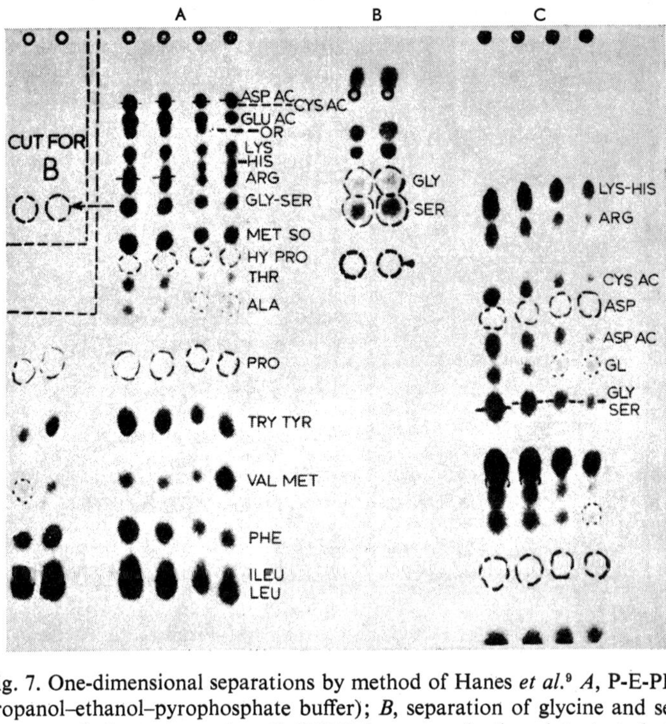

Fig. 7. One-dimensional separations by method of Hanes *et al.*[9] *A*, P-E-PP (*n*-propanol–ethanol–pyrophosphate buffer); *B*, separation of glycine and serine by reversed chromatography; *C*, P-E-T (*n*-propanol–ethanol–tartrate buffer). Loadings of each amino acid present (*left to right*): 0.4, 0.3, 0.2, 0.1 μmole. See Figs. 2 and 8 for abbreviations.

The patterns of separation obtained with these systems are shown in Fig. 7, together with that obtained using a complementary system worked out for dealing with glutamine and asparagine. This third system contains 62.5 vol. of *n*-propanol–37.5 vol. of ethanol–25 vol. of aqueous tartrate buffer (P–E–T system). 0.10 *M* tartrate buffer, pH 3.40 ± 0.03, is prepared from 60.0 g of tartaric acid in 3800 ml of water and approximately 84 ml of 5 *N* NaOH. The final volume is adjusted to 4 l with water and the P–E–T solvent prepared from 750 ml of buffer solution and 1250 ml *n*-propanol. Although this solvent contains no ethanol the latter is present in the preconditioning medium which contains 6 ml of water and 10 ml of ethanol. The

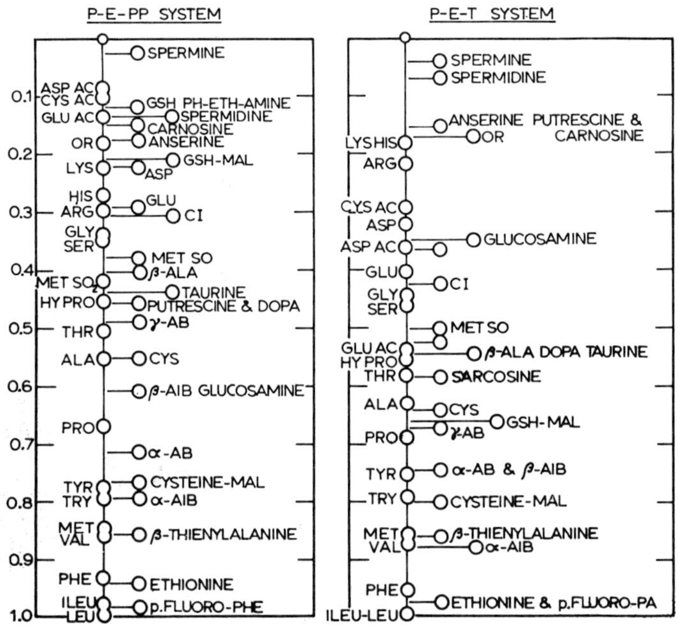

Fig. 8. Distribution of amino acids and derivatives in relation to leucine in the P–E–PP and P–E–T systems[9]. *α*-AB, *α*-aminobutyric acid; *β*-AB, *β*-aminobutyric acid; *α*-AIB, *α*-aminoisobutyric acid; *β*-AIB, *β*-aminoisobutyric acid; *β*-ALA, *β*-alanine; ASP, asparagine; CYSTEINE-MAL, cysteine-maleimide; *p*-FLUORO-PA, *p*-fluorophenylalanine; GLU, glutamine; GSH, glutathione; GSH-MAL, glutathione-maleimide; MET SO_2, methionine sulphone; MET SO, methionine sulphoxide. See Fig. 2 for other abbreviations.

optimum flow rate is such that glycine or serine travels 14–17 in. in 40 h at 34°.

The distribution of amino acids and other amino compounds of interest is shown in Fig. 8. In 18 cases, percentage recoveries of ninhydrin colour lay between 95–100 after development with a modified reagent (see p. 19). Reverse chromatography of an excised portion of a P–E–PP chromatogram gave colour values for glycine and serine 82% and 75% respectively of the direct values.

The amino acid spots are detected with a dip reagent (*see below*) colour development being kept to the minimum necessary to delineate the zones. The individual spots are cut out and weighed in order to calculate the paper blank value. Ammonia is removed from the excised spots by wetting with a solution containing 18.6 g of boric acid, 65 ml of 6 *N* aqueous potassium hydroxide solution and 935 ml

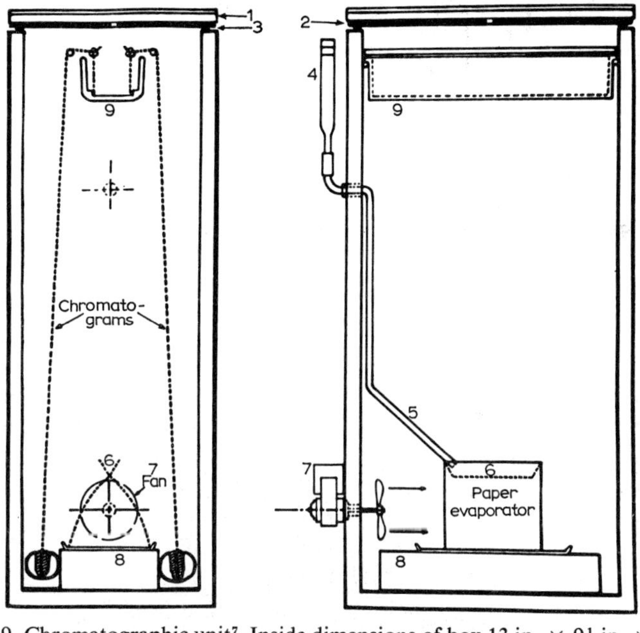

Fig. 9. Chromatographic unit[7]. Inside dimensions of box 13 in. × 9¼ in. ×28¼ in. high. 1, Weighted cover; 2, glass lid with central hole; 3, neoprene foam gasket; 4,5,6, receiver tube, delivery tube and paper evaporator; 7, fan; 8, combined support for evaporator and positioner for bottom clamps of chromatogram; 9, porcelain solvent trough (adjustable in height).

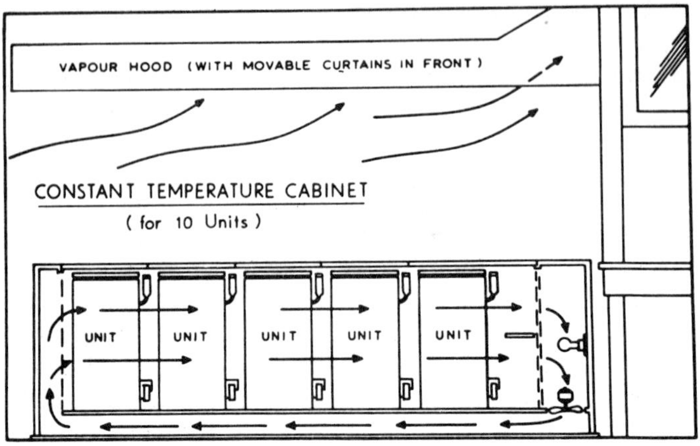

Fig. 10. Arrangement for chromatography at constant temperature[7].

Fig. 11. Cabinet for chromatography at two different temperatures installed at Twyford Laboratories.

of methanol. The spots are cut into strips, dropped into boiling tubes flushed out with nitrogen, and treated with 3 ml of a modified ninhydrin reagent (see below) at 100° for 20 min. Paper blanks cut from each chromatogram are treated similarly and a number of reagent blank tubes are included in each analysis. The coloured solutions are diluted to convenient volumes with 50 vol. % aqueous ethanol and the extinction values measured at 570 mμ or 440 mμ in the case of proline and hydroxyproline.

The chromatographic units used by the Toronto school are of special interest. Solvent flow is regulated by a filter-paper wick sewn to the top of the chromatogram and a special paper evaporator and small fan ensure rapid equilibration of the system (Fig. 9).

One of the best arrangements for temperature control of the units and for ventilation was constructed by Hanes and Isherwood quite early in the development of chromatography. A diagrammatic illustration of the cabinet is shown in Fig. 10. A similar chamber modified to include glazed front panels and hinged glazed lids is used in the Twyford Laboratories (Fig. 11). The two sections of the cabinet can be operated at different temperatures and the suction from the hood can be localised over a particular lid by opening the corresponding vent.

The solvent system *tert.*-amyl alcohol–acetic acid–water (20:1:20) separates amino acid isomers and discriminates between the straight-chain and branched-chain forms of some aliphatic amino acids. R_{Leucine} values for several groups of isomers determined by Gray *et al.*[31] are shown in Table 2.

DETECTION OF AMINO ACIDS

Ninhydrin

Suitable solutions for spraying are (a) 0.1% ninhydrin in *n*-butanol saturated with 0.5 *N* citrate buffer, pH 5.5 and (b) 0.1% ninhydrin in ethyl alcohol alone or containing 5% collidine. A useful dip treatment described by Hanes contains 125 mg of ninhydrin in 25 ml of glacial acetic acid, 20 ml of collidine and 955 ml of ethyl acetate. The solution must be freshly made up and a trace of hydrindantin is added for use with phenol chromatograms. In alcohol solution containing collidine as the free base the amino acid spots show up

References p. 38

TABLE 2

SEPARATION OF AMINO ACID ISOMERS ON PAPER USING THE UPPER PHASE OF FRESHLY PREPARED *tert.*-AMYL ALCOHOL–ACETIC ACID–WATER (20:1:20)[31]

Amino acid isomer	$R_{Leucine}$ *at 20°C*	*Time in days taken for separation of each group of isomers on Whatman No. 3 MM paper*[a]
β-Methyl-L-aspartic acid (*threo* form)[c]	0.20	
β-Methyl-DL-aspartic acid (*erythro* form)[c]	0.23	20
β-Hydroxy-DL-aspartic acid (*threo* form)	0.03	
β-Hydroxy-DL-aspartic acid (*erythro* form)[c]	0.06	12
γ-Methyl-DL-glutamic acid (natural form)	0.41	
Allo-γ-methyl-DL-glutamic acid	0.48	4
L-Isoleucine (*erythro* form)[c]	0.88	
D-Alloisoleucine (*threo* form)[c]	0.81	4
Cis-α-(methylcyclopropyl)glycine	0.56	
Trans-α-(methylcyclopropyl)glycine	0.67	3
4-Hydroxy-L-proline (*erythro* form)[c]	0.23	
Allo-4-hydroxy-D-proline (*threo* form)[c]	0.19	4
DL-Valine	0.49	
DL-Norvaline	0.66	1
L-Isoleucine	0.88	
L-Leucine	1.00	2
L-Norleucine	1.14	
Homoisoleucine[b] (2-amino-4-methyl hexanoic acid)	1.39	
Homoleucine[b] (2-amino-5-methyl hexanoic acid)	1.50	2
Homonorleucine[b] (2-amino-heptanoic acid)	1.59	

[a] Samples (10γ) of the compounds were applied to the origin of 44 cm-long descending chromatograms as streaks 2.0 × 0.3 cm. The amino acids were finally revealed with ninhydrin.

[b] These identifications are tentative.

[c] The relationship between the two asymmetric centres, where this is known, is expressed by the use of the terms *threo* and *erythro* as proposed by Vickery[32].

in different colours depending on the temperature and time of development. A 5% solution of cyclohexylamine or dicyclohexylamine in alcohol has also been proposed for this purpose[33].

For *quantitative development* of the ninhydrin colour Hanes[10] has used the following reagent. 7.6 g of ninhydrin is dissolved in 1140 ml of redistilled methyl-cellosolve and to this is added 610 ml of water and 250 ml of 4 *N* acetate buffer, pH 5.5 (Moore and Stein, see p. 79). A vacuum is applied to remove dissolved gases and the mixture allowed to stand at 5° under nitrogen for 12 h. Finely powdered hydrindantin (1.2 g) is then added and the bottle shaken at intervals. The reagent, which is dispensed under nitrogen, can be checked by adding 2 ml to 1 ml 8 *M* sodium hydroxide. A stable blue colour develops, the extinction of which at 570 mμ should be in the range 0.35 to 0.9. More hydrindantin must be added if the extinction is lower than this value.

To render the spots permanent for photographic purposes the papers can be sprayed with 1% copper nitrate in ethyl alcohol (Kawerau and Wieland[34]). The spots then appear pink on a green background. Many amino acids and peptides will fluoresce under ultra-violet light if the paper is first heated to 90° for 20 min[35]. The type of paper has some influence on the fluorescence and contact with phenolic vapours must be avoided for satisfactory spots.

Jepson and Smith[36] have described a multiple dip procedure whereby maximum information is given with sequences such as ninhydrin (or isatin), followed by Ehrlich reagent, followed by Sakaguchi reagent or diazo reagent (*see below*). Ninhydrin can be applied over isatin.

Trinitro-benzene-1-sulphonic acid[37]

The sodium salt of 2,4,6-trinitro-benzene-1-sulphonic acid (TNBS) is prepared[38] by adding finely ground sodium bisulphite (20 g) to a suspension of picryl chloride (20 g) in absolute ethanol (240 ml), and boiling the mixture under reflux for 6 h. The precipitate is recrystallized twice from water and once from 50% ethanol; yield 35%. The free acid, obtained by recrystallization of the sodium salt from 5 *N* HCl, has a melting point of 183°.

Papers to be treated with the reagent are first made slightly alkaline by spraying with a solution containing aqueous saturated borax and methanol (1:3). The dried paper is dipped into a freshly prepared solution of TBNS Na (2 mg) in dry acetone (100 ml) and hung in a wellventilated dark area. Within 20 min, amino acids and some peptides appear as bright yellow spots on a white background (sensitivity, 0.01 μmole or less of amino acid). The reagent has been used to detect ^{14}C-labeled amino acid spots on paper without removing radioactivity from the carboxyl group[38].

Proline and hydroxyproline

The presence of proline or hydroxyproline is best confirmed using the isatin reagent. Its use is based on studies of Grassmann and Arnim[39] and was first proposed as a spray by Acher, Fromageot and Jutisz[40]. The paper is treated with a solution containing 100 mg of isatin, 50 ml of *n*-butanol and 5 ml of glacial acetic acid and then heated at 40° for 5 min. Proline gives an intense blue colour (sensitivity 0.2 μg) and hydroxyproline gives a pale blue colour (sensitivity 0.8 μg). Phenylalanine and tyrosine give blue-green spots. In the more recent method of Barrollier *et al.*[41] two reagents are employed and prepared as follows.

(a) 1 g of isatin
1.5 g of zinc acetate
1 ml of pyridine
100 ml of *iso*propanol

(b) 1 g of isatin
1.5 g of zinc acetate
1 ml of glacial acetic acid
95 ml of *iso*propanol
5 ml of water

The zinc acetate and isatin are dissolved in the *iso*propanol by warming in a water-bath at 80° after which the solution must be kept cool. After spraying the chromatogram with either solution the paper is heated to 80–85° for 30 min. Excess isatin may be washed off quickly with water. The pyridine reagent gives a greater variety of colours with amino acids (see Table 3) but the acid reagent provides more intense colours (sensitivity for proline 3×15^{-8} mole/cm^2). A 0.2% solution of isatin in *n*-propanol has been used for the detection of *prolyl peptides* in which the proline residue has a free amino group (Monier and Jutisz[42]).

A further interesting reagent also studied by the same authors is the following:

0.125 g of sodium 1,2-naphthaquinone-4-sulphonate
0.5 g of sodium acetate
0.125 ml of quinoline
10 ml of water
90 ml of *sec.*-butanol

The salts are dissolved in water, the quinoline and *sec.*-butanol are then added, and the mixture filtered after a few minutes. The reagent will not keep. The chromatogram is sprayed and allowed to stand at room temperature for one hour. Excess reagent is washed off the paper with a 20% solution of zinc acetate in 95% alcohol. Colours produced are rose with proline (3×10^{-8} mole/cm^2) and hydroxyproline (10^{-7} mole/cm^2), a characteristic yellow-green with glycine and glycylglycylglycine (10^{-7} mole/cm^2), intense blue with lysine, arginine and ornithine, and various shades of blue and grey with other amino acids.

Tyrosine

A specific reagent for tyrosine is a 0.1% solution of α-nitroso-β-naphthol in 75% alcohol. The chromatogram is sprayed with the reagent, dried, sprayed again with 10% aqueous nitrous acid and finally heated to 90° for 3 min. Tyrosine and tyrosine peptides appear red on a pale green background (sensitivity, 1 μg), (Acher and Crocker[43]).

Tryptophan

Tryptophan and other compounds containing the indole nucleus give purple or violet colours with the Ehrlich reagent. A typical spray reagent is 0.5 g of *p*-dimethylaminobenzaldehyde in 100 ml 95% alcohol containing 2 ml of concentrated hydrochloric acid. After some hours tryptophan appears as a violet spot, kynurenine appears orange and citrulline yellow (Reddi and Kodicek[44]). (Sensitivity 1 μg tryptophan, 5 μg kynurenine.) The reagent is not completely specific for indoles, and has even been proposed for confirming the presence of hydroxyproline after treatment with isatin. Hydroxyproline alone undergoes the colour change blue to purple (Jepson and Smith[36]). Ureides and aromatic amines produce various shades of yellow with the reagent.

References p. 38

TABLE 3

COLOURS DEVELOPED WITH PYRIDINE–ISATIN REAGENT[41]

Amino acid	*Colour*	*Concentration (mole/cm^2)*
α-Alanine	Red	10^{-6}
α-Alanine	Pale pink	10^{-7}
β-Alanine	Dark red	10^{-6}
β-Alanine	Lavender	10^{-7}
Arginine	Deep pink	10^{-6}
Aspartic acid	Bright red	10^{-6}
Cystine	Yellow-brown	10^{-6}
Cystine	Faint yellow	10^{-7}
Glutamic acid	Bright red	10^{-6}
Glycine	Orange-red	10^{-6}
Histidine	Orange-red	10^{-6}
Leucine	Red	10^{-6}
Leucine	Orange-red	3×10^{-7}
Isoleucine	Bright red	10^{-6}
Lysine	Red	
Lysine	Green-yellow	10^{-7}
Methionine	Pink	10^{-6}
Ornithine	Deep Pink	10^{-6}
Hydroxyproline	Blue-green	10^{-6}
Hydroxyproline	Faint blue	3×10^{-8}
Phenylalanine	Red-brown	10^{-6}
Phenylalanine	Yellow	10^{-7}
Proline	Intense blue	10^{-6}
Proline	Pale blue	3×10^{-8}
Serine	Pink	10^{-6}
Threonine	Pink	10^{-6}
Tryptophan	Red-brown	10^{-6}
Tryptophan	Light-brown	10^{-7}
Tyrosine	Light-brown	10^{-6}
Tyrosine	Faint yellow	10^{-7}
Valine	Red	10^{-6}

Histidine

The classical reagent for histidine and other imidazoles is the Pauly reagent prepared from a solution of 1 g of sulphanilic acid in 100 ml *N* hydrochloric acid and an equal volume of 0.7% aqueous sodium nitrite. The paper is sprayed, dried in a stream of air and sprayed again lightly with 10% sodium carbonate solution. Various shades of red and orange are given by histidine, thiol histidine, ergothionine and other imidazoles which are not N-substituted. Sanger and Tuppy[45] have described a modification which avoids the use of aqueous solutions. A mixture of equal volumes of 1% (w/v) *p*-anisidine in ethanolic 0.11 *N* hydrochloric acid and 10% (v/v) amyl nitrite in ethanol is allowed to stand for a few minutes and sprayed on to the paper. Colour is developed with ammonia vapour or by spraying with ethanolic potassium hydroxide solution. Histidine and histidine peptides give a dark-red spot (sensitivity 0.1 μg), tyrosine gives only a faint orange colour. Care must be taken to remove all traces of phenol from the paper before using the reagent.

Sluyterman[46] has described a quantitative method based on the Pauly reaction. A standard solution is prepared by dissolving 1 mg of histidine hydrochloride in 2.5 ml of water containing 0.25% ethylenediamine tetracetate. A 2-ml aliquot of this solution is pipetted into a flask after which 0.2 ml of 1% sulphanilic acid in 1.4 *N* hydrochloric acid, and 0.2 ml of 5% nitric acid are added. 20% sodium carbonate solution (0.6 ml) is added after 10 min with gentle swirling. Exactly 2 min later 2 ml of 75% alcohol are added and the solution mixed well. The extinction values are read at 530 mμ.

Newman and Turnbull[47] discovered that histidine and histamine at pH 11–12 give a violet colour with potassium tri-iodide. The colour development can be checked by the addition of sodium dithionite and the extinction measured at 560 mμ. The reaction is specific for histidine and histamine (sensitivity 1 μg/ml). Tryptophan and tyrosine do not form colours but interfere with the histidine colour formation.

Arginine

Arginine and monosubstituted guanidines in general give red colours with the Sakaguchi reagent prepared as follows[43]: (a) 0.01% of

References p. 38

α-naphthol in 95% alcohol containing 5% urea; (b) 5% sodium hypobromite solution (2 g bromine in 100 ml of 5% sodium hydroxide).

Five pellets of sodium hydroxide are added to 10 ml of solution (a) and the clear solution used to spray the chromatogram. After drying at room temperature the paper is sprayed lightly with solution (b). (Sensitivity 0.1 μg arginine.) Solutions (a) and (b) will keep for a few weeks at 0°. In a recent modification N-bromosuccinimide is substituted for hypobromite[48].

Glycine and taurine

The chromatogram is sprayed with a 0.2% solution of *o*-phthalaldehyde in acetone and heated at 100° for 2 min. It is then dipped in a 1% solution of potassium hydroxide in 95% ethanol and heated for a further 10 min at 100°. Glycine gives a green colour and taurine a red colour (sensitivity 1 μg[49]).

Sulphur-containing amino acids

There are many tests for the sulphur-containing amino acids and considerable care is needed for their successful application on the microgram scale. Cystine, cysteine and methionine undergo partial oxidation, the particular products depending on the circumstances. Before chromatography it is usual to oxidise these amino acids deliberately to convert cystine to cysteic acid and methionine to methionine sulphoxide or sulphone. It is possible to treat a mixture of amino acids as a spot on paper with hydrogen peroxide (30 vol.) or bromine water. However, numerous artifacts[50] may easily arise and it is advisable to treat the sample to be analysed with performic acid prior to spotting on paper. (See p. 99). It is reported that several proteins when heated at 100° with an aqueous suspension of cadmium hydroxide at pH 9.5 release hydrogen sulphide quantitatively from cysteine and cystine[51].

Substituted maleimides

A simple reaction is that which takes place between cysteine and substituted maleimides.

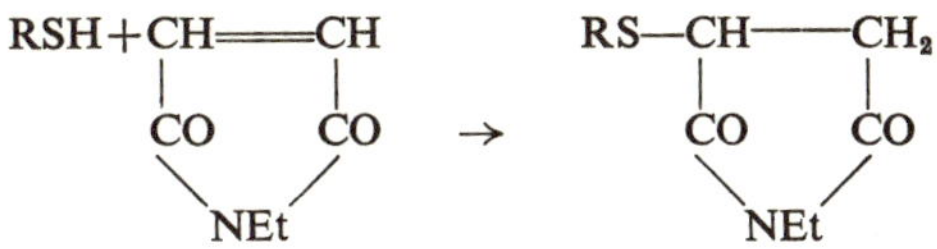

N-ethylmaleimide[52] (NEM) was introduced by Hanes *et al.*[53] as a useful reagent for studying reactions involving cysteine by paper chromatography. An aqueous solution containing cysteine is treated with 1/7.5 *M* NEM or 0.005 *M* NEM can be added to a spot on paper. The compound N-(4-dimethylamino-3,5-dinitrophenyl) maleimide has recently been described by Witter and Tuppy[54] as a coloured sulphydryl reagent. The preparation of the derivative of cysteine is as follows. 610 mg of the DDPM reagent in 20 ml acetone is mixed with 351 mg cysteine hydrochloride in 20 ml of deoxygenated water. The pH of the mixture is adjusted to 5 by the addition of solid sodium bicarbonate and the mixture left at room temperature for 90 min. The solid material left after evaporation of the solution is washed with water and the orange derivative may be purified by precipitation with ether from acetic acid solution. The maleimides are unstable to alkali.

Iodine-azide

Compounds containing the group C=S or C—SH catalyse the reaction $2NaN_3 + I_2 = 2NaI + 3N_2$. A suitable solvent for spraying chromatograms contains 1.5% sodium azide in 0.01 *M* potassium in ethanol. Thiol compounds like cysteine cause rapid evolution of nitrogen with simultaneous bleaching of the brown background. Cystine reacts in about 15 min, methionine after 60 min. Cysteic acid does not give the reaction.

Tetrazolium salts

The paper is sprayed with 0.5% 2,3,5-triphenyltetrazolium chloride in chloroform, dried and then sprayed again with a 10% solution of sodium ethoxide or sodium carbonate in alcohol. Cysteine gives a red spot (sensitivity 0.5 μg).

Nitroprusside

A deep red colour is produced when sodium nitroprusside is added to a thiol compound. For the detection of cystine or peptides containing cystine the chromatogram is first blotted with a sheet of

References p. 38

filter paper wetted lightly with a solution containing 0.5 g of sodium cyanide in 2 ml of water diluted to a volume of 25 ml with methanol. The paper is then sprayed lightly but rapidly with a nitroprusside solution prepared by dissolving 250 mg of sodium nitroprusside in 1 ml of dilute sulphuric acid and then adding 2 ml of concentrated ammonium hydroxide and methanol to a final volume of 25 ml. The solution should be filtered before use (sensitivity 0.5 μg cysteine).

Polychromatic techniques

As many as 20 amino acids, in amounts as low as 1.2 γ, can be completely differentiated after chromatography for 6 h by spraying with *ninhydrin–cupric nitrate*. The characteristic coloured complexes are listed in Table 4. The ninhydrin–cupric nitrate indicator of Moffat and Lytle[55] consists of two solutions, *A* and *B*, combined in a ratio of 25 to 1.5 just before using. Solution *A* contains 0.2% ninhydrin in absolute ethanol (50 ml), glacial acetic acid (10 ml) and 2,4,6-collidine (2 ml); solution *B* is 1% cupric nitrate trihydrate in absolute ethanol. Following a run in butanol–acetic acid–water the paper is sprayed with the indicator solution and heated at 105° for 1½–2 min.

Coloured spots against a light blue background are obtained with *diethylamine-ninhydrin*[56]. After chromatography in butanol–acetic acid–water the *wet* paper is exposed to the vapour of diethylamine for 2 min. The sheet is dried at 75–80°, sprayed with 0.25% ninhydrin in *n*-butanol, and heated for 10 min at 75–80°. The colours which develop are shown in Table 5.

SEPARATION OF PEPTIDES

Useful tables of R_F values of peptides in various solvents have been published by Consden[1], Knight[57], and by Moore and Baker[58]. A short list of R_F values of peptides taken from Moore and Baker is given in Table 8 together with the R_F values of amino acids in the same solvents. The compositions of the solvent systems are given in Table 6. It is possible, however, to arrive at a reasonably accurate figure for the R_F of a given small peptide by using the expression derived by Pardee[59] relating the R_F values of peptides with those of their constituent amino acids. This interesting extension of the theory

Text continued p. 30

TABLE 4

COLOURS OF AMINO ACIDS AFTER USE OF NINHYDRIN-CU$(NO_3)_2$ INDICATOR[55]

Compound	Colour
Cystine	Grey
Lysine	Reddish brown, pink ring forms upon standing
Histidine	Light brown with dark brown ring, inside a yellow ring
Asparagine	Golden
Arginine	Dark purple
Serine	Greenish brown, red ring forms upon standing
Aspartic acid	Light blue (if removed from the oven too soon, it will be light green)
Glycine	Orange brown with bright orange ring
Threonine	Greenish brown, changes to purplish brown on standing
Glutamic acid	Purple, fades slightly upon standing
Alanine	Dark purple
Proline	Light green with yellow ring
Cysteine	Grey
Tyrosine	Light brown
Valine	Purple
Methionine	Greyish purple with yellow ring
Tryptophan	Brown with bright blue ring (ring fades rapidly)
Isoleucine	Light blue
Phenylalanine	Greenish yellow
Leucine	Light purple with yellow ring

TABLE 5

COLOURS OF AMINO ACIDS AFTER USE OF DIETHYLAMINE-NINHYDRIN INDICATOR[56]

Compound	Colour
Aspartic acid	Green-blue
Glycine	Red
Histidine	Brown
Hydroxyproline	Salmon
Lysine	Grey-blue
Phenylalanine	Yellow-brown
Proline	Yellow
Serine	Dark brown
Threonine	Blue
Tryptophan	Brown
Tyrosine	Yellow-brown
Others	Purple

TABLE 6

COMPOSITION OF SOLVENT SYSTEMS[58]

Solvent (composition see below)	*A*	*B*	*C*	*D*	*E*
Distilled water	15	50	25	25	
88% Formic acid	15				
Glacial acetic acid		10	10	5	
Methyl alcohol (abs.)					
Ethyl alcohol (abs.)					
n-Propyl alcohol					
n-Butyl alcohol		40	65	12.5	
tert.-Butyl alcohol	70				
tert.-Amyl alcohol					
Methyl ethyl ketone					
Acetone					
Liquified phenol				25	[a]
m-Cresol					
Pyridine					
0.5% Aqueous urea					
10% Trisodium citrate					[a]
Conc. ammonium hydroxide					
Phase[b]	M	U	M	U	U

A Formix
B 4:1:5 Butanol–acetic acid
C BuOH–acetic acid
D PBA
E Phenol (citrate)
F Phenol (H_2O)(NaCN and NH_3 atmos.)
G *m*-Cresol

[a] Liquified phenol saturated aqueous 10% trisodium citrate.
[b] M: miscible; U: upper layer used.

TABLE 6

CONPOSITION OF SOLVENT SYSTEMS[58]

F	*G*	*H*	*I*	*J*	*K*	*L*	*M*	*N*	*O*
c	d	23	30	40	20		20	30	10
							80		
		77		10					
			70						60
				50					
					40				
								35	
					40				
						60			
c									
	d								
							4	35	
						40			
									30
U	U	M	M	M	M	M	M	M	M

H 77% Ethanol
I 70% Propanol
J BuOH–EtOH
K TBK[e]
L Acetone–urea
M Pyridine-mix
N Pyridine–amyl alcohol
O Propanol–NH_3

[c] Liquified phenol saturated with water with addition of 4 to 6 crystals of sodium cyanide; a beaker of 0.3% NH_4OH present in the tank during each run.

[d] *m*-Cresol saturated with water; a beaker of 0.03% NH_4OH present in the tank during each run.

[e] A beaker of 5% aqueous diethylamine (Eastman No. 92) present in the tank during each run.

of paper chromatography as developed by Martin and Synge[60,61] is outlined below.

The relation of the partition coefficient of a solute to its rate of movement in a system can be expressed by

$$R = A/A_L + \alpha A_S$$

where A = cross-sectional area of paper + water + solvent
A_L = cross-sectional area of solvent phase
A_S = cross-sectional area of water phase
α = partition coefficient

Now
$$R_F = \frac{\text{distance travelled by solute zone}}{\text{distance travelled by solvent front}}$$
$$= RA_L/A = A_L/A_L + \alpha A_S$$

Or
$$\alpha = A_L/A_S(1/R_F - 1) \tag{1}$$

Assuming ideal conditions the work, $\Delta F°$ required to move 1 mole of solute from the aqueous phase to the solvent phase under standard conditions is given by

$$\Delta F° = RT\ln\alpha = RT\ln A_L/A_S(1/R_F - 1)$$

The total work $\Delta F°_{(p)}$ for a peptide of n residues can be divided into the work required to transfer n amino acid residues, $\Sigma\ \Delta F°_{R(p)}$; $n - 1$ peptide bonds, $(n - 1)\Delta F°_{\text{CONH}}$; and the terminal amino and carboxyl groups $\Delta F°_{(p)\text{COO}^-} + \Delta F°_{(p)\text{NH}_3{}^+}$

$$\Delta F°_{(p)} = RT\ln A_L/A_S(1/R_{F(p)} - 1)$$
$$= (n - 1)\Delta F°_{\text{CONH}} + \Delta F°_{(p)\text{COO}^-}\Delta F°_{(p)\text{NH}_3{}^+} + \Sigma\Delta F°_{R(p)}$$

Similarly to transfer n free amino acids which make up the peptide

$$\Delta F°_{(a)} = \Sigma RT\ln A_L/A_S(1/R_{F(a)} - 1)$$
$$= n\Delta F°_{(a)\text{COO}^-} + n\Delta F°_{(a)\text{NH}_3{}^+} + T\Delta F°_{R(a)}$$

Assuming $\Delta F°_{R(p)} = \Delta F°_{R(a)}$ then the following expression is possible

$$RT\ln(1/R_{F(p)}-1) = (n-1)A+B+\Sigma\ RT\ln(1/R_{F(a)}-1) \tag{2}$$

where $R_{F(p)}$ and $R_{F(a)}$ are the R_F values of peptide and amino acids respectively, A and B are constants, R is the molar gas constant and T the absolute temperature.

The constants A and B are evaluated from a plot of $RT \ln (1/R_{F(p)} - 1) - \Sigma RT \ln (1/R_{F(a)} - 1)$ against the number of peptide bonds, $(n - 1)$, where A is the slope and B the intercept.

TABLE 7

VALUES OF CONSTANTS A AND B IN VARIOUS SOLVENTS

Solvent	*A* *(Cal/mole)*	*B* *(Cal/mole)*
A. Formix	—14.0	—112.6
B. Butanol–acetic acid (4:1:5)	—695.9	—349.7
E. Phenol citrate	—322.8	157.4
F. Phenol–(H_2O)(NaCN + NH_3 atmos.)	—172.5	27.8
G. *m*-Cresol	—1187.2	303.6
H. 77% Ethanol	—456.2	187.2
K. TBK	—828.4	—232.9
M. Pyridine mix	—172.3	125.0
O. Propanol–NH_3	—190.5	—103.4

The validity of equation (2) above has been checked by Moore and Baker for 46 peptides in 9 different solvent systems. In the majority of cases the difference between the observed and calculated R_F values was less than ± 0.05. Peptides that occasionally varied by more than 0.10 from the calculated values often contained sarcosine, glutamic acid, aspartic acid, lysine or phenylalanine. The values of constants A and B used in the calculations for each solvent are listed in Table 7. An interesting finding is that there is no difference in the R_F values for groups of peptides in which the sequence of amino acids is altered. For example, alanylglycylglycine, glycyalanylglycine, and glycylglycylalanine, all have identical R_F values. An interesting use of equation (2) in sequence studies on phosphoglucomutase was reported by Milstein and Sanger[62].

Detection of peptides

When sprayed with even high concentrations of ninhydrin, peptides often show up very feebly. Cyclic peptides, highly acidic peptides

References p. 38 *Text continued p. 36*

TABLE 8

EXPERIMENTAL R_F VALUES OF PEPTIDES AND AMINO ACIDS[58]

(Whatman No. 4 paper)

	A Formix	*B* BuOH-Acetic 4:1:5	*C* BuOH: Acetic 6.7:1:2.5	*D* PBA	*E* Phenol citrate	*F* Phenol H_2O (NaCN + NH_3 atmos.)	*G* m-Cresol (NH_3 atmos.)	*H* 77% EtOH	*I* 70% PrOH	*J* BuOH EtOH	*K* TBK	*L* Acetone-urea	*M* Pyridine mix	*N* Pyridine amyl	*O* Propanol-NH_3
Glycylglycine	0.32	0.22	0.12	0.27	0.37	0.61	0.15	0.18	0.21	0.05	0.19	0.59	0.46	0.20	0.41
Glycylsarcosine	0.42	0.17	0.11	0.41	0.68	0.80	0.44	0.33	0.30	0.05	0.18	0.73	0.49	0.21	0.43
DL-Alanylglycine	0.47	0.23	0.16	0.41	0.53	0.69	0.25	0.32	0.28	0.06	0.29	0.68	0.63	0.29	0.49
DL-Leucylglycine	0.75	0.54	0.53	0.66	0.68	0.86	0.56	0.61	0.63	0.23	0.60	0.83	0.75	0.59	0.74
L-γ-Glutamylglycine	0.35	0.15	0.12	0.23	0.18	0.32	0.04	0.16	0.18	0.02	0.05	0.63	0.52	0.16	0.25
Glycyl-L-alanine	0.47	0.30	0.21	0.44	0.50	0.71	0.25	0.27	0.28	0.07	0.28	0.67	0.60	0.27	0.48
DL-Alanyl-L-alanine	0.61	0.39	0.31	0.55	0.68	0.78	0.31	0.42	0.34	0.10	0.36	0.66	0.71	0.34	0.57
D-Alanyl-D-alanine	0.57	0.36	0.28	0.50	0.65	0.76	0.41	0.40	0.43	0.09	0.31	0.74	0.63	0.32	0.57
α-L-Glutamyl-L-alanine	0.53	0.25	0.25	0.39	0.32	0.48	0.03	0.24	0.10	0.04	0.08	0.63	0.54	0.16	0.28
γ-L-Glutamyl-β-alanine	0.40	0.20	0.19	0.38	0.33	0.42	0.04	0.22	0.21	0.03	0.05	0.66	0.58	0.20	0.28
Glycyl-L-serine	0.27	0.11	0.08	0.15	0.29	0.52	0.10	0.18	0.26	0.03	0.17	0.66	0.45	0.21	0.41
Glycyl-α-aminoisobutyric acid	0.54	0.40	0.31	0.56	0.61	0.80	0.41	0.37	0.42	0.09	0.35	0.69	0.61	0.30	0.56
Glycyl-α-amino-n-butyric acid	0.56	0.39	0.30	0.51	0.62	0.82	0.38	0.41	0.44	0.10	9.34	0.74	0.66	0.32	0.58
L-alanyl-L-α-amino-n-butyric acid	0.71	0.50	0.45	0.61	0.78	0.86	0.53	0.55	0.48	0.12	0.37	0.80	0.72	0.44	0.65
Glycyl-L-threonine	0.35	0.24	0.12	0.20	0.41	0.66	0.16	0.24	0.32	0.04	0.21	0.71	0.50	0.26	0.48

Glycyl-D-valine	0.63	0.43	0.41	0.54	0.64	0.82	0.45	0.46	0.48	0.13	0.43	0.74	0.68	0.43	0.58
Glycyldehydrovaline	0.53	0.40	0.33	0.54	0.65	0.85	0.49	0.29	0.35	0.08	0.32	0.83	0.49	0.29	0.58
Glycyl-D-norvaline	0.64	0.49	0.45	0.54	0.65	0.84	0.51	0.47	0.51	0.13	0.45	0.74	0.65	0.44	0.64
L-alanyl-L-norvaline	0.79	0.60	0.58	0.69	0.84	0.88	0.64	0.62	0.58	0.21	0.49	0.80	0.74	0.53	0.71
Glycyl-L-leucine	0.73	0.61	0.57	0.66	0.70	0.86	0.57	0.52	0.59	0.23	0.54	0.77	0.70	0.55	0.71
Glycyldehydroleucine	0.65	0.58	0.55	0.63	0.70	0.87	0.55	0.43	0.52	0.23	0.54	0.87	0.61	0.48	0.65
α-Glutamyl-L-leucine	0.70	0.64	0.60	0.65	0.52	0.63	0.20	0.68	0.66	0.19	0.20	0.80	0.73	0.35	0.50
Glycyl-L-isoleucine	0.74	0.63	0.55	0.64	0.71	0.86	0.57	0.52	0.58	0.20	0.53	0.80	0.71	0.54	0.70
Glycyl-L-alloisoleucine	0.71	0.59	0.54	0.63	0.72	0.83	0.56	0.53	0.57	0.20	0.54	0.79	0.70	0.52	0.71
Glycyl-L-norleucine	0.72	0.65	0.58	0.65	0.73	0.86	0.60	0.57	0.63	0.21	0.58	0.83	0.69	0.57	0.74
Glycyldehydrophenyl-alanine	0.59	0.52	0.43	0.58	0.67	0.85	0.54	0.36	0.52	0.24	0.56	0.89	0.55	0.55	0.70
DL-Alanyl-DL-phenyl-alanine	0.76	0.62	0.55	0.72	0.76	0.87	0.62	0.52	0.59	0.30	0.56	0.83	0.72	0.60	0.75
α-L-Glutamyl-L-phenyl alanine	0.73	0.62	0.56	0.62	0.51	0.64	0.18	0.35	0.31	0.09	0.20	0.84	0.68	0.37	0.43
L-Cystinyl-L-cystine	0.38	0.32	0.21	0.35	0.46	0.68	0.20	0.34	0.38	0.18	0.42	0.75	0.53	0.43	0.56
L-Cystinyl-D-cystine	0.08	0.11	0.03	0.17	0.08	0.37	0.02	0.02	0.12	0.02	0.08	0.30	0.18	0.09	0.30
D-Cystinyl-L-cystine	0.08	0.10	0.03	0.15	0.16	0.34	0.02	0.00	0.11	0.01	0.15	0.41	0.22	0.09	0.31
Glycyl-D-methionine	0.64	0.41	0.31	0.49	0.71	0.81	0.45	0.38	0.47	0.12	0.41	0.80	0.59	0.46	0.64
Glycyl-L-tryptophan	0.57	0.52	0.42	0.62	0.68	0.83	0.47	0.34	0.43	0.19	0.51	0.73	0.59	0.57	0.65
α-L-Aspartyl-D-histidine	0.12	0.07	0.04	0.24	0.38	0.45	0.05	0.04	0.08	0.01	0.06	0.41	0.27	0.12	0.23
Glycyl-L-lysine-H_2SO_2	0.12	0.05	0.07	0.13	0.35	0.83	0.51	0.07	0.08	0.02	0.13	0.39	0.14	0.09	0.41
Glycyl-L-aspartic acid	0.16	0.12	0.05	0.14	0.35	0.56	0.11	0.12	0.23	0.05	0.12	0.60	0.32	0.18	0.33
Glycyl-L-glutamic acid	0.34	0.31	0.11	0.18	0.17	0.29	0.03	0.14	0.20	0.03	0.07	0.62	0.43	0.14	0.24
α-L-Glutamyl-L-glutamic acid	0.41	0.25	0.21	0.24	0.07	0.10	0.00	0.15	0.17	0.02	0.03	0.63	0.50	0.08	0.18
Glycyl-D-tyrosine	0.50	0.41	0.32	0.41	0.56	0.74	0.27	0.37	0.44	0.09	0.36	0.81	0.59	0.53	0.52
Glycylglycylglycine	0.24	0.15	0.09	0.22	0.44	0.72	0.22	0.18	0.26	0.03	0.21	0.69	0.40	0.21	0.34
D-Alanylglycylglycine	0.39	0.20	0.13	0.33	0.62	0.79	0.34	0.28	0.35	0.03	0.27	0.76	0.54	0.30	0.45

References p. 38

TABLE 8 (continued)

EXPERIMENTAL R_F VALUES OF PEPTIDES AND AMINO ACIDS[58]

(Whatman No. 4 paper)

	A Formix	*B* BuOH-Acetic 4:1:5	*C* BuOH: Acetic 6.7:1:2.5	*D* PBA	*E* Phenol citrate	*F* Phenol H_2O (NaCN + NH_3 atmos.)	*G* m-Cresol (NH_3 atmos.)	*H* 77% EtOH	*I* 70% PrOH	*J* BuOH EtOH	*K* TBK	*L* Acetone-urea	*M* Pyridine mix	*N* Pyridine amyl	*O* Propanol-NH_3
DL-Leucylglycylglycine	0.65	0.50	0.44	0.67	0.82	0.85	0.60	0.59	0.59	0.16	0.54	0.87	0.74	0.57	0.72
Glycyl-DL-alanylglycine	0.35	0.23	0.13	0.36	0.59	0.79	0.33	0.28	0.36	0.04	0.23	0.73	0.52	0.30	0.47
Glycyl-DL-leucylglycine	0.65	0.58	0.47	0.63	0.75	0.88	0.61	0.57	0.60	0.17	0.55	0.85	0.72	0.56	0.73
Glycyl-DL-phenylalanylglycine	0.63	0.47	0.38	0.61	0.76	0.84	0.59	0.44	0.54	0.14	0.49	0.85	0.65	0.59	0.66
Glycylglycyl-L-alanine	0.38	0.22	0.15	0.38	0.61	0.80	0.37	0.28	0.31	0.05	0.27	0.70	0.52	0.28	0.46
Glycylglycyl-DL-leucine	0.65	0.54	0.47	0.66	0.82	0.86	0.67	0.55	0.58	0.16	0.51	0.86	0.72	0.51	0.68
Glycylglycyl-DL-phenylalanine	0.65	0.49	0.41	0.53	0.82	0.88	0.58	0.42	0.51	0.11	0.50	0.85	0.65	0.54	0.64
Glycylglycyl-L-cystine	0.07	0.08	0.07	0.09	0.19	0.52	0.05	0.04	0.15	0.01	0.05	0.51	0.15	0.10	0.22
DL-Alanyl-DL-alanyl-L-cystine	0.16	0.08	0.08	0.16	0.33	0.65	0.09	0.09	0.13	0.01	0.09	0.57	0.29	0.16	0.34
Glycylglycylglycylglycine	0.18	0.12	0.06	0.22	0.58	0.75	0.24	0.17	0.24	0.02	0.16	0.64	0.37	0.21	0.32
Glycine	0.41	0.17	0.13	0.22	0.40	0.50	0.14	0.30	0.29	0.08	0.17	0.54	0.41	0.24	0.38
Sarcosine	0.47	0.18	0.09	0.52	0.77	0.77	0.52	0.42	0.35	0.15	0.25	0.69	0.56	0.36	0.56
Alanine	0.51	0.28	0.20	0.39	0.56	0.60	0.28	0.40	0.36	0.13	0.24	0.66	0.59	0.36	0.50
α-Amino-*n*-butyric acid	0.69	0.34	0.30	0.48	0.71	0.75	0.47	0.51	0.46	0.18	0.29	0.68	0.65	0.41	0.54
Norvaline	0.80	0.49	0.44	0.62	0.79	0.81	0.60	0.61	0.60	0.29	0.39	0.74	0.71	0.51	0.64
δ-Aminovaleric acid	0.65	0.46	0.35	0.62	0.81	0.81	0.59	0.48	0.38	0.11	0.20	0.62	0.63	0.28	0.47
Norleucuie	0.87	0.63	0.58	0.70	0.83	0.85	0.71	0.67	0.74	0.43	0.52	0.80	0.75	0.61	0.72
α-amino-*n*-heptylic acid	0.90	0.71	0.68	0.77	0.85	0.88	0.77	0.70	0.83	0.54	0.62	0.85	0.77	0.68	0.83

Valine	0.73	0.45	0.37	0.58	0.75	0.78	0.57	0.58	0.60	0.24	0.38	0.74	0.73	0.46	0.67
Leucine	0.78	0.61	0.59	0.70	0.78	0.82	0.67	0.61	0.61	0.41	0.51	0.78	0.75	0.61	0.74
Isoleucine	0.82	0.59	0.52	0.68	0.81	0.85	0.68	0.64	0.68	0.37	0.47	0.80	0.77	0.56	0.73
Serine	0.37	0.16	0.12	0.19	0.32	0.45	0.10	0.32	0.29	0.08	0.28	0.60	0.44	0.28	0.39
Threonine	0.44	0.17	0.15	0.26	0.48	0.54	0.21	0.42	0.38	0.13	0.48	0.71	0.58	0.38	0.47
Phenylalanine	0.76	0.53	0.49	0.68	0.84	0.85	0.72	0.57	0.61	0.34	0.54	0.77	0.69	0.63	0.73
Tyrosine	0.59	0.24	0.30	0.35	0.59	0.64	0.31	0.45	0.44	0.24	0.34	0.79	0.61	0.59	0.56
Diiodotyrosine	0.68	0.62	0.59	0.67	0.76	0.63	0.36	0.51	0.50	0.34	0.31	0.77	0.64	0.66	0.42
Tryptophan	0.59	0.43	0.35	0.50	0.70	0.75	0.53	0.39	0.46	0.17	0.51	0.65	0.47	0.55	0.64
Proline	0.58	0.27	0.23	0.65	0.88	0.84	0.75	0.49	0.42	0.13	0.25	0.69	0.62	0.37	0.53
Hydroxyproline	0.39	0.17	0.13	0.36	0.66	0.68	0.42	0.40	0.33	0.08	0.18	0.61	0.52	0.32	0.42
Histidine	0.16	0.10	0.08	0.37	0.61	0.72	0.42	0.17	0.23	0.13	0.22	0.51	0.38	0.27	0.43
Arginine	0.25	0.10	0.08	0.34	0.59	0.93	0.61	0.12	0.19	0.05	0.05	0.18	0.17	0.17	0.36
Ornithine	0.16	0.09	0.07	0.21	0.37	0.76	0.40	0.14	0.13	0.04	0.09	0.18	0.13	0.15	0.32
Lysine	0.22	0.08	0.06	0.25	0.54	0.87	0.46	0.12	0.16	0.03	0.10	0.19	0.20	0.14	0.36
Aspartic acid	0.31	0.16	0.09	0.10	0.08	0.19	0.00	0.20	0.13	0.02	0.05	0.53	0.44	0.13	0.19
Glutamic acid	0.42	0.17	0.17	0.24	0.14	0.26	0.03	0.22	0.20	0.03	0.05	0.64	0.54	0.17	0.25
Methionine	0.69	0.44	0.36	0.57	0.77	0.77	0.82	0.56	0.51	0.25	0.39	0.73	0.68	0.52	0.65
Cysteine-HCl	0.07	0.08	0.03	0.08	0.25	0.32	0.06	0.12	0.13	0.02	0.05	0.31	0.22	0.22	0.25
Cystine	0.07	0.05	0.03	0.06	0.18	0.39	0.05	0.11	0.11	0.00	0.05	0.30	0.30	0.11	0.19

References p. 38

and those in which the terminal amino group is acylated or substituted, *e.g.* N-acetyl, N-methyl, give very little or no colour with ninhydrin. In these cases exposure to chlorine gas to form the labile N—Cl compounds has been proposed by Rydon and Smith[63]. The paper is made into a cylinder and dropped into a tall measuring jar full of the gas. It is taken out after 30 min and placed in a current of dry air. After removal of all the chlorine gas in this way the paper is sprayed with a 1% aqueous solution of soluble starch containing 1% potassium iodide. A better background is obtained if the paper can be left overnight in a large desiccator containing sodium hydroxide pellets. With care the method can be very sensitive and will detect peptides of any chain length.

O-Tolidine (saturated solution in 2% acetic acid, 1 part) and potassium iodide (0.05 *N* solution, 1 part) may be used instead of starch-iodide[64]. The dried paper is sprayed with 0.05 *M* sodium tetraborate solution, exposed to chlorine vapour, and then waved in the air for 2–3 min. A corner of the paper is tested with the tolidine iodide reagent for background which can be reduced, if excessive, by fleeting exposure to ammonia vapour[65,66]. A 1% solution of *tert.*-butyl hypochlorite[67] in cyclohexane is a convenient substitute for chlorine gas. Papers sprayed with the reagent are dried and aerated at room temperature for 1 h. Application of starch-iodide reveals the peptides as blue-black spots against a white or light blue background. 0.1 μg of leucylglycine in an area of 1 cm^2 is easily detected.

Thin layer chromatography

A discussion of paper chromatography would be incomplete without some reference to the important new method of analysis whereby separations are accomplished in thin layers of adsorbents spread on glass plates. Although the principle of thin layer chromatography was described as early as 1938 by Izmailov and Shraiber[68] the practical problem of producing very thin layers (250 μ thick) was not solved until Stahl[69] produced a suitable device in 1958. A spreader with ancillary equipment is available from Desaga GmbH, Heidelberg (Agents in United Kingdom, Camlab Ltd., Cambridge). Stahl[70] has given a comprehensive account of the subject and Randerath[71] has provided a laboratory manual of great practical value.

Attempts to exploit the method for separations of organic compounds utilizing thin layers of powdered cellulose[72–75] have met with limited success but microcrystalline cellulose ('Avirin', American Viscose Co., Marcus Hook, Penn.) is a promising new material. In a preliminary study by Wolfrom *et al.*[76] microcrystalline cellulose (100 g) and water (430 ml) were blended for 30–45 sec and the slurry layered on glass plates (20 × 20 × 0.4 cm) by means of a **Desaga** applicator at the optimum thickness of 1.0 mm. After drying the plates overnight or at 80° a fine film was obtained which adhered strongly to the glass. Good separations of amino acids and sugars were achieved in three hours or less of development time. In the solvent system butanol–acetic acid–water (3:1:1) the R_F values for valine, isoleucine and leucine were 0.51, 0.61 and 0.64 respectively.

Amino acids can be separated by adsorption chromatography in thin layers of Silica Gel G 'Merck'. The plates must be air-dried

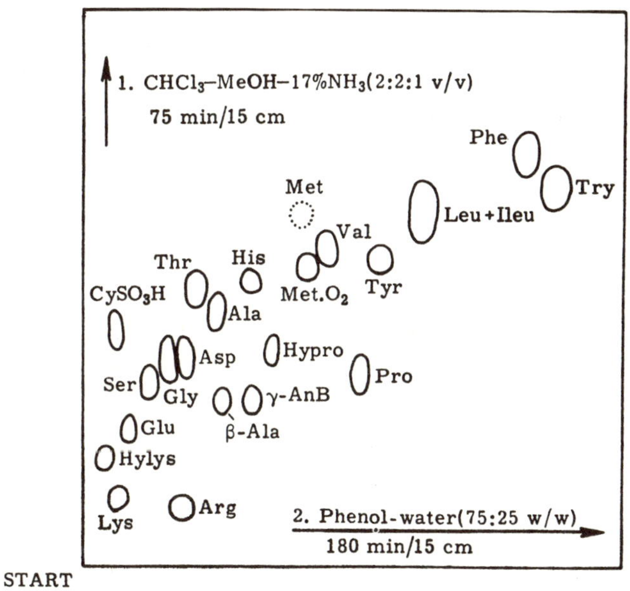

Fig. 12. Two-dimensional separation of a mixture of 20 amino acids + β-alanine + γ-amino-*n*-butyric acid (γ-AB) oxidized with performic acid. Load per amino acid: 0.5 μg in a total of 0.5 μl of 0.1 *N* HCl. Fahmy *et al.*[77].

References p. 38

and best results are obtained with 1–5 μg quantities of each amino acid. Fig. 12 shows a two-dimensional separation of a mixture of 22 amino acids obtained by Fahmy *et al.*[77]. Oxidation of the sample with performic acid is advisable otherwise cysteine and cystine give several secondary spots due to partial oxidation on the medium. The solvent systems used were chloroform–methanol–17% aqueous ammonia (2:2:1 v/v) and, in the second direction, phenol–water (75:25 w/w).

REFERENCES

1 R. Consden, A. H. Gordon and A. J. P. Martin, *Biochem. J.*, 38 (1944) 224.
2 C. E. Dent, *Biochem. J.*, 43 (1948) 169.
3 G. L. Kowkabany and H. G. Cassidy, *Anal. Chem.*, 22 (1950) 817.
4 R. Consden and A. H. Gordon, *Nature*, 162 (1948) 180.
5 R. G. Westall, *Biochem. J.*, 42 (1949) 249.
6 E. F. McFarren, *Anal. Chem.*, 23 (1951) 168.
7 C. S. Hanes, *Can. J. Biochem. Physiol.*, 39 (1961) 119.
8 E. H. M. Wade, A. T. Matheson and C. S. Hanes, *Can. J. Biochem. Physiol.*, 39 (1961) 141.
9 C. S. Hanes, C. K. Harris, M. A. Moscarello and E. Tigane, *Can. J. Biochem. Physiol.*, 39 (1961) 163.
10 A. T. Matheson, E. Tigane and C. S. Hanes, *Can. J. Biochem. Physiol.*, 39 (1961) 417.
11 E. Tigane, E. H. M. Wade, J. T. Wong and C. S. Hanes, *Can. J. Biochem. Physiol.*, 39 (1961) 427.
12 C. K. Harris, E. Tigane and C. S. Hanes, *Can. J. Biochem. Physiol.*, 39 (1961) 439.
13 O. J. Draper and A. L. Pollard, *Science*, 109 (1949) 448.
14 S. M. Partridge, *Biochem. J.*, 42 (1948) 238.
15 A. L. Levy and D. Chung, *Anal. Chem.*, 25 (1953) 396.
16 L. B. Rockland and J. C. Underwood, *Anal. Chem.*, 26 (1954) 1557.
17 K. Dus, R. G. Bartsch and M. D. Kamen, *J. Biol. Chem.*, 237 (1962) 3083.
18 R. R. Redfield and E. S. Guzman Barron, *Arch. Biochem. Biophys.*, 35 (1952) 443.
19 L. Fowden, *Biochem. J.*, 48 (1951) 327.
20 L. Fowden, *Biochem. J.*, 50 (1952) 355.
21 J. F. Roland and A. M. Gross, *Anal. Chem.*, 26 (1954) 502.
22 B. Keil, in: A. Neuberger (Ed.), *Symposium on Protein Structure*, Methuen & Co. Ltd., London, 1958, p. 90.
23 I. Dubrovskaia, N. N. Ostrovskaia and A. E. Flubokina, *Biokhimiya*, 23 (1958) 489.
24 N. Subramanian and M. V. L. Rao, *J. Sci. Ind. Res.* (*India*), 14C (1955) 56.
25 A. E. Pasieka, H. J. Morton and J. F. Morgan, *J. Natl. Cancer Inst.*, 16 (1956) 995.

[26] A. E. Pasieka, H. J. Morton and J. F. Morgan, *Can. J. Biochem. Physiol.*, 38 (1958) 171.
[27] R. J. Block and D. Bolling, in: *Amino Acid Composition of Proteins and Foods*, 2nd Ed., C. C. Thomas, Springfield, Illinois, 1950.
[28] R. J. Block, *Science*, 108 (1948) 608.
[29] W. H. Stein and S. Moore, *J. Biol. Chem.*, 178 (1949) 79.
[30] G. E. Connell, G. H. Dixon and C. S. Hanes, *Can. J. Biochem. Physiol.* 33 (1955) 416.
[31] D. O. Gray, J. Blake, D. H. Brown and L. Fowden, *J. Chromatog.*, 13 (1964) 276.
[32] H. B. Vickery, *J. Org. Chem.*, 28 (1963) 291.
[33] T. L. Hardy, D. O. Holland and J. H. C. Nayler, *Anal. Chem.*, 27 (1955) 971.
[34] E. Kawerau and T. Wieland, *Nature*, 168 (1951) 77.
[35] A. J. Woiwod, *Nature*, 166 (1950) 272.
[36] J. B. Jepson and I. Smith, *Nature*, 172 (1953) 1100.
[37] T. Okuyama and K. Satake, *J. Biochem. (Tokyo)*, 47 (1960) 454.
[38] C. F. Baxter and I. Senoner, *Anal. Biochem.* 7 (1964) 55.
[39] W. Grassmann and K. V. Arnim, *Ann.*, 519 (1935) 192.
[40] R. Acher, C. Fromageot and M. Jutisz, *Biochim. Biophys. Acta*, 5 (1950) 81.
[41] J. Barrollier, J. Heilman and E. Watzke, *Z. Physiol. Chem.*, 304 (1956) 21.
[42] R. Monier and M. Jutisz, *Biochim. Biophys. Acta*, 14 (1954) 551.
[43] R. Acher and C. Crocker, *Biochim. Biophys. Acta*, 9 (1952) 704.
[44] K. K. Reddi and E. Kodicek, *Biochem. J.*, 53 (1953) 286.
[45] F. Sanger and H. Tuppy, *Biochem. J.*, 49 (1951) 463.
[46] L. A. Ae. Sluyterman, *Biochim. Biophys. Acta*, 38 (1960) 218.
[47] P. M. Newman and J. H. Turnbull, *Biochem. J.*, 74 (1960) 379.
[48] Kazuo Satake and J. Murray Luck, *Bull. Soc. Chim. Biol.*, 40 (1958) 1743.
[49] G. Curzon, *Nature*, 173 (1954) 314.
[50] M. B. van Halteren, *Nature*, 168 (1951) 1090.
[51] F. M. Robbins and J. A. Fioriti, *Nature*, 200 (1963) 577.
[52] E. Friedmann, D. H. Marrian and I. Simon-Reuss, *Brit. J. Pharmacol.*, 4 (1949) 105.
[53] C. S. Hanes, F. J. R. Hird and F. A. Isherwood, *Nature*, 166 (1950) 288.
[54] A. Witter and H. Tuppy, *Biochim. Biophys. Acta*, 45 (1960) 429.
[55] E. D. Moffat and R. I. Lytle, *Anal. Chem.* 31 (1959) 926.
[56] R. Circo and B. A. Freeman, *Anal. Chem.* 35 (1963) 262.
[57] C. A. Knight, *J. Biol. Chem.*, 190 (1951) 753.
[58] T. B. Moore and C. G. Baker, *J. Chromatog.*, 1 (1958) 513.
[59] A. B. Pardee, *J. Biol. Chem.*, 190 (1951) 757.
[60] A. J. P. Martin and R. L. M. Synge, *Biochem. J.*, 35 (1941) 1358.
[61] A. J. P. Martin, *Ann. N.Y. Acad. Sci.*, 49 (1948) 249.
[62] C. Milstein and F. Sanger, *Biochem. J.*, 79 (1961) 456.
[63] H. N. Rydon and P. W. G. Smith, *Nature*, 169 (1952) 922.
[64] F. Reindel and W. Hoppe, *Chem. Ber.* 87 (1954) 1103.
[65] H. Zahn and E. Rexroth, *Z. Anal. Chem.* 148 (1955) 181.
[66] J. O'Donnell and E. O. P. Thompson, *Australian J. Biol. Sci.*, 17 (1964) 271.

[67] R. H. Mazur, B. W. Ellis and P. S. Cammarata, *J. Biol. Chem.*, 237 (1962) 1619.
[68] N. A. Izmailov and M. S. Shraiber, *Farmatsiya (Sofia)*, 3 (1938) 1.
[69] E. Stahl, *Chemiker Ztg.*, 82 (1958) 323.
[70] E. Stahl, *Dünnschicht-Chromatographie*, Springer Verlag, Berlin, 1962.
[71] K. Randerath, *Thin-layer Chromatography* (translated by D. D. Libman), Academic Press, New York, 1963.
[72] K. Randerath and H. Struck, *J. Chromatog.*, 6 (1961) 365.
[73] K. Randerath, *Biochem. Biophys. Res. Commun.*, 6 (1961) 452.
[74] A. Schweiger, *J. Chromatog.*, 9 (1962) 374.
[75] P. Wollenweber, *J. Chromatog.*, 9 (1962) 369.
[76] M. L. Wolfrom, D. L. Patin and R. M. de Lederkremer, *Chem. Ind. (London)*, (1964) 1065. See also: J. G. Heathcote and K. Jones, *Biochem. J.*, 97 (1965) 15P.
[77] A. R. Fahmy, A. Niederwasser, G. Pataki and M. Brenner, *Helv. Chim. Acta*, 44 (1961) 2022.

Chapter 2

HIGH-VOLTAGE PAPER ELECTROPHORESIS

The simple but elegant technique of high-voltage electrophoresis developed from the work of Wieland and Fischer in 1948[1]. Although some useful separations had been obtained using a potential gradient of 5–20 V/cm it became clear that the conditions were quite unsatisfactory for resolution of a complex mixture of amino acids or peptides. The high diffusion rates of low molecular weight compounds set a limit to the analytical value of the method and workers began to experiment with potential gradients of 50–200 V/cm. The dramatic sharpening of the zones at these voltages may be seen in the illustrations on pages 51 and 52.

APPARATUS

Separations are carried out with the paper strip held vertically in a large volume of organic solvent or placed horizontally between two cooled metal or glass plates.

Vertical strip method of Michl[2]

The apparatus (Fig. 13) consists very simply of a rectangular or cylindrical tank, A, containing at the bottom a slurry of buffer and cellulose powder, and a small upper electrode vessel, B, also containing buffer slurry. The electrodes, E, are platinum wires sealed into glass tubes which are open at the lower end and provided with a terminal at the upper end. The paper strip is moistened with buffer solution and partially dried between filter papers until the ratio of the weight of air-dried strip to the weight of retained buffer is about 1:1. Electrophoretic separations are carried out with the paper hanging between the two electrode chambers and the whole system filled with an inert organic liquid to conduct the heat away from the strip.

In the modification of Ryle *et al.*[3] (Fig. 14) a tall buffer chamber rests in the bottom of the other larger vessel and the coolant is toluene. The electrodes are platinum wires arranged as in the original

References p. 60

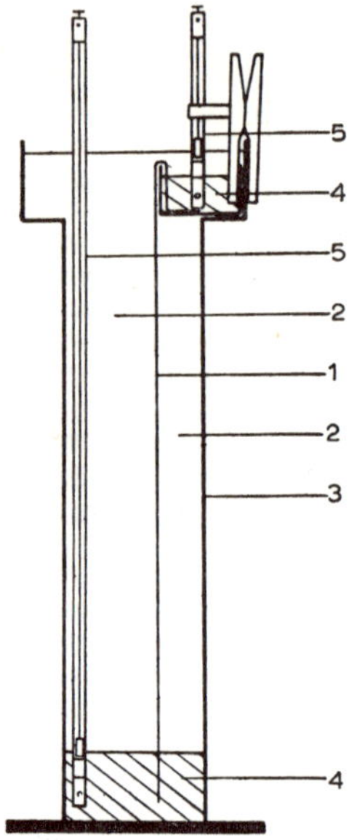

Fig. 13. High-voltage electrophoresis apparatus of Michl[4]. 1, Filter paper strip; 2, organic solvent; 3, glass vessel; 4, buffer slurry with cellulose paper; 5, electrodes (Ag/AgCl); 6, clip.

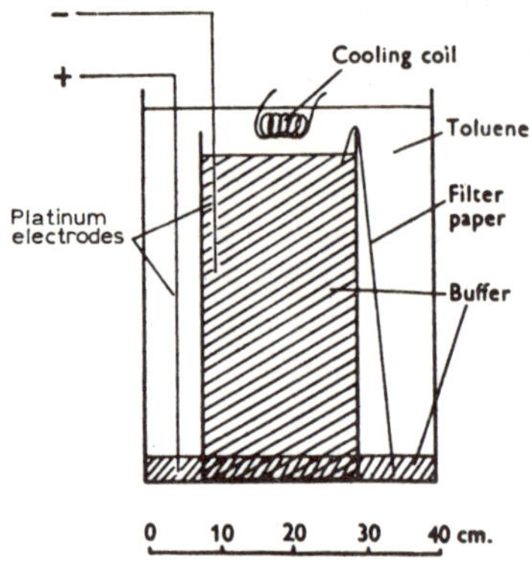

Fig. 14. Michl apparatus modified by Ryle *et al.*[3]

Michl apparatus. Suitably shaped glass rods prevent the paper from touching the sides of vessels and a cooling coil immersed in the toluene is necessary at high currents. In the arrangement where the buffer solution in the cathode vessel is higher than that in the anode vessel the electroendosmotic flow is to a certain extent balanced by the gravitational flow. The sample is applied as a thin line to within 1 cm of each edge and allowed to dry. The paper, which can be up to 20 cm wide, is then laid on a glass plate with a glass rod underneath the line of application. It is wetted with buffer to within about 1 cm of the sample and the buffer allowed to flow in slowly from both sides thus concentrating the zone. Excess buffer is removed by blotting between filter papers before placing the strip in the apparatus. A simple apparatus constructed from Perspex for runs carried out in Varsol (Esso White Spirit 100) has been described by Katz *et al.*[5]. An efficient cooling system is used.

Ideally the solvent should be completely immiscible with both the sample and the buffer used. Among those that have been recommended are toluene, cumene, heptane, chlorobenzene, white spirit and carbon tetrachloride. Toluene appears to be the solvent of choice for many workers, its only drawbacks being high inflammability and high toxicity. The equipment in this case must be assembled in a fume-cupboard and strict attention paid to the very real fire hazard. Electrical connections must be rigid and secure to avoid sparking, and switching operations done away from the apparatus. When non-inflammable solvents are used it is still necessary to ensure that electrical connections are adequate and the connecting cables heavily insulated. Some relevant physical properties of solvents compiled by Michl[4] are given in Table 9. During a run convection currents transfer the generated heat from the strip to the walls of the tank and if necessary additional external cooling can be applied. It is possible that 0.2 W/cm^2 of the area of the strip can be removed in this way.

TABLE 9

PROPERTIES OF SOLVENTS

Compound	*Thermal conductivity in cal.cm^{-1} degree^{-1}sec^{-1}*	*Solubility in water (g/100 ml)*	*Vapour pressure in mm Hg at 20°C*	*Explosion limits in mixture with air*
Chlorobenzene	0.00030	0.049 (30°)	9	—
Heptane	0.00034	0.0052 (16°)	35	1–6%
Carbon tetrachloride	0.00028	0.077 (25°)	91	—
Toluene	0.00036	0.063 (25°)	22	1.3–7%

Strip between cooled plates

The need for more efficient cooling prompted the design of apparatus in which the paper strip is laid between cooled, electrically insulated plates. The heat dissipating capacity of water cooled plates is about ten times that of liquid heat exchangers. In the apparatus of Michl[4,6] (Fig. 15), the paper strip is laid on a glass plate, which in turn is placed on a water-cooled plate. A heavy glass plate covers the strip, the ends of which are connected by means of cellophane

References p. 60

membranes and a filter paper wick to the buffer container. The function of the cellophane membrane is to prevent too much imbibition of buffer into the partially dried paper strip. The buffer container is connected to the electrode chamber by means of an inverted tube plugged at both ends with glass wool or cellulose.

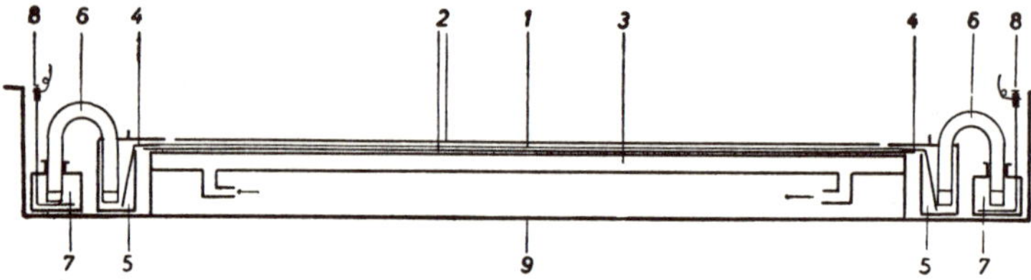

Fig. 15. Apparatus of Michl with water-cooled glass plate. 1, Filter paper strip; 2, glass plates; 3, condenser; 4, cellophane membrane; 5, buffer container; 6, bridge; 7, electrode chamber; 8, electrode ($Cu/CuCl_2$); 9, case.

Dissipation of 1 W/cm^2 of filter paper is possible in the apparatus designed by Gross[7,8] (Fig. 16A and B). The cooling plates are made of aluminium alloy, $NP_5/6M$ (British Aluminium Co. Ltd.) and their overall dimensions are 22 × 13 × 1 in. Deep grooves ¼ in. wide and ⅜ in. deep form the channelling system permitting water flow of 12 litre/min. The channelling in the two plates is arranged counter-current fashion to smooth out slight temperature differences (Fig. 17). It is highly important to remove such local gradients as a difference of 1° causes a 3% change in migration velocity[7]. The surface of the plates is precision-ground to a tolerance of 0.001 in. over a whole area. A pressure control device rests on top of the upper plate and consists of an inflatable plastic air cushion approximately 20 × 13 in. (Pneumatic Tent Co. Ltd., Dorking, England) contained in a casing made of steel sheet, ¼ in. thick, and joined through an inlet tube to a compressed air supply. An important feature of the technique is the use of thin polythene sheeting (0.01 in.) as the insulating material. One or two layers will usually give adequate insulation and at the same time provide sufficient heat conduction. Occasional breakdown of insulation will occur at very high voltages in the form of pinholes but experience in the use of the apparatus is the best guide to spotting this trouble.

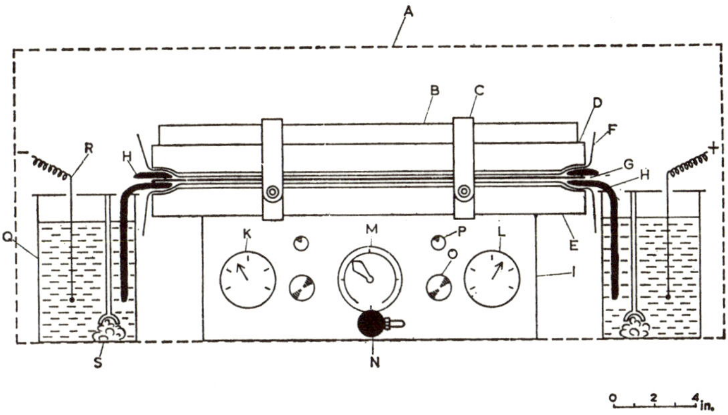

Fig. 16A. Apparatus of Gross with cooled metal plates[8]. A, safety cage; B, pneumatic pressure device; C, retaining bracket; D, top cooling plate; E, bottom cooling plate; F, insulating film; G, paper strip; H, thick paper pad; I, instrument panel; K, air pressure gauge; L, water flow gauge; M, timing device; N, knob for plate locking device; O, pressure reducing valve; P, neon indicator light; Q, buffer vessel; R, platinum electrode; S, cotton wool plug.

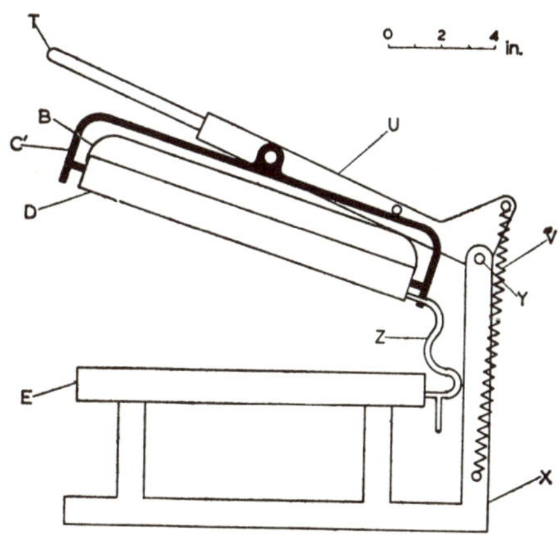

Fig. 16B. Diagrammatic side view of the lifting device. T, handle; U, lever; V, steel spring; X, stand; Y, pivoting pin; Z, heavy rubber tubing; C′, levelling bracket.

References p. 60

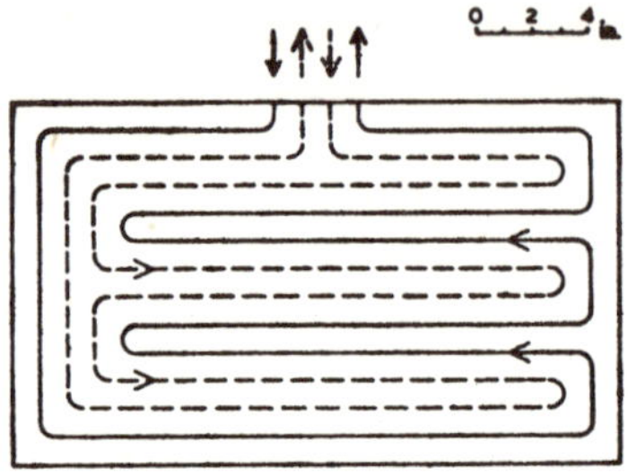

Fig. 17. Channelling system in cooling plate; arrows indicate direction of flow.[8]

Contact between the paper strip and the buffer compartment is made by means of pads of wet filter paper enclosed in cellophane sleeves. The buffer vessels are constructed of $\frac{1}{4}$ in. polythene sheet and hold about 4.5 l of electrolyte solution. Communication between the two halves is afforded by a narrow horizontal slit, $\frac{1}{2}$ in. wide, plugged with cotton wool. The electrodes are 6 in. long platinum wire, 18 SWG. This type of apparatus is available commercially (Locarte Co., London; Shandon Scientific Co., London).

The circuit diagram of a power pack suitable for use with this apparatus is shown in Fig. 18. It is capable of supplying 500 mA at 5 kV although current consumption in excess of 350 mA presents difficulties in heat dissipation.

Fig. 19 shows the circuit diagram of another power unit designed specifically for the vertical strip, toluene coolant apparatus. The unit will supply up to 2500 V at 50 mA and incorporates a mercury switch safety device. The latter is incorporated into the lid of the wire-mesh box containing the tank assembly and is connected into the primary circuit of the supply. High voltage cannot be applied while the lid is open and when the supply is switched on the delay circuit takes 30 sec to close before it is possible to switch on the E.H.T.

SEPARATION OF AMINO ACIDS

Useful amino acid separations have been achieved with both types of apparatus. The promising results of Kickhöfen and Westphal[9] in 1952 was a stimulus to other German workers and led to the extensive investigations of Heilmeyer *et al.*[10] on the non-protein ni-

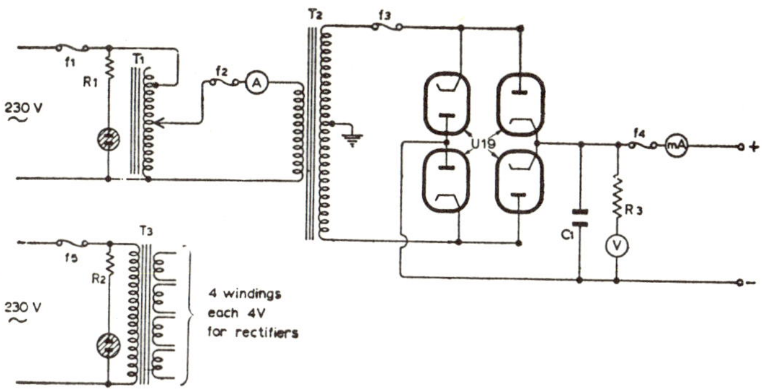

Fig. 18. Circuit diagram for a 5 kV power unit[8]. Meters: A = 0–10A, A.C. 50$\sim$; mA = 0–500 mA, m/c, V = 0–5000. Resistances: R_1 = 220 kΩ; R_2 = 220 kΩ; R_3 = 5 MΩ 10 W. Transformers: T_1 = 100 R Variac; T_2 = 230 V in, 2250 0 2250 V, 500 mA out; T_3 = 230 V in, 4 V 4 A, four times, high insulation. Condenser: C_1 = 3 mF, 8000 V, D.C. Fuses: f_1 and f_2 = 10 A; f_3 = 500 mA; f_4 = 500 mA; f_5 = 1 A.

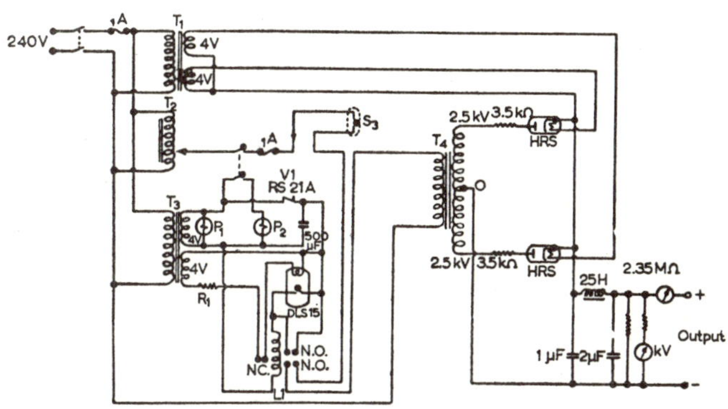

Fig. 19. Circuit diagram for a 2.5 kV, 50 mA power unit. Designed by J. S. Chilvers, Institute of Animal Physiology, Babraham, Cambs. Tl, Rectifier filament transformer insulated to 2,500 V; T2, Variable transformer controlling output voltage (Berco Regavolt, 0–250 V); T3, transformer supplying delay switch DLS 15 and relay L1; T4, E.H.T. transformer, 2,500 V 50 mA; R1, resistor hand wound to give 30 sec delay with DLS 15; L1, 50-Ohm relay 2 pairs contacts normally opened (N.O.), one pair contacts normally closed (N.C.); P1, pilot lamp indicating mains switching; P2, pilot lamp indicating E.H.T. switching; S3, mercury tilting switch incorporated in case lid.

References p. 60

trogen of human serum. There followed the interesting qualitative separation of amino acids from protein hydrolysates by Spengler and Knedel[11]. In these experiments a liquid heat exchanger was used and the results (strips A and B) are compared in Fig. 20 with those obtained in the horizontal plate apparatus of Gross[12] (strips C and D). The results of Spengler and Knedel differ considerably from those of Gross and other workers; the separations shown in Fig. 22 are in fact those usually obtained in laboratories today. The effect of a buffer at high pH is shown in Fig. 21 (strip E). The conditions for the separations are as follows:

A. Pyridine–acetic acid–water buffer, pH 3.6; Whatman No. 1 paper 4.5 × 60 cm; 5600 V during 50 min at 2°; hexane coolant. 200–250 μg of protein hydrolysate.

B. Formic acid, pH 1.9. Other conditions as in A.

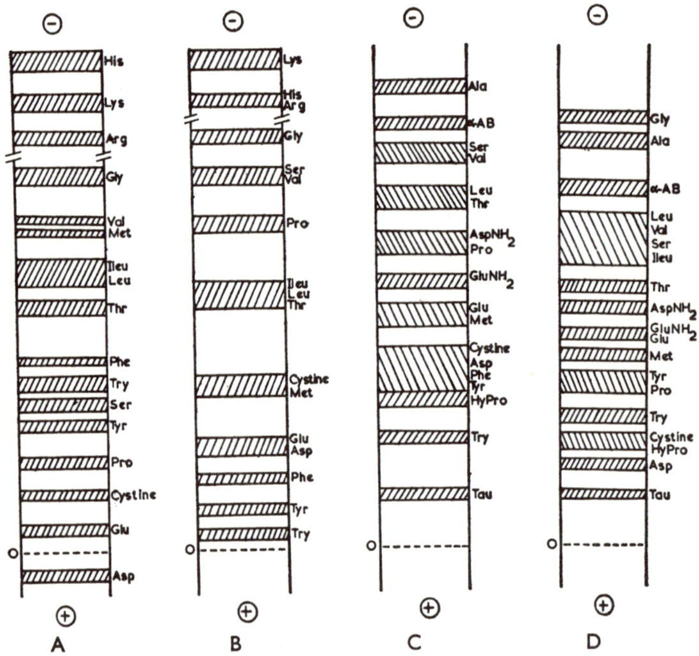

Fig. 20. Qualitative separations of amino acids by electrophoresis. A and B at pH 3.6 and pH 1.9 using hexane heat exchanger[11]. C and D at pH 2.0 and pH 2.9 using cooled metal plates. (Diagrams constructed from results of Gross[12].)

C. Whatman No. 3 MM paper, 4 × 22½ in.; 1.5 *M* formic acid + 2 *M* acetic acid (1:1) buffer, pH 2.0; 90 V/cm, 6 mA/cm, 1¼ h; 2–10 μg of each amino acid. Plate cooling with tap water.
D. Whatman No. 3 MM paper, 5 × 22½ in.; acetate buffer, pH 2.9 (454 ml 2 *M* acetic acid, 6 ml 2 *M* sodium acetate, 540 ml water); 110 V/cm, 2.4 mA/cm, 1½ h.
E. 0.05 *M* sodium borate buffer, pH 9.2; 90 V/cm, 8 mA/cm, 1 h.

Until quite recently separations of amino acids have been on a qualitative basis. There seems little doubt now, however, that fairly

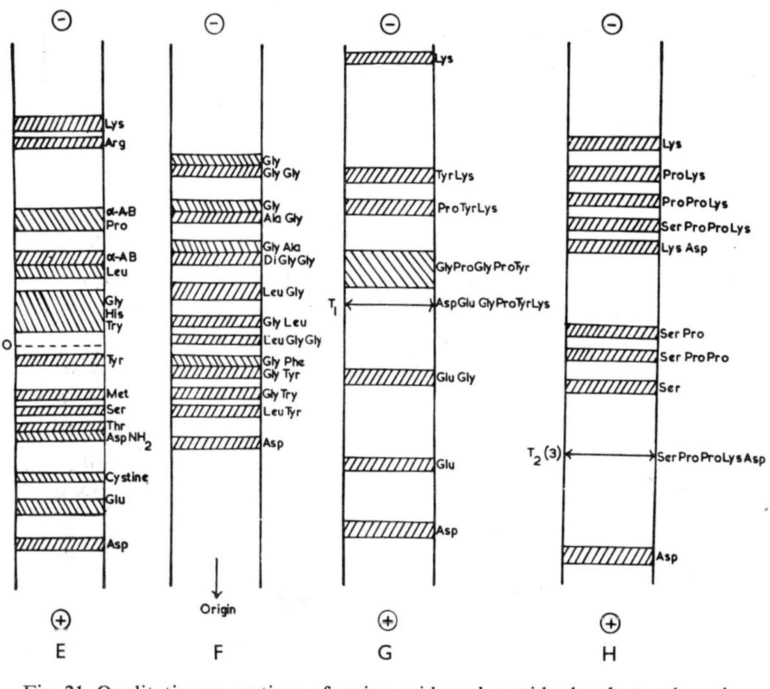

Fig. 21. Qualitative separations of amino acids and peptides by electrophoresis. E, separation of amino acids at pH 9.2 (diagram constructed from results of Gross[12]); F, separation of peptides at pH 2.0 (constructed from results of Gross[12]); G, partial acid hydrolysate of peptide Tl (12 *N* HCl, 37°, 72 h) derived from β-melanocyte-stimulating hormone, run at pH 6.5 (Harris and Roos[21]); H, partial acid hydrolysate of peptide T2 (3) (12 *N* HCl, 100°, 1 h) derived from β-melanocyte-stimulating hormone, run at pH 3.5 (Harris and Roos[21]).

References p. 60

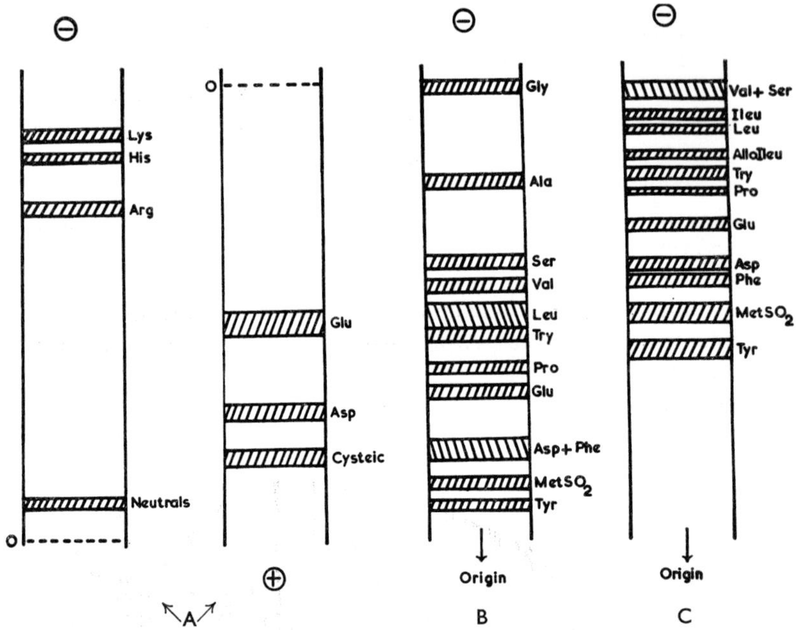

Fig. 22. Quantitative separations of amino acids by electrophoresis. A, pH 5.2; B, pH 1.85; C, pH 1.81 (from results of Atfield and Morris[13]).

quantitative results can be expected from high voltage paper electrophoresis. Preliminary investigations by Atfield and Morris[13] have already indicated that recoveries of amino acids from a hydrolysate of a performic acid oxidized protein (200 μg) are mostly within $\pm 5\%$ and always within $\pm 10\%$. The essential part of the apparatus used consists of two metal plates, 15×105 cm maintained at $2 \pm 1°$; during 8 hours operation the moisture content of the strip was 92 $\pm 5\%$ at 300 W dissipation. Three separate electrophoretic runs are required to separate 18 amino acids; a diagrammatic representation of the developed strips are shown in Fig. 22. The experimental conditions are as follows:

A. Pyridine–acetate buffer, pH 5.2 (2% pyridine in 1% aqueous acetic acid), Whatman 3 MM paper, 75 V/cm for 200 min. The two halves of the run are illustrated. The neutral amino acids are displaced only slightly form the origin.

B. Formic–acetic acid, pH 1.85 (2.5% formic acid, 7.8% acetic acid), 100 V/cm for 270 min. Phenylalanine + aspartic acid, leucine + isoleucine, unresolved pairs.
C. Cadmium–formic–acetic acid, pH 1.81 (2% formic acid, 20% acetic acid, 2–4 m*M* cadmium acetate), 100 V/cm for 360 min. Valine + serine, unresolved pair.

After electrophoresis the strips are dried for 15 min at 80°, dipped in a cadmium–ninhydrin reagent (based on that of Heilmann *et al.*[14]) allowed to air-dry and stored for 18–20 h in a dry ammonia-free atmosphere. The red pigment is eluted with sulphur-free methanol and the extinction values read at 500 mμ (325 mμ for proline). The reagent of Heilmann is prepared from cadmium acetate (100 mg), water (10 ml), acetic acid (5 ml), acetone (100 ml) and ninhydrin (1 g) in this order.

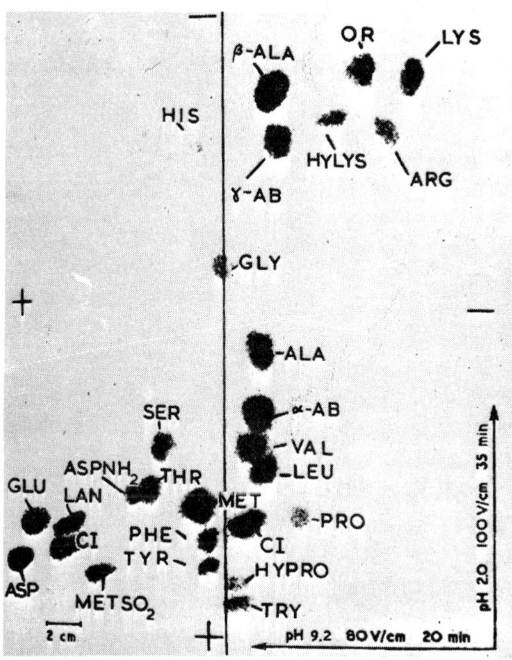

Fig. 23. Two-dimensional separation of amino acids by electrophoresis (from Gross[15]).

References p. 60

Blackburn and Lee[16] have drawn attention to the probability that the mobilities of the various amino acids have different temperature coefficients. With a formic acid–acetic acid buffer, pH 1.85, optimum separations were achieved when the temperature of the water emerging from the plates was 11°. In the apparatus of Pasieka[17] a compressor and expansion valve utilizing direct expansion of Freon 12 is used to control the temperature of the lower aluminium plate.

From an examination of E and C in Figs. 20 and 21 it would appear possible to gain greater resolution from electrophoresis in two directions at two different pH values. If this is done the results are comparable to those obtained by two-dimensional paper chromatography. Figs. 23 and 24 show the patterns obtained by Gross[15]. An initial run at pH 2.0 followed by a short run at pH 9.2 gave the

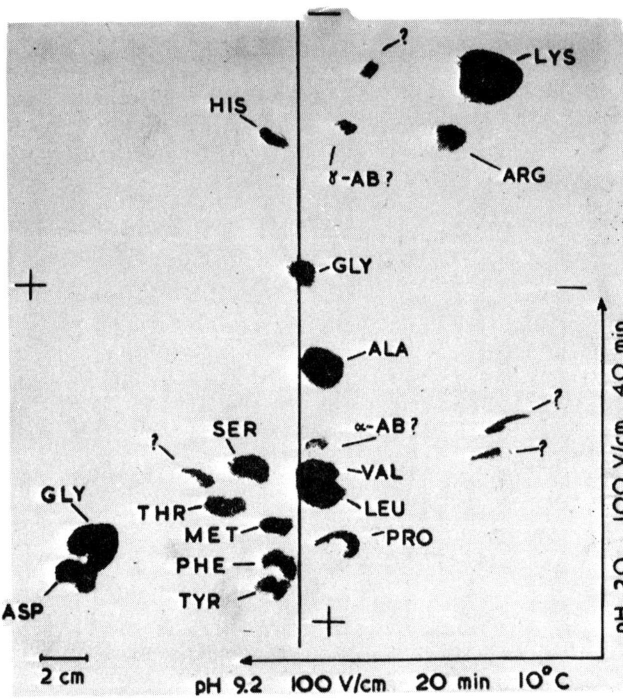

Fig. 24. Two-dimensional separation of amino acids in a casein hydrolysate (from Gross[15]).

remarkable achievement of 28 amino acids completely resolved in less than an hour. After the first run in 0.75 *M* formic acid the paper is dried for 10 min at 90° and for 1 h in a stream of cold air to remove most of the formic acid, trimmed and sprayed lightly from an atomizer with a 0.05 *M* sodium borate solution of pH 9.2. The paper is then ready for the run in the second dimension.

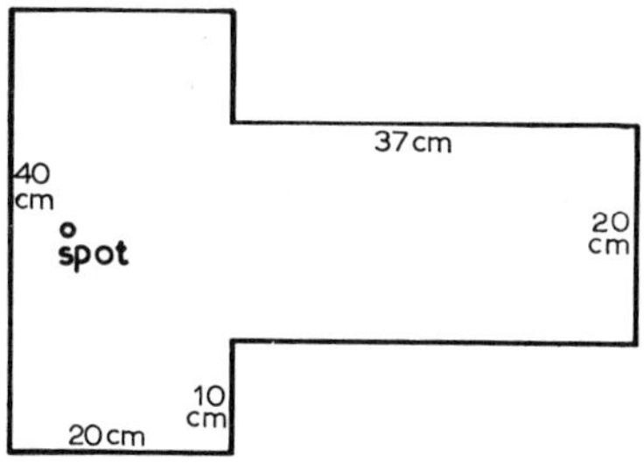

Fig. 25. T-shaped paper sheet for electrophoresis followed by chromatography.

Until the apparatus for two-dimensional electrophoresis of Gross becomes available commercially a combination of *electrophoresis with chromatography* will give satisfactory separations. Dixon *et al.*[18] have described an electrophoresis run at pH 3.6 followed by paper chromatography in *n*-propanol–pyrophosphate buffer. A T-shaped piece of Whatman No. 3 paper of the dimensions shown in Fig. 25 is rolled up as far as the horizontal bar of the T. The latter is wetted with pyridine–acetic acid buffer, pH 3.6, from the two ends and submitted to high voltage electrophoresis for 20 min at 1500 V. Dixon *et al.* did this in a Michl liquid coolant apparatus but electrophoresis between pressure plates is equally effective. The ends of the horizontal bar in contact with the wicks in the latter case must be trimmed before proceeding with chromatography. At pH 3.6 the bases are in one spot, the neutral amino acids in another, and the acidic amino acids, glutamic, aspartic and cysteic are resolved. If the P–E–PP system of Hanes is to be used for chromatography the strip must first be dried and then wetted from the ends with 0.05 *M* pyrophosphate–sodium chloride buffer, pH 8.9 (see p. 12).

In the method of Richmond and Hartley[19] electrophoresis is carried out at pH 2.2 (4% v/v formic acid, 0.3% pyridine) followed

Referensec p.60

by chromatography in *n*-butanol-*n*-butyl acetate–acetic acid–water (19:1:5:25). A peptide hydrolysate is applied in 1 cm bands a quarter of the way up from the bottom, anode, end of a 58 × 20 cm Whatman No. 3 or 3 MM paper. The papers are wetted with pH 2.2 buffer and subjected to electrophoresis under toluene at 2000 V for 70 min. Cysteic acid migrates 8 cm towards the anode, lysine 36 cm towards the cathode, while histidine and arginine overlap at about 34 cm from the origin. The strip 12–30 cm from the origin is cut out and sewn to the top of another sheet of paper and the amino acids

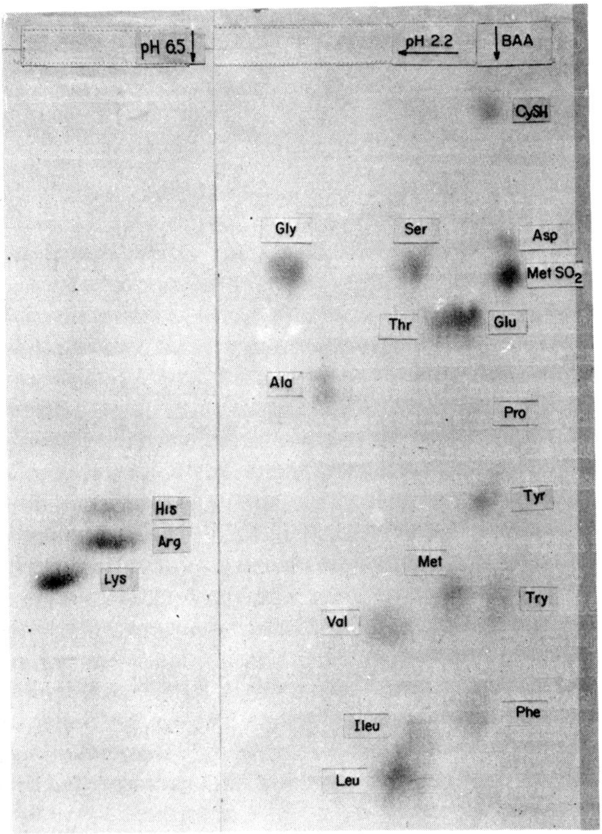

Fig. 26. Separation of amino acids by electrophoresis at pH 2.2 followed by chromatography (from Richmond and Hartley[19]).

contained in the strip separated in the butanol–butyl acetate solvent in a run of 16–24 h (Fig. 26). If necessary the bases may be separated further by electrophoresis at pH 6.5.

Improved separations of the pairs threonine–glutamic acid, and leucine-isoleucine are obtained in the procedure of Rothman and Higa[20]. Conditions are: electrophoresis in formic–acetic acid buffer, pH 1.8 (25 ml of 88% formic acid plus 87 ml of glacial acetic acid per litre of solution) at 62 V/cm for about 80 min, followed by chromatography using *sec.*-butanol–3% aqueous ammonia (5:2). Lys, Arg, His, Gly and Ala are resolved in the first direction and are not submitted to chromatography. Carboxymethylcysteine, methionine sulphoxide and methionine sulphone are also separated by this method.

SEPARATION OF PEPTIDES

One-dimensional or two-dimensional electrophoresis in combination with paper chromatography constitute the most powerful method yet devised for the study of complex mixtures of peptides on a small scale. To illustrate the wide range of problems current-

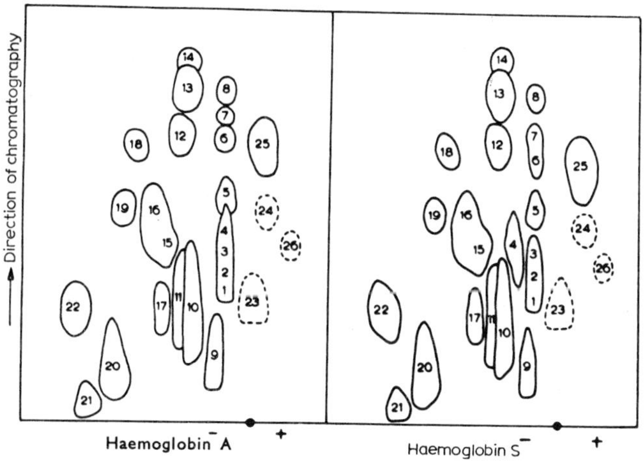

Fig. 27. Peptide maps of trypsin digests of haemoglobin A and S. Dotted lines indicate peptides which only became visible after heating the chromatogram (from Ingram[22]).

ly being attacked with the aid of these techniques experimental details are given for schemes developed during some of the more extensive investigations.

The approximate positions relative to aspartic acid of a number of small peptides (5–15 μg) separated at pH 2.0 in a potential gradient of 90 V/cm are shown in Fig. 21 (strip F). Small peptides derived from larger peptides of 6–10 residues can usually be separated cleanly and rapidly in a single direct electrophoresis. Thus, to confirm the proposed structure for β-melanocyte stimulating hormone peptides obtained by Harris and Roos[21] from enzymic hydrolysates were split further by partial acid hydrolysis. Fragments from two of these peptides gave the electrophoretic sparations shown in

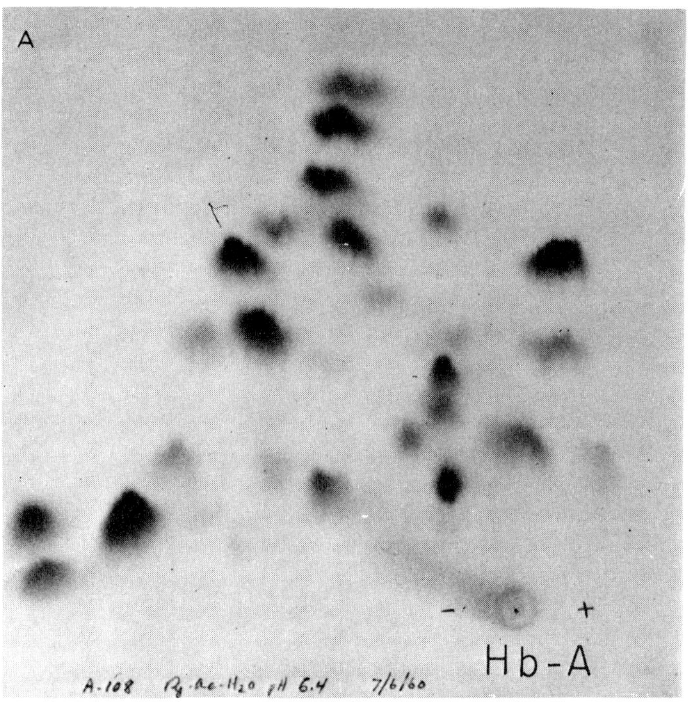

Fig. 28A. Photograph of the peptide map of haemoglobin A; electrophoresis at pH 6.4 for 2.5 h (1 kV), chromatography in isoamylalcohol–pyridine–water (35:35:30). Baglioni[23].

Fig. 21 (strip G and H). Conditions employed were for peptide Tl, pyridine–acetic acid–water (10:0.4:90, by vol), pH 6.5, 35–40 V/cm for 2 h; and for peptide T2 (3), pyridine–acetic acid–water (1:10:189) pH 3.5, 35–40 V/cm for 2 h. The heat exchanger was toluene.

An example of the use of paper chromatography in conjunction with electrophoresis of peptides is taken from the studies of Ingram[22] on haemoglobin. Trypsin digests of haemoglobin A and haemoglobin S are compared after electrophoresis in Whatman No. 3 MM paper in pyridine–acetic acid–water, pH 6.5, 19 V/cm for 150 min, followed by chromatography in the system. *n*-butanol – acetic acid – water (3 : 1 : 1) for 24 h. The patterns obtained (Fig. 27) are almost indentical but show one important difference. The zone marked number 4 in the haemoglobin A pattern has clearly mi-

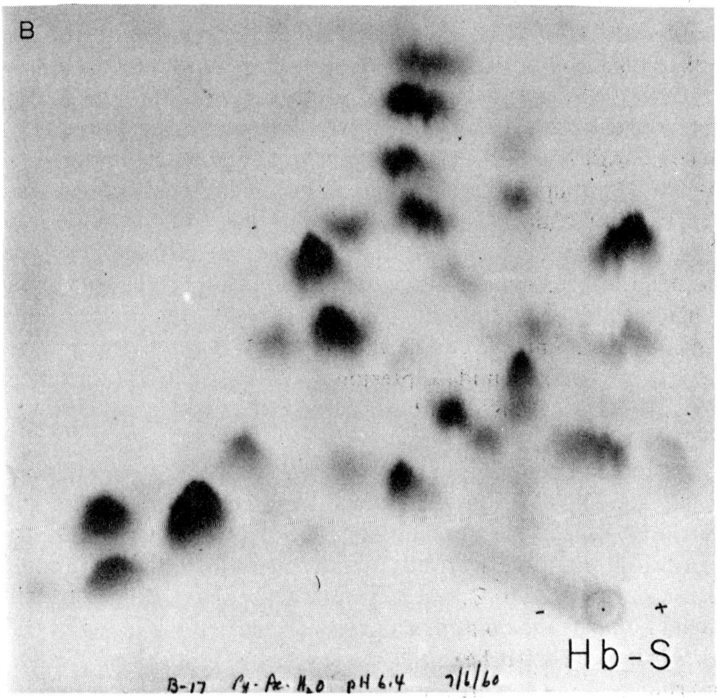

Fig. 28B. Photograph of the peptide map of haemoglobin S. See also legend of Fig. 28A. Baglioni[23].

References p. 60

grated nearer to the cathode in the haemoglobin S pattern. Ingram has shown that the glutamic acid residue of the former haemoglobin is replaced by a valine residue in the latter.

More compact spots were obtained by Baglioni[23] when the solvent[24] pyridine–isoamyl alcohol–water (30:30:35) was used for the chromatographic separation. From photographs of the peptide maps shown in Figs. 28A and B it will be seen that the original spot 15 has resolved into two peptides, and spots 1, 2, 7 and 8 no longer appear. Each of the spots 16, 17, 20 and 21 is now known to contain two peptides.

Two further ingenious applications of paper electrophoresis in protein structure are worth mentioning. The first concerns the means of identification of the C-terminal peptide of the γ-chain of human foetal haemoglobin by Naughton and Hagopian[25]. These workers found it possible to digest peptides *in situ* on paper sprayed with carboxypeptidase (0.01 mg/ml) dissolved in ammonium carbonate buffer, pH 8.8. Before use the treated strip was hung at room temperature for 18 h in a jar lined with paper wet with the same buffer, and then dried at 100° for 30 min to remove buffer. The γ-chain was digested with trypsin and the insoluble core material removed. A sample of the supernatant containing the soluble peptides was then run in two directions at pH 6.4 to give a diagonal distribution of peptides across the paper. Another sample was digested *in situ* on the strip before running in the second direction. Three spots only remained along the diagonal. These were free lysine, Val.Lys, and a peptide containing tyrosine and histidine. Val.Lys was not attacked by carboxypeptidase presumably because it is a dipeptide, and Tyr.His was thus identified as the C-terminal peptide of the γ-chain.

The other interesting application is a method for determining disulphide bridges in proteins which was described by Brown and Hartley[26]. An enzyme digest of the protein was submitted to electrophoresis in one direction and a guide strip was then cut from the paper and exposed for 2 h to performic acid vapour. This converted each cystine peptide to a pair of cysteic acid peptides. After drying *in vacuo* over NaOH the strip was stitched to a full sheet of paper and run in the second direction under the same conditions as before. Most of the peptides appeared on a 45° diagonal across the paper; each cystine peptide yielded a parallel pair of cysteic acid peptides lying off this diagonal. The method was used for allocating the disulphide bridges of chymotrypsinogen A[26] and lysozyme[27].

Two new solvent systems for chromatography, in conjunction with electrophoresis, are methylcellosolve–formic acid–*sec.*-butanol–water (2:1:8:3)[28] and *n*-butanol–*sec.*-butanol–acetic acid–water–iso-amyl alcohol–ethyl acetate (1:1:1:1:0.5:0.5). The latter was used in the preparation of peptide maps from enzyme digests of nucleohistones[29].

Radioactive peptides

In the extensive studies being conducted by the Cambridge school on the amino acid sequence around the reactive serine residue of

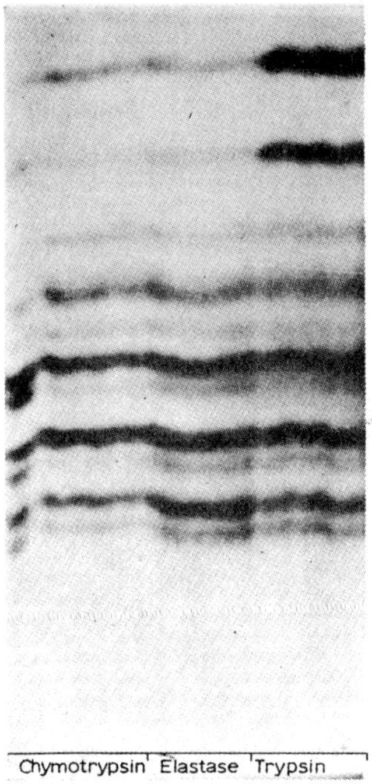

Fig. 29. Comparison of partial hydrolysates of di-isopropoxyl [^{32}P] phosphinyl-enzymes. Reproduction of radioautograph obtained on electrophoresis at pH 3.5 (from Naughton *et al.*[30]).

References p. 60

certain proteolytic enzymes use is made of radioautography for detection of the significant peptides[19]. The di-isopropoxyl(^{32}P)-phosphinyl enzyme is prepared from the active enzyme (25 mg) dissolved in 5 ml of 0.1 *M* phosphate buffer, pH 8.0, and 0.63 ml of 0.02 *M* (^{32}P) DFP (containing between 1 and 30 μC of ^{32}P/mg) in propan-2-ol. After incubation at room temperature for 3 h the mixture is dialyzed against two changes of 5 l of 0.01 *M*-acetic acid at 2° and freeze-dried. Separations of the labelled phosphoserine peptides obtained from partial hydrolysates are done on Whatman No. 52 filter paper in pyridine–acetic acid buffer of pH 3.5 and 6.5. The relative migration rates of the peptides are very sensitive to slight differences in pH and the values obtained will therefore depend on the buffer content of the toluene when the method of Michl is used. Addition of 0.3% (v/v) of acetic acid and 8% (v/v) of pyridine to the toluene coolant is recommended for the pH 3.5 and 6.5 runs respectively. Fig. 29 is a reproduction of the radioautograph obtained on electrophoresis of partial acid hydrolysates of di-isopropoxyl (^{32}P)-phosphinyl-enzymes at pH 3.5, 40 V/cm, 2–5 h. Exposure times for the preparation of radioautographs depend on the specific activity of the peptides. Using Ilford Industrial G X-ray film Naughton *et al.*[30] found that whereas 24 h was sufficient exposure for bands containing 1 μmC of ^{32}P, 10–15 days were necessary when the bands contained only 0.1 μmC. Not more than 0.1 mg of peptide fraction is applied to the paper as a thin 2 cm line. For two-dimensional electrophoresis a useful technique demonstrated by Milstein and Sanger[31] is to carry out the first run normally on a paper strip and then sew the strip with a sewing machine to a larger sheet for the second run. The paper can then be wetted with the buffer to be used in the second dimension in such a way that an even flow from both sides is obtained with consequent sharpening of the bands.

REFERENCES

1 T. Wieland, *Angew. Chem.*, 60 (1948) 313.
2 H. Michl, *Monatssch. Chem.*, 82 (1951) 489.
3 A. P. Ryle, F. Sanger, L. F. Smith and R. Kitai, *Biochem. J.*, 60 (1955) 541.
4 H. Michl, *Chromatog.*, 1 (1959) 11.
5 A. M. Katz, W. J. Dreyer and C. B. Anfinsen, *J. Biol. Chem.*, 234 (1959) 1897.

[6] H. Michl, H. Kuhn and H. Bühn, *Fachliche Mitt. Oesterr. Tabakregie*, (1956) 13–16; *C.A.*, 51 (1957) 674.
[7] D. Gross, *Nature*, 172 (1953) 908.
[8] D. Gross, *J. Chromatog.* 5 (1961) 194.
[9] B. Kickhöfen and O. Westphal, *Z. Naturforsch.*, 7b (1952) 659.
[10] L. Heilmeyer, R. Clotten, I. Sano, A. Sturm, Jr. and A. Lipp, *Klin. Wochschr.*, 32 (1954) 831.
[11] G. H. Spengler and M. Knedel, *Klin. Wochschr.*, 34 (1956) 389.
[12] D. Gross, *Nature*, 176 (1955) 72.
[13] G. N. Atfield and C. J. O. R. Morris, *Biochem. J.*, 81 (1961) 606.
[14] J. Heilmann, J. Barollier and E. Watzke, *Z. Physiol. Chem.*, 309 (1957) 219.
[15] D. Gross, *Nature*, 184 (1959) 1298.
[16] S. Blackburn and G. R. Lee, *Biochem. J.*, 87 (1963) 1P.
[17] A. E. Pasieka, *Can. J. Biochem. Physiol.*, 39 (1961) 1313.
[18] G. H. Dixon, D. L. Kauffman and H. Neurath, *J. Biol. Chem.*, 233 (1958) 1373.
[19] V. Richmond and B. S. Hartley, *Nature*, 184 (1959) 1869.
[20] F. Rothman and A. Higa, *Anal. Biochem.*, 3 (1962) 173.
[21] J. I. Harris and P. Roos, *Biochem. J.*, 71 (1959) 434.
[22] V. M. Ingram, *Biochim. Biophys. Acta*, 28 (1958) 539.
[23] C. Baglioni, *Biochim. Biophys. Acta*, 48 (1961) 392.
[24] H. G. Wittmann and G. Braunitzer, *Virology*, 9 (1959) 726.
[25] M. A. Naughton and H. Hagopian, *Anal. Biochem.*, 3 (1962) 276
[26] J. R. Brown and B. S. Hartley, *Biochem. J.*, 89 (1963) 59P.
[27] J. R. Brown, *Biochem. J.*, 92 (1964) 13P.
[28] J. E. Euger and W. E. Savige, *Photochem. Photobiol.*, 2 (1963) 25.
[29] L. S. Hnilica, C. W. Taylor and H. Busch, in: J. Bonner and P. Ts'o (Eds.), *The Nucleohistones*, Holden-Day, Inc., San Francisco, 1964, p. 84.
[30] M. A. Naughton, F. Sanger, B. S. Hartley and D. C. Shaw, *Biochem. J.*, 77 (1960) 149.
[31] C. Milstein and F. Sanger, *Biochem. J.*, 79 (1961) 456.

Chapter 3

ION-EXCHANGE CHROMATOGRAPHY OF AMINO ACIDS AND PEPTIDES

The ion-exchange resins commonly used in protein chemistry for the separation of amino acids, peptides, proteins and related substances are derived from polystyrene or polymethacrylic acid. Styrene itself produces linear polymers, but copolymerization with the cross-linking agent divinyl-benzene gives materials containing a three-dimensional network of benzene rings. Sulphonation of these materials leads to cation exchangers of the following structure.

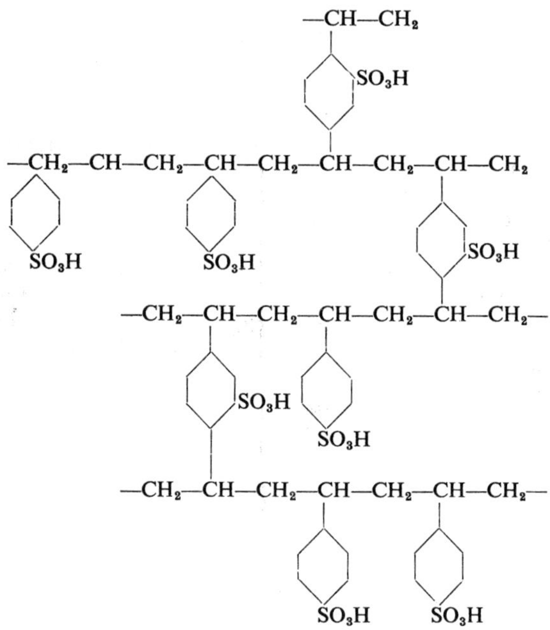

Commercial materials have a capacity of about 5 mequivs. per g which corresponds to slightly less than one sulphonic acid group per benzene ring.

Anion exchangers are prepared from the same cross-linked polystyrene network by introduction of $—CH_2Cl$ groups into the benzene rings at the para position and then treating with an amine. Primary and secondary amines yield weakly basic resins, tertiary amines yield strongly basic resins containing quaternary ammonium groups. A product obtained by reaction with trimethylamine has the structure

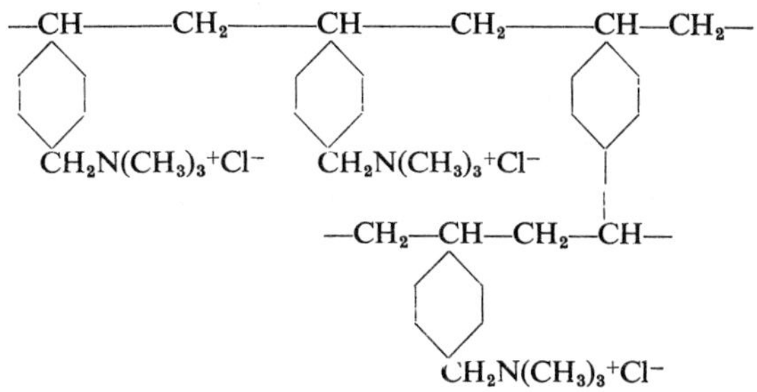

Weakly acidic resins are prepared by the copolymerization of methacrylic acid with a crosslinking agent. The resulting polymer has the structure

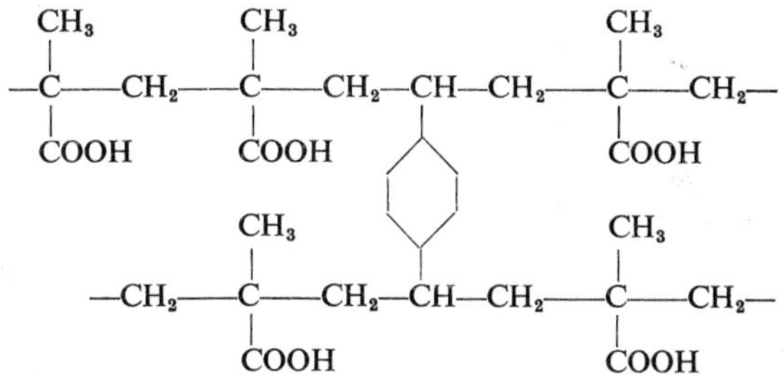

Highly cross-linked resins form a matrix impermeable to all but the smallest solute molecules, but low cross-linked resins permit some penetration of larger molecules. A description of resins available commercially is given in Table 10. The separation of all the common amino acids by ion-exchange resins has been achieved by displacement development and by elution. These two different techniques are discussed separately.

References p. 107 *Text continued p. 66*

TABLE 10

CHARACTERISTICS OF ION-EXCHANGE RESINS

Resin	*Type*		*Approx. capacity*		*Manufacturer*
			mequivs./g (dry)	*mequivs./ml resin in water*	
Dowex 50-X12[a]	Strongly acidic cation exchanger	SO_3^-	5.0	2.3	Dow-Chemical Co., Michigan, U.S.A.
Dowex 50-X8	Strongly acidic cation exchanger	SO_3^-	5.1	1.7	Dow-Chemical Co.
Dowex 50-X4	Strongly acidic cation exchanger	SO_3^-	5.2	1.2	Dow-Chemical Co.
Dowex 50-X2	Strongly acidic cation exchanger	SO_3^-	5.2	0.7	Dow-Chemical Co.
Amberlite IR-120	Strongly acidic cation exchanger	SO_3^-	4.25	1.9	Rohm & Haas Co., Philadelphia, U.S.A.
Zeo-Karb 225	Strongly acidic cation exchanger	SO_3^-	4.8	2.1	The Permutit Co. Ltd., London, England
Dowex 1-X10	Strongly basic anion exchanger	$—N^+(CH_3)_3$	3.0	1.5	Dow-Chemical Co.
Dowex 1-X8	Strongly basic anion exchanger	$—N^+(CH_3)_3$	3.2	1.4	Dow-Chemical Co.
Dowex 1-X4	Strongly basic anion exchanger	$—N^+(CH_3)_3$	3.5	1.2	Dow-Chemical Co.
Dowex 1-X2	Strongly basic anion exchanger	$—N^+(CH_3)_3$	3.5	0.8	Dow-Chemical Co.

Dowex 2-X10	Strongly basic anion exchanger	—$N^+(CH_3)_2(C_2H_4OH)$	3.0	1.5	Dow-Chemical Co.
Dowex 2-X8	Strongly basic anion exchanger	—$N^+(CH_3)_2(C_2H_4OH)$	3.2	1.4	Dow-Chemical Co.
Dowex 2-X4	Strongly basic anion exchanger	—$N^+(CH_3)_2(C_2H_4OH)$	3.2	1.2	Dow-Chemical Co.
Amberlite IRA-400	Strongly basic anion exchanger	—$N^+(CH_3)_3$	2.5	1.0	Rohm & Haas Co.
De-Acidite FF	Strongly basic anion exchanger	—$N^+(CH_3)_3$	3.8	1.5	The Permutit Co.
Dowex 3	Weakly basic	polyamine, beads	6	3	Dow-Chemical Co.
Amberlite IR-4B	Weakly basic	polyamine, granular	10	3	Rohm & Haas Co.
Amberlite IR-45	Weakly basic	polyamine, beads	5	2	Rohm & Haas Co.
De-Acidite E	Weakly basic	aliphatic amine, granular	9	3	The Permutit Co.
De-Acidite G	Weakly basic	aliphatic amine, beads	3.5	1.3	The Permutit Co.
Amberlite IRC-50	Weakly acidic	acrylic resin, beads	10.0	3.5	Rohm & Haas Co.
Amberlite XE64	Weakly acidic	acrylic resin, powder	10.0		Rohm & Haas Co.
Zeo-Karb 226	Weakly acidic	acrylic resin, beads	10.0	3.5	The Permutit Co.
Retardation 11 A8	Strongly basic+ weakly acidic	—$N^+(CH_3)_3$ +$RCHCOO^-$			Bio-Rad Lab., Calif. (U.S.A.)

[a] X followed by a number denotes % cross-linking agent.

References p. 107

AMINO ACID ANALYSIS BY DISPLACEMENT DEVELOPMENT

In this technique amino acids are adsorbed on a column of ion-exchange resins in amounts which will saturate part of the resin. The column is then developed with a competitive ion which has a strong affinity for the resin so the zone of amino acids is displaced continuously. The amino acids separate into further zones by mutual displacement as a result of their different affinities for the resin. The technique was developed by Partridge and co-workers during the period 1949–1952 and a survey of their results appeared in a communication by Partridge and Brimley[1].

As a simple picture of the displacement process involving neutral amino acids the following scheme is helpful.

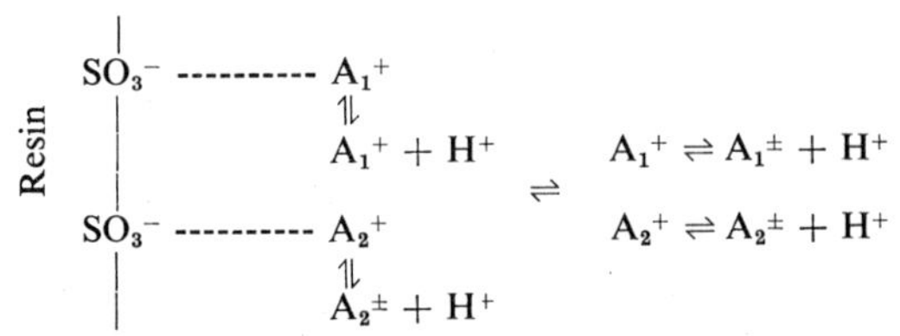

An amino acid is adsorbed in the cationic form A^+ and this form is in reversible equilibrium with hydrogen ions and uncharged molecules in the surface layer. The system at the surface of the resin is again in equilibrium with the solution remote from it, but as the resin is charged, the concentration of hydrogen ions at the interface is different from that in the ambient solution. Displacement of the weaker base A_2 will depend mainly upon the partial suppression of the cationic form of A_2 by the stronger base A_1 but will also depend on the relative affinities of the resin for the cations A_1^+ and A_2^+. Factors which contribute to the total adsorption energy are the valency and degree of hydration of the ion, molecular size and aromaticity. That the order of displacement is largely determined by the basic or acidic strengths of the solutes is clearly illustrated by the data in Table 11. With the exception of the amino acids proline, methionine, cystine, phenylalanine and tyrosine, the experimentally determined sequence of displacement of the listed solutes from the polystyrene resins is in the order of pK values.

In the method of Partridge and Brimley multiple columns of

TABLE 11

THE SEQUENCE OF DISPLACEMENT OF A RANGE OF SOLUTES FROM TWO RESINS AT ROOM TEMPERATURE[1]

(The solutes are arranged in sequence of displacement from the resins. Relevant pK values are from Cohn and Edsall[76], and Edsall and Wyman[77].)

Sulphonated cross-linked polystyrene (4.5% divinylbenzene)		
Aspartic acid	pK_1	1.995
Hydroxyproline	pK_1	1.815
Threonine	pK_1	2.096
Serine	pK_1	2.186
Glutamic acid	pK_1	2.19
Proline	pK_1	1.970
Glycine	pK_1	2.350
Alanine	pK_1	2.348
Valine	pK_1	2.32
Methionine	pK_1	2.28
Leucine	pK_1	2.36
Cystine	pK_1	2.26
Creatine	pK_1	2.62
Phenylalanine	pK_1	1.83
β-Alanine	pK_1	3.551
Trimethylamine oxide	pK	4.5 approx.
Creatinine	pK	4.8 approx.
Histidine	pK_1	1.82
	pK_2(-Im)	6.00
Methylhistidine	pK_2(-Im)	6.48
Carnosine	pK_2	6.83
Anserine	pK_2	7.04
Hydroxylysine	—	—
Lysine	pK_2	9.18
Ammonia	pK	9.245
Arginine	pK_2	9.01
	pK_3(-Guan)	12.5 approx.
Methylamine	pK	10.615
Lysine	pK_3(-ε)	10.79
Proline	pK_2	10.640
β-Alanine	pK_2	10.235
Alanine	pK_2	9.867
Valine	pK_2	9.62
Leucine	pK_2	9.60
Glycine	pK_2	9.780
Carnosine	pK_3	9.51

TABLE 11 (continued)

Sulphonated cross-linked polystyrene (45% divinylbenzene)		
Threonine }	pK_2	9.100
Serine }	pK_2	9.205
Histidine	pK_3	9.17
Methionine	pK_2	9.21
Methylhistidine	pK_3	8.85
Cystine	pK_3	7.85
Tyrosine	pK_2	9.11
	pK_3(—OH)	10.95
Acetic acid	pK	4.756
Glutamic acid	pK_2	4.25
Aspartic acid	pK_2	3.910
Hydrochloric acid	—	—

diminishing diameter were used. Tyrosine and phenylalanine were first removed from the protein hydrolysate by adsorption on activated charcoal. The remaining solution was passed through a fore-column packed with a sulphonated polystyrene resin when arginine, lysine, most of the histidine and small amounts of leucine, methionine and glucosamine and cystine were adsorbed. The effluent from this column was passed through three coupled columns packed with Zeo-Karb 215, a sulphonated phenol-formaldehyde type resin. The amino acids were displaced from the latter with 0.15 *N* ammonium hydroxide, and from the fore-column with 0.075 *N* sodium hydroxide. This primary fractionation into groups of amino acids is shown schematically in Fig. 30.

Further fractionation was effected in the following ways. Glutamic acid was removed from Band II by adsorption on De-Acidite B and subsequent elution with 0.5 *N* hydrochloric acid. Serine and threonine were separated on Zeo-Karb 215 with 0.1 *N* ammonia in 50% aqueous acetone. Glycine and alanine from Band III were separated on Dowex 2 and the proline and valine of Band IV were resolved on Zeo-Karb 215 at 55° (Partridge and Brimley[2,3]). Methionine and the leucines were similarly separated either on Zeo-Karb 215 at 55° or on Dowex 2. Fractionation of Band VI was done on a sulphonated polystyrene column by development with 0.1 *N* sodium chloride. This removed the neutral amino acids and gluco-

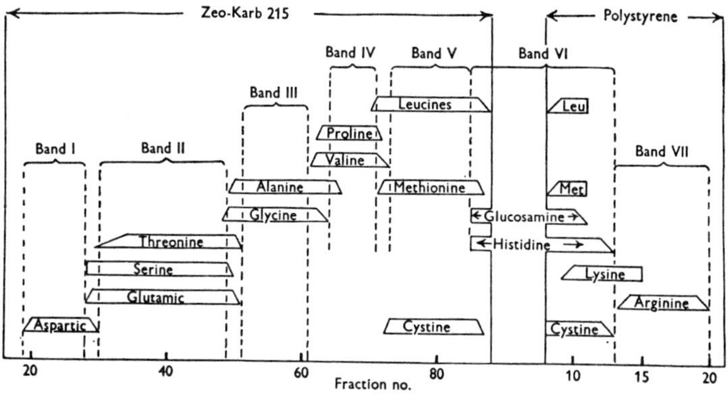

Fig. 30. Primary fractionation of the hydrolysis products of 280 g of commercial egg albumin. The top section of the column was packed with sulphonated polystyrene and the lower sections with Zeo-Karb 215. (From Partridge and Brimley[1]).

samine. The basic amino acids histidine and lysine remaining on the column were displaced with 0.1 N sodium hydroxide. These bases were finally added to the lysine and arginine of Band VII and the mixture resolved on a column of sulphonated polystyrene with 0.1 N sodium hydroxide.

Displacement development is particularly suitable for large scale operation. Partridge and Brimley isolated 16 amino acids from 280 g of commercial egg albumin with an overall recovery of 54% of the dry weight of protein taken.

AMINO ACID ANALYSIS BY ELUTION (MOORE AND STEIN)

High resolution with near quantitative recovery of solutes has made elution from ion-exchange resins the most powerful method for amino acid analysis available today. A simple picture of the elution of two cationic species by sodium ions is shown in Fig. 31. In contrast to displacement analysis only small amounts of solutes are adsorbed on the resin. Initially, the leading boundaries of A_1^+ and A_2^+ are sharp as Na^+ ions are displaced. The trailing boundaries, where Na^+ ions are displacing A_1^+ and A_2^+, are diffuse and soon overtake the leading boundaries. The points of maximum concentration of A_1^+ and A_2^+ decrease as the zones travel down the column and as a result of the diffusion of both leading and trailing bounda-

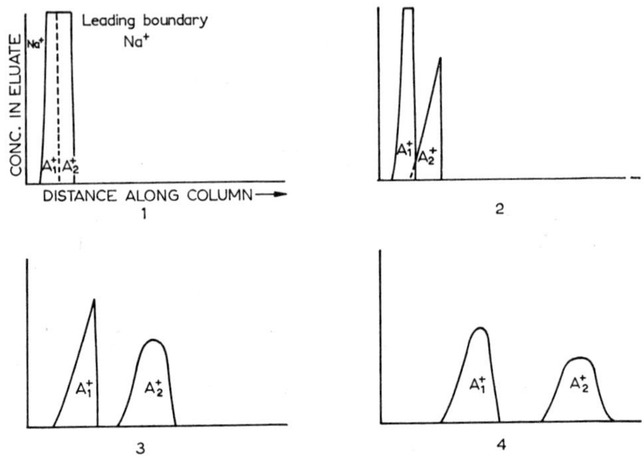

Fig. 31. Formation of elution zones as two cationic species travel down a column of resin in the sodium form.

ries symmetrical 'peaks' are eventually eluted from the column. A more accurate description of the process would also take into account mutual displacement effects of the different ionic species. Factors governing the resolution of solutes, apart from the relative affinities, are the pH and ionic strength of the eluent, the temperature of the system and the characteristics of the resin. Further discussion will be found on p. 91.

Hold-up volume, elution volume and effluent volume are terms commonly used in describing chromatographic procedures. They are defined as follows:

Hold-up volume (V^1) = Volume of solvent which is held between the particles of solid.

Elution volume (V) = Total volume of effluent less the hold-up volume.

Total effluent volume ($V + V^1$) = Volume of liquid flowing from the column from the time of application of solute to maximum concentration of solute emerging from column.

The expression $R_F = V^1/(V + V^1)$ may be compared with a similar expression given on p. 30.

Amino acid analysis by ion exchange methods as we know it today, was developed over a number of years by the Rockefeller Institute group. In 1950 Stein and Moore[4] demonstrated that a mixture of 18 amino acids could be resolved almost completely on a 55-cm column of Dowex 50-X8 with hydrochloric acid (Fig. 32). The full elegance of the elution method was revealed, however,

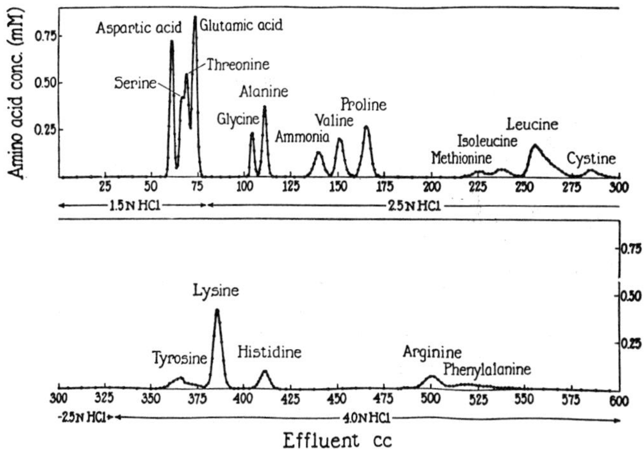

Fig. 32. Fractionation of a known mixture of amino acids on Dowex 50-X8. Solvents, 1.5 *N*, 2.5 *N* and 4.0 *N* HCl. Column dimensions, 0.9 × 55 cm. (From Stein and Moore[4].)

in the procedure[5] of 1951 when buffers of progressively increasing pH from 3.4 to 11 were used. Separation of amino acids and related compounds from a synthetic mixture containing 32 components on a 100-cm column of Dowex 50-X8 was achieved (Fig. 33). There followed further modifications[6] which made use of the resin Dowex 50-X4 and introduced the principle of gradient elution. A synthetic mixture containing 50 components were resolved successfully on a 150-cm column.

In the latest procedure the bead-form of the Dowex resins has been abandoned in favour of the 'micropowder' form of the 8% cross-linked resin Amberlite IR-120. Two columns are employed, one for the neutral and acidic amino acids (150 cm) and one for the basic amino acids (15 cm). A complete amino acid analysis is possible in 48 hours using a fraction collector or in 22 hours using automatic recording equipment. The following details of the method are taken from Moore, Spackman and Stein[7], and Spackman,

Stein and Moore[8] except where an alternative reference is given. It should be noted, however, that a procedure similar to that of Moore and Stein has been worked out in the U.K. by Campbell *et al.*[9] using Zeo-Karb 225-WR. 1.55.

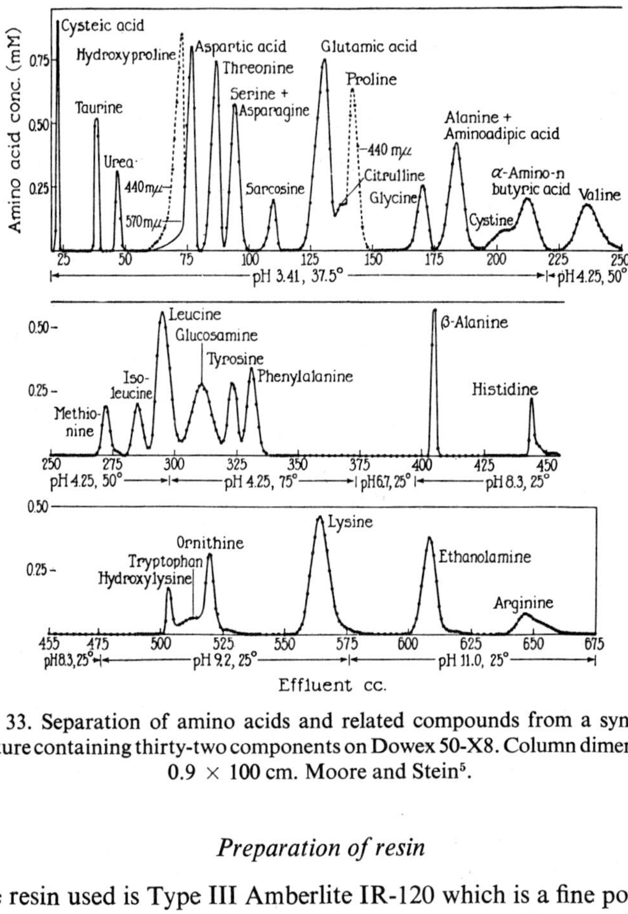

Fig. 33. Separation of amino acids and related compounds from a synthetic mixture containing thirty-two components on Dowex 50-X8. Column dimensions 0.9 × 100 cm. Moore and Stein[5].

Preparation of resin

The resin used is Type III Amberlite IR-120 which is a fine powder of 400–600 mesh (Product of Rohm and Haas Co., marketed through Fisher Scientific Co., New York, and through the British Drug Houses, England). The most important preliminary is the fractionation of the commercial product into particles of appropriate size as this will influence the performance of the column. The presence of very fine particles will make the column too tight and give slow flow rates. On the other hand the resolving power of the resin is a function of the surface area of resin available for exchange and too many coarse particles in the preparation will adversely affect the resolution. The recommended particle sizes are given in Table 12.

TABLE 12

RECOMMENDED PARTICLE SIZES OF AMBERLITE IR-120

Use of resin	*Particle diameter*
Automatic recorder, 15- and 50-cm columns	25 — 30μ
Automatic recorder, 150-cm column	40 $\pm$ 7μ
Fraction collector, 15- and 50-cm columns	40 $\pm$ 7μ
Fraction collector, 150-cm column	56 $\pm$ 9μ

Formerly the required particle size was obtained by passing the wet resin through sieves of appropriate mesh. Hydraulic separation by the method of Hamilton[10] has been found to be more convient and permits reproducible results. The equipment is illustrated in Fig. 34. A number of laboratories use a multiple-column hydraulic method enabling several fractions to be collected in one operation, and where distilled water is a scarce commodity a closed circuit system may be operated. Apparatus constructed along these lines will effectively fractionate micro-bead ion-exchange resin, powdered resin, Sephadex and Celite[13].

Before fractionation the very fine particles are removed from the resin by suspension in water. One pound of the dry sodium salt of the resin is suspended in about 10 l of water. If purchased in the H^+ form the resin is converted to the sodium salt by washing with 2 *N* NaOH on a Büchner funnel until the filtrate is alkaline. The suspension in water is allowed to settle for about 6 h. The supernatant liquid which contains the very fine particles is withdrawn by suction and the operation repeated four or five times. The product is then washed slowly on a Büchner funnel with 2 l of 4 *N* hydrochloric acid, followed by 500 ml of water. The moist resin is suspended in 2 l of 2 *N* sodium hydroxide and the mixture heated on the steam-bath for 1 h. The resin is finally collected on a funnel and washed with water until neutral. Even after this treatment commercial

References p. 107

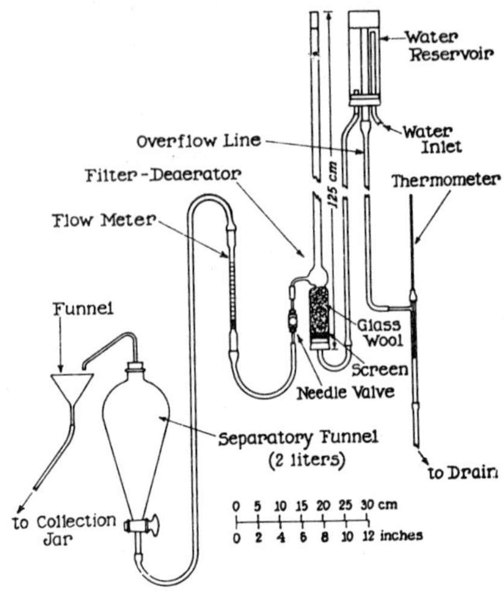

Fig. 34. Equipment used for separation of resin particles by hydraulic method of Hamilton[10].

resins have been shown to contain traces of some amino acids, but fortunately the amounts are too small to influence the results of amino acid analysis of protein hydrolysates[11].

For the fractionation, 400 ml or less of the settled resin is suspended in water to give a total volume of about 1 litre and transferred to the 2-litre separating funnel. With the needle valve fully open and the tube to the bottom of the separating funnel disconnected the flow-rate and temperature are checked. The tube is then attached to the stem of the funnel and a flow of about 600 ml per min maintained through the funnel. When the funnel is about 3/4 full the flow is reduced to 1·10 ml per min and maintained at this rate until no particles leave the outlet tube. The flow rate is then increased to 280 ml per min to give particles of average diameter 40 μ. After this particles of about 56 μ diameter can be collected at a flow rate of 580 ml per min. It is important to re-cycle the fractions twice before use. (A suitable flow meter is the Type 124 glass-tube Rotameter calibrated 0–1,000 ml obtainable from the Rotameter Manufacturing Co. Ltd., Croydon, England.) For comparison purposes the wet screening data are roughly 200–300 mesh for 56 μ particles and not less than 400 mesh for 40 μ particles. The fractionated resin is best stored in 0.2 *N* sodium hydroxide in polythene bottles at 4°.

Preparation of columns

Glass columns of 0.9–1.0 cm diameter are of length 165 cm or 25 cm above a sintered plate, and are fitted with jackets. The 150 cm column to be used with fraction collectors is poured from resin containing particles of 50 μ average diameter. It is first washed with 1 litre of 4 *N* hydrochloric acid per 200 to 300 ml of settled resin, then with water, 2 *N* sodium hydroxide and water again. The sodium salt is washed with 0.2 *N* buffer at pH 4.25 (Table 13) until the filtrate is about pH 4. A slurry of the resin is made in this buffer so that 200 ml of supernatant liquid is present per 100 ml of settled resin. All bubbles should be removed from the slurry by gentle stirring after it has been allowed to stand for an hour or two. The column is packed in sections from five portions of slurry of about 50 ml. The first portion is allowed to settle under gravity for a few minutes and then under air pressure of 15–30 cm of mercury until no further drop in the height of the surface of the resin occurs. When the level of the buffer has fallen to the resin surface a second portion of slurry is added. The other portions are added to bring the height of the column to a few centimetres above the 150 cm mark. An extension tube of about 40 cm is necessary to contain the final portions. Before use a new 150 cm column should be washed with 0.2 *N* sodium hydroxide and equilibrated with boiled buffer at pH 3.25 at room temperature. The flow rate is checked at 50° and should be about 12 ml per h at 20–40 cm pressure.

The 15-cm column is poured similarly in one section from a slurry of resin in 0.35 *N* buffer at pH 5.28. The resin of 40μ particles can be used for this purpose.

Buffers

The compositions of the buffers are listed in Table 13. They are stored at 4° after addition of 0.1% phenol to prevent the growth of micro-organisms. Buffer for immediate use is brought to the boil and layered with mineral oil. For uniform drop size BRIJ35 (a polyoxyethylene lauryl alcohol)[12] is added to the buffers just

References p. 107

TABLE 13

PREPARATION OF SODIUM CITRATE BUFFERS[7]

pH	*Sodium ion (N)*	*Citric acid H_2O (g)*	*NaOH, 97% (g)*	*HCl conc. (ml)*	*Final vol. (l)*	*Phenol added (g)*	*Fraction collector*		*Automatic recorder*	
							TG(a)	*BRIJ(b)*	*TG(a)*	*BRIJ(b)*
3.25 ± 0.01 (c)	0.20	840	330	426	40	40	10	5]	5	[2
4.25 ± 0.02	0.20	840	330	188	40	40	10	5	5	2
5.28 ± 0.02	0.35	491	288	136	20	20	—	5	–	2
4.26 ± 0.02	0.38	532	312	307	20	20	—	5	–	2
6.50 ± 0.05	0.38	133	78	6.3	5	5	—	5	–	2

(a) Amount of thiodiglycol added per litre when methionine is present.
(b) Amount of BRIJ35 detergent solution added per litre.
(c) If adjustment of pH is necessary, addition of 1 ml of 50% NaOH or 2 ml of concentrated HCl causes a change of about 0.01 pH unit.

before use. The BRIJ solution is prepared by dissolving 50 g of solid detergent in 100 ml of hot water[6].

Operation of 150-cm column

Buffer at pH 3.25 containing BRIJ35 (and if necessary thiodiglycol) is passed through the column, and water at 50° circulated through the jacket. The sample solution should be of pH 2.0 to 2.5 and the optimum load is 0.5 to 1.0 μmole per amino acid or 1 to 2 mg of hydrolyzed protein in a volume of 2 ml. The solution is applied to the top of the column by means of a bent tip pipette and washed in with three aliquots of buffer at pH 2.2. A few centimetres of pH 3.25 buffer are added, the column is mounted over a fraction collector and the effluent is collected in 2-ml fractions (Fig. 35).

Elution with buffer at pH 3.25 is discontinued and buffer at pH 4.25 substituted at a point which must be determined by trial.

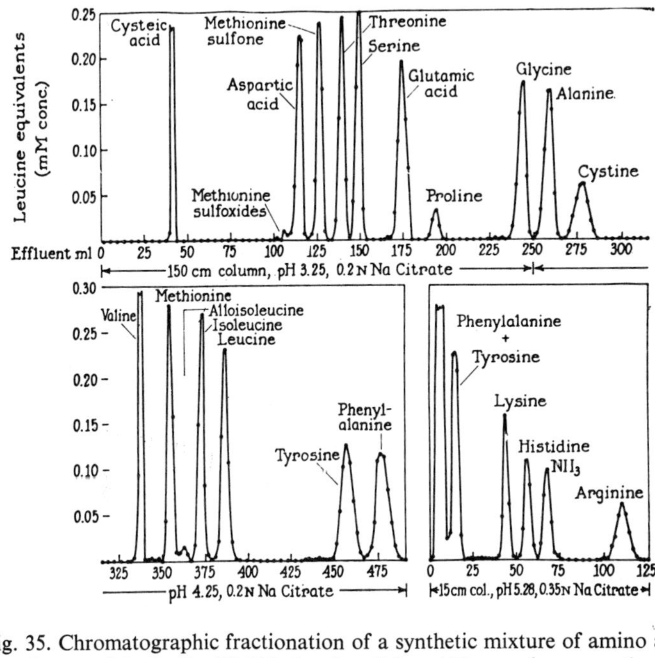

Fig. 35. Chromatographic fractionation of a synthetic mixture of amino acids on columns of Amberlite IR-120. Load on 0.9 × 150 cm column was 1 μmole of each amino acid (0.5 μmole of cystine). (From Moore *et al.*[7])

References p. 107

It will be seen from Fig. 35 that the cystine peak is well separated from alanine and the valine peak is sharpened by the pH 4.25 buffer front. This usually obtains when the buffer change is made at an effluent volume 2.15 times that at which the aspartic acid peak has emerged, or at about 260–280 ml. The diameter of the tube will influence the rate of elution of all the amino acids. The elution volume of cystine is dependent on the pH of the initial buffer, and the exact pH value required to separate cystine from alanine should be noted so that subsequent lots of the buffer may be suitably adjusted. The effluent volume for phenylalanine should not exceed 500 ml.

To regenerate the 150-cm column for re-use the basic amino acids retained by the resin are eluted with 0.2 *N* sodium hydroxide (free from carbon dioxide and containing 5 ml of BRIJ35 solution per litre). The column is then re-equilibrated with buffer at pH 3.25.

Operation of 15-cm column

The basic amino acids are determined by elution at 50° with buffer at pH 5.28 and a rate of 25–30 ml per hour (Fig. 35). The sample

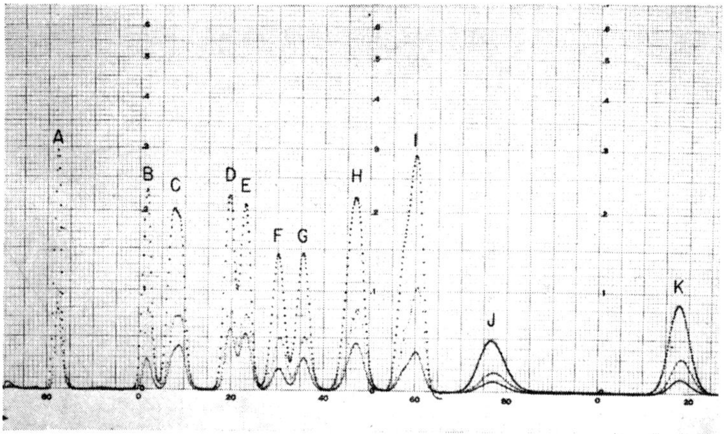

Fig. 36. Resolution of basic amino acids on a 50-cm column of Amberlite IR-120 at 50° with 0.70 N sodium citrate buffer, pH 5.28, as eluent. A, α-aminobutyric acid; B, glucosamine; C, galactosamine + hydroxylysine; D, ornithine; E, lysine; F, isomer of galactosamine; G, histidine; H, methyl-histidine; I, ammonia, ethanolamine and carnosine; J, tryptophan; K, arginine. Kominz[19].

need only be half the size of that applied to the 150-cm column and the analysis is complete after about 70 2-ml fractions have been collected. The position of histidine is sensitive to pH and the correct value to place the peak between those of lysine and ammonia must be determined. It is not usually necessary to regenerate the column but, if required, this can be accomplished with 10 ml of 0.35 *N* sodium hydroxide containing BRIJ35, after which 80 ml of the pH 5.28 buffer is run through the column. A 30-cm column may be used for better resolution of the bases. A 6 hour procedure (flow rate 30–40 ml/h) which will resolve hexosamines, α-aminobutyric acid, ornithine, methyl-histidines, as well as lysine, histidine, ammonia and arginine has been described by Kominz[19]. The sample is loaded on a 50-cm column and developed at 50° with 0.70 *N* sodium citrate buffer, pH 5.28 (see Fig. 36).

Many other related compounds in blood plasma, urine and tissue extracts can be separated on a 50-cm column and a 150-cm column operated at 30° and 50°. The elution schedules are given in Fig. 37 A and B.

Estimation of amino acids by ninhydrin

Ninhydrin reagents developed for the estimation of amino acids have been described by Moore and Stein[14], Cocking and Yemm[16], Hanes and co-workers (see p. 19), Troll and Cannon[17], and Heilmann *et. al.* (see p. 51). The following details are from Moore and Stein[15].

Hydrindantin is prepared by adding 80 g of ascorbic acid in 400 ml of water at 40°, to 80 g of ninhydrin in 2 l of water at 90°. The mixture is left for 30 min and then cooled to room temperature. The hydrindantin is filtered off, washed well with water and dried over P_2O_5 in a vacuum desiccator protected from light.

4 *N* Sodium acetate buffer (pH 5.5) is prepared from 2720 g of sodium acetate (trihydrate) and 500 ml of glacial acetic acid in a total volume of 5 l. The undiluted solution should be at pH 5.51 ± 0.03. If final adjustment of the pH is necessary, 5 g of NaOH correspond to about 0.04 pH unit.

The ninhydrin reagent is made up by adding 250 ml of the buffer at pH 5.5 to a solution of 20 g ninhydrin and 3 g of hydrindantin in 750 ml of methyl Cellosolve (monomethyl ether of ethylene glycol;

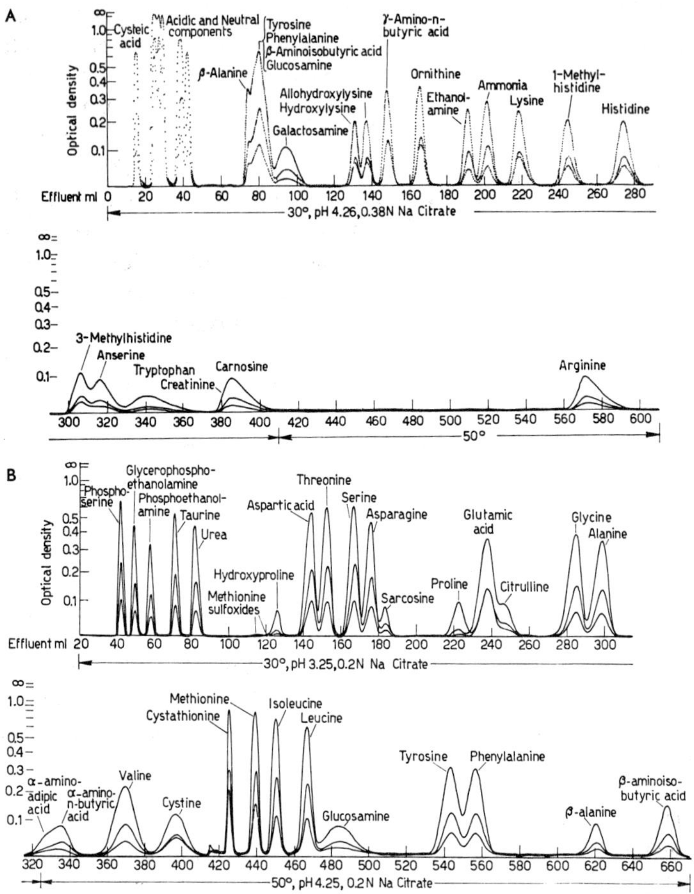

Fig. 37. Chromatographic analysis of mixtures of amino acids and related compounds most commonly encountered in physiological fluids. *A*. 50-cm column at 30° and 50°; *B*. 150-cm column of Amberlite IR-120 at 30° and 50°. Glutamine has the same rate of travel as asparagine; cysteic acid and phosphoserine are not separated; galactosamine, if present, emerges with tyrosine and phenylalanine in A; homocysteine precedes but partially overlaps β-alanine; felinine is slightly more retarded than citrulline, but overlaps it (from Spackman *et al.*[8]).

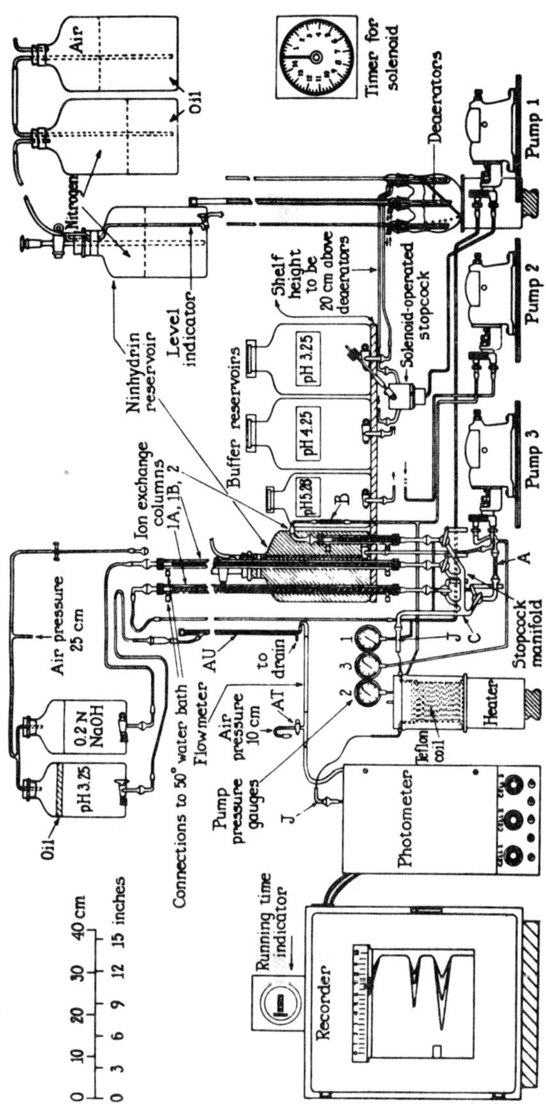

Fig. 38. Automatic recording apparatus used in chromatographic analysis of mixtures of amino acids (from Spackman *et al.*[8]). Influent buffers are drawn from the reservoir bottles through the deaerating equipment and pumped through the ion-exchange columns at a constant rate. Ninhydrin reagent is pumped into the influent stream at C and the effluent-ninhydrin mixture is carried in a Teflon coil through a boiling water bath to develop the colours formed when ninhydrin reacts with amino acids. The extinction value of the coloured solution is measured continuously by a photometer assembly and registered on a recorder.

distilled over ferrous sulphate and shown to be peroxide free). Methyl Cellosolve will remain peroxide-free for at least 4 months if stored over a zinc–copper couple (6 strips of zinc foil, 2 × 15 cm, previously immersed for 1 min in acidified 5% copper sulphate solution, for 2 l of solvent)[18]. The ninhydrin solution is stored in a dark dispenser bottle connected to a nitrogen reservoir as shown at the top of Fig. 38. It is not very stable at room temperature and not more than 2 weeks' supply should be prepared at once.

For analysis of 1 ml or 2 ml effluent fractions, 1 ml of the ninhydrin solution is added, the capped tubes are shaken for a few seconds and then heated in a covered boiling water-bath for exactly 15 min. It is convenient to analyze racks of 50 fractions at the same time. After development of the purple colour the fractions are diluted with 5 ml of ethanol–water (1:1), cooled in front of an electric fan and shaken for 30 sec to oxidize most of the residual hydrindantin. The extinction values are read at 570 mμ and 440 mμ for proline and hydroxyproline. The same coloured product, diketohydrindylidene-diketohydrindamine, is formed in the ninhydrin reaction from all the α-amino acids, except cysteine[14], according to the reaction shown on page 86.

With proline and hydroxyproline yellowish red products are formed which have an absorption maximum at 440 mμ. The colour yields from different amino acids vary slightly and these values calculated on a molar basis relative to leucine are listed in Table 14. The colour yields are relatively independent of the initial pH of the sample provided the substance analyzed is dissolved in 0.2 *N* buffers in the pH range 3 to 7.

The amino acid composition of chymotrypsinogen B (Table 15) as determined by Kassell and Laskowski[20] is a recent example of an analysis carried out using an amino acid analyzer. The samples of protein (15–20 mg) were hydrolyzed with 4 ml of redistilled constant boiling hydrochloric acid in Pyrex tubes sealed in a high vacuum and placed in boiling toluene (110.8°) for 7, 11, 16, 30 and 70 h. The HCl was removed from each hydrolysate by evaporation in a vacuum desiccator.

It has been recently pointed out by Ikawa and Snell[21] that during evaporation of a hydrolysate in a vacuum desiccator a compound of glutamic acid and serine may be formed which chromatographs between alanine and cystine. In the analysis of Kassell and Laskow-

TABLE 14

COLOUR YIELDS FROM AMINO ACIDS AND RELATED COMPOUNDS ON MOLAR BASIS RELATIVE TO LEUCINE[15]

Determined on 2-ml samples of 0.1 mM solutions in the buffers (pH 2.2 to 5) in which the compounds emerge in the Dowex 50-X4 chromatographic procedure; heating time 15 minutes; read at 570 mμ. The urea and creatinine were 3.0 mM.

Compound	*Colour yield*	*Compound*	*Colour yield*
Aspartic acid	0.94	β-NH_2-isobutyric acid	0.44
Threonine	0.94	Carnosine	0.93
Serine	0.95	Citrulline	1.04
Proline (440 mμ)	0.225	Creatinine	0.027
Glutamic acid	0.99	Cysteic acid	0.99
Glycine	0.95	Diaminopimelic acid (per 2 NH_2 groups)	1.24
Alanine	0.97		
Valine	0.97	Ethanolamine	0.91
Half cystine	0.55	Felinine	0.95
Methionine	1.02	Glutamine	0.99
Isoleucine	1.00	Glucosamine	1.03
Leucine	1.00	Glutathione (oxidized, half)	0.93
Tyrosine	1.00	Glycerophosphoethanolamine	0.50
Phenylalanine	1.00	Hydroxylysine	1.12
Ammonia	0.97	Hydroxyproline (440 mμ)	0.077
Lysine	1.10	Methionine sulphone	1.02
Histidine	1.02	Methionine sulphoxide	0.98
Tryptophan	0.94	1-Methylhistidine	0.88
Arginine	1.01	3-Methylhistidine	0.86
α-NH_2-adipic acid	0.96	Ornithine	1.12
β-Alanine	0.50	Phosphoethanolamine	0.43
Anserine	0.78	Sarcosine	0.28
Asparagine	0.95	Taurine	0.88
α-NH_2-n-butyric acid	1.02	Urea	0.0314
γ-NH_2-butyric acid	1.01		

References p. 107

TABLE 15

AMINO ACID COMPOSITION OF CHYMOTRYPSINOGEN B[20]

(Total N is 16.2%)

Amino acid	Protein: As amino acid (% dry wt.)	Protein: As amino acid residue (% dry wt.)	N (% of total N)	Calculated No. of residues per 100 g protein × 10³	Calculated No. of residues per molecule, mol wt. 24,000, with standard deviation	α-Chymotrypsinogen No. of residues per molecule, mol wt., 25,100[22]
Alanine[a]	7.60	6.06	7.37	85.3	20.5 ± 0.5	21.7
Arginine[b]	3.58	3.21	7.11	20.6	4.9 ± 0.1	4.0
Aspartic acid[c]	10.42	9.01	6.77	78.3	18.8 ± 0.4	21.8
Half-cystine[d]	3.91	3.33	2.81	32.6	7.8 ± 0.4	10.0
Glutamic acid[a]	9.67	8.49	5.68	65.8	15.8 ± 0.6	14.2
Glycine[a]	6.32	4.80	7.28	84.2	20.2 ± 0.6	23.3
Histidine[e]	1.21	1.07	2.02	7.8	1.9 ± 0.1	1.9
Isoleucine[f,g]	4.35	3.75	2.87	33.1	8.0[h]	9.9
Leucine[b]	9.28	8.01	6.12	70.8	17.0 ± 0.5	18.8
Lysine[b]	6.26	5.49	7.40	42.9	10.3 ± 0.2	13.2
Methionine[d]	2.37	2.08	1.37	15.9	3.8 ± 0.2	1.9
Phenylalanine[b]	4.39	3.92	2.30	26.6	6.4 ± 0.2	6.5
Proline[i]	5.70	4.81	4.29	49.5	11.9 ± 0.3	8.7
Serine[j]	8.11	6.72	6.67	77.2	18.5 ± 0.5	30.1
Threonine[j]	10.00	8.49	7.26	84.0	20.2 ± 0.2	23.0
Tryptophan[k]	5.24	4.78	4.44	25.7	6.2 ± 0.2	7.0
Tyrosine[l]	2.28	2.05	1.09	12.6	3.0 ± 0.1	4.1
Valine[f]	10.34	8.75	7.63	88.3	21.2[h]	22.4
Amide groups[m]			5.45	63.0	15.1 ± 0.4	24.1
Total		94.9	95.9		219[n] = 23,086	

[b] Average of 30- and 70-hour hydrolysates. For arginine, the value given is the average of the amino acid analyzer data (3.53%) and of a colorimetric method (3.63%).

[c] Taken from hydrolysate of oxidized protein.

[d] Average of amino acid analyzer data and a colorimetric method.

[e] Average of 16-, 30- and 70-hour hydrolysates.

[f] 70-hour hydrolysate (3 determinations).

[g] Includes allo-isoleucine.

[h] Insufficient number of determinations to calculate standard deviation.

[i] Average of 5.76% (amino acid analyzer), 5.55% (naphthoquinonesulphonate method), and 5.78% (acid-ninhydrin method).

[j] Intercept for zero time calculated for a second order reaction.

[k] Average of 5.27% (alkaline–stannite hydrolysis, colorimetric method), and 5.21% (alkaline hydrolysis, amino acid analyzer).

[l] Average of 2.18% (acid hydrolysis, amino acid analyzer) and 2.38% (alkaline stannite hydrolysis, colorimetric method).

[m] Average of 14.9 residues/24,000 by acid hydrolysis and 15.2 residues/24,000 by alkaline hydrolysis. Two preparations gave the same results.

[n] Lysine and phenylalanine are counted as the next highest residues. By quantitative paper chromatography, 10.9 residues of lysine and 6.4 residues of phenylalanine were found. Phenylalanine is more likely to be low than high in both methods.

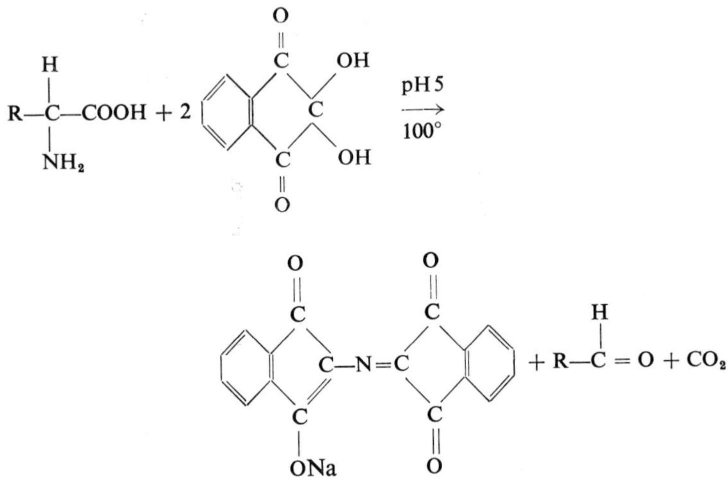

ski the values for glutamic acid and serine might be low for this reason.

From the analytical data the authors were able to divide the amino acids into three groups:

(1) Those giving constant values between 11 and 70 h of hydrolysis: aspartic acid, glutamic acid, proline, alanine, methionine, glycine and tyrosine. The arithmetic mean was used.

(2) Those for which values increased with time of hydrolysis: lysine, histidine, arginine, valine, isoleucine, leucine and phenylalanine. The average of the constant part of the curve was taken.

(3) Those whose values decreased with time of hydrolysis: threonine and serine. Second order reaction curves were drawn.

Two further points of interest about Table 15 are the additional results of independent methods of analysis and the comparison with the analytical data for α-chymotrypsinogen of Wilcox *et al.*[22]. A valuable and up-to-date reference work on primary protein constitution giving tables of amino acid composition of many purified proteins was compiled by Tristram and Smith[23] in 1963.

Hydrolysis of proteins may also be carried out by simple refluxing with glass-distilled constant-boiling HCl or with formic acid–HCl (1:1) for 22–72 h (5 ml/mg protein). Under these conditions the recoveries of serine and threonine are usually taken to be 90% and

95% respectively. Conversion of serine to *O*-sulphoserine has been observed[24] after acid hydrolysis in the presence of sulphate ion. Appreciable quantities of sulphate or phosphate should be removed from protein solutions before hydrolysis. The behaviour of protein-bound cystine during acid hydrolysis has been investigated by Fletcher and Robson[25]; bis-(2-amino-2-carboxyethyl) trisulphide appears to be the main decomposition product, and since this is largely decomposed to cystine and other products under standard conditions of ion-exchange chromatography it would not normally be detected[26]. Tryptophan is destroyed under the conditions of hydrolysis by acids and this amino acid is usually estimated by a spectrophotometric method or by ninhydrin after hydrolysis of the protein by alkali (see Ch. 11). Chlorotyrosines and bromotyrosines are possible transformation products of tyrosine during acid hydrolysis[27].

Cation exchange resins in the H^+ form have been used for the hydrolysis of proteins particularly in studies on mucoproteins. During acid hydrolysis monosaccharides may be distinguished by their different rates of liberation, and the function of the resin is to bind amino sugars and amino acids thus minimizing sugar–amino acid interactions. The following two examples illustrate the procedure. Mucoprotein (50 mg), water (2.5 ml) and prepared ion-exchange resin (Amberlite IR-120; 1 g) were heated in a closed tube with a rubber cap at 95–100° for 24–30 h[28]. Luteinizing hormone (2 mg) was hydrolyzed in 1 ml of a suspension of resin in 0.25 *N*

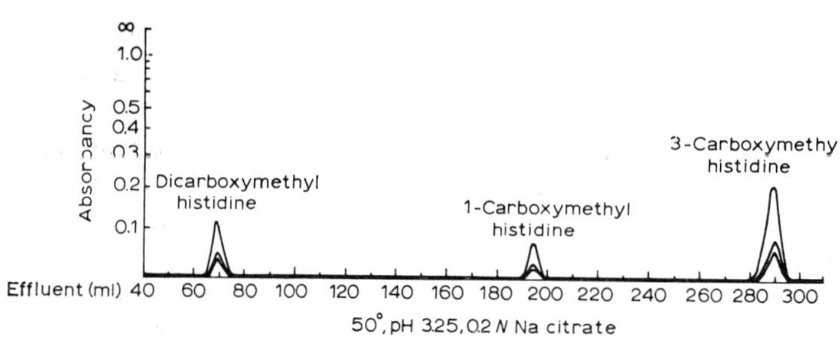

Fig. 39. Chromatographic separation of carboxymethyl derivatives of histidine. The 150-cm column of Amberlite IR-120 on the amino acid analyzer was used. Crestfield *et al.*[31].

HCl (Dowex 50 × 12, 200–400 mesh; 1 mg moist resin per 3 ml dilute acid). Hexosamines were eluted from the resin with 2 *N* HCl[29,30].

Carboxymethyl amino acids arising from alkylation reactions used by protein chemists (see p. 124) have been chromatographed on Amberlite IR-120. The S-carboxymethylcysteine peak emerges in front of aspartic acid; the positions of the three histidine derivatives[31] are shown in Fig. 39.

Desalting of amino acid solutions may be accomplished in a variety of ways (see p. 268), and ion-exchange resins have been used for this purpose in the past with varying success. In the conventional two-stage process the neutral and acidic amino acids are adsorbed on Dowex 2 and eluted with acetic acid, and the basic amino acids and tryptophan adsorbed on Dowex 50 and eluted with HCl. Alternatively, the amino acids are adsorbed on Dowex 2 in the OH^- form, to remove cations, and then eluted with 1 *N* HCl. The chloride is removed with the bicarbonate form of the resin[32].

A novel type of resin, AG11A8, is available which possesses both cation and anion exchange sites, and by a process of ion retardation desalting is possible in one operation[33]. The resin shows a high preference for retarding chloride and its accompanying cation, but it is not suitable for removing citrate and acetate. Adsorbed ions, except for hydrogen ions, can be eluted from the resin with water.

Isolation of pure amino acids

It is possible to isolate amino acids by elution methods but not on the scale indicated above for displacement development. Hirs *et al.*[34] used four columns of ion-exchange resins for the isolation of amino acids present in an acid hydrolysate obtained from 2.5 g of bovine serum albumin. Elution was effected with ammonium formate or acetate buffers in the pH range 3 to 7 and the buffers were removed by sublimation at 40°. All the amino acids with the exception of methionine were obtained analytically pure in an average yield of 60%. About 50% of the cystine was racemized during hydrolysis. Volatile acids have also been used. Hirs *et al.*[35] described a procedure where aspartic acid, glutamic acid and tyrosine were isolated by elution from Dowex 1-X8 acetate with 0.5 *N* acetic acid and the remaining amino acids eluted from Dowex 50-X4 with 1 to 4 *N* hydrochloric acid (Fig. 40 A and B).

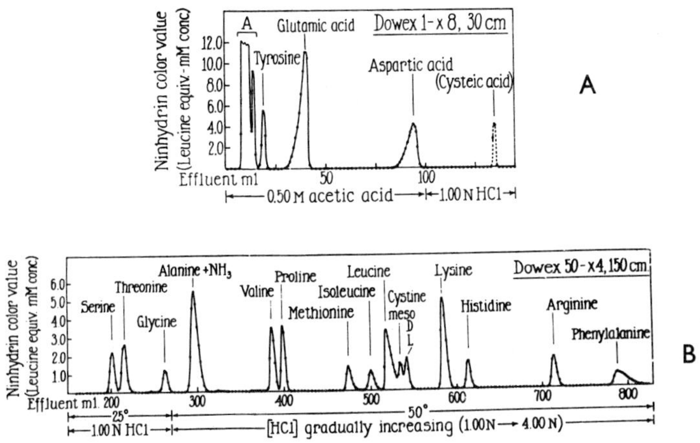

Fig. 40 A and B. Elution of amino acids from Dowex 1-X8 with 0.5 *N* acetic acid at 25° and from Dowex 50-X4 with 1 to 4 *N* HCl at 25° to 50°. The sample added to the second column was the mixture of the 14 amino acids comprising peak A obtained from the first column (from Hirs *et al.*[35]).

Automatic recording apparatus

Much labour is involved during the course of structural work on proteins and it is inevitable that an element of automation should enter into this field. An automatic recording apparatus for chromatographic amino acid analysis was constructed in 1956 by Spackman *et al.*[36], and full details were published in 1958 by the Rockefeller Institute group[8].

The buffers are pumped through the column at a flow rate of 30–40 ml per h and ninhydrin solution is introduced at a constant rate into the effluent flowing from the column. Colour development is accomplished by pumping the mixture through a coil of Teflon capillary tubing immersed in a boiling water-bath. The intensity of the colour formed when amino acids are present is measured by continuous photometry at 570 mμ and at 440 mμ for proline and hydroxyproline. The diagram of the apparatus in Fig. 38 (p. 81) is selfexplanatory. Among the problems to be solved for the successful design and operation of such equipment are (a) maintenance of constant flow rates of buffer and ninhydrin solution, (b) avoidance of mixing of the effluent after the solution leaves the column, (c) control of bubble formation in the heating coil and (d) avoidance of precipitation of chemicals as the solution cools after leaving the heating coil. The latter difficulty has been met by employing a ninhydrin reagent containing methyl Cellosolve (3 l), 4 *M* sodium acetate (1 l), ninhydrin (80 g) and stannous chloride ($SnCl_2 \cdot 2H_2O$, 1.600 g). A more recent method[37] uses sodium propionate (201.8 g), propionic acid (93 ml), methyl Cellosolve (500 ml), ninhydrin (20 g) and water (250 ml). The more favourable solubility of ninhydrin in this medium should eliminate clogging of capillary lines and the reagent need not be protected from light or air. Cyanide of final concentration 25×10^{-6} *M* is also included, not in the reagent itself but in the column buffers. The use of cyanide in this connexion has been discussed by Rosen[38].

An important feature of the performance of automatic apparatus is the remarkably constant base line of $\pm$ 0.001 optical density unit before and after a peak. Useful results are obtained with as little as 1/10 μmole of solute and in view of the excellent reproducibility of effluent volumes the identification of components of a mixture present in low concentration is feasible.

Further instrumentation has brought additional valuable aids *e.g.* a stream divider diverts a fraction of the effluent from a preparative column for analysis while the main portion is channeled off to an external fraction collector, an expander unit increases the sensitivity of the instrument, an electronic integrator reduces time spent on calculations. Automatic equipment is available commercially. A model based on the original specification of the Rockefeller Institute group is illustrated in Fig. 41 (similar equipment is made in the U.K. by Evans Electroselenium, London). Another type of commercial apparatus (Bender and Hobein, Munich) employs as a pumping device three glass syringes fitted with precision ground pistons with sealing rings. The design of the equipment is based on the method of Hannig[39].

Interest in the automatic method is growing and pioneers in the art are striving for improved resolution, greater accuracy and higher speeds of operation. As a result of a fresh look at the underlying mechanisms of column behaviour a series of papers giving data of

Fig. 41. Beckman/Spinco amino acid analyzer. (By permission of Beckman Instruments, Inc., Spinco Division, California.)

considerable practical importance has appeared from Hamilton and co-workers[40–43]. Some of their findings are summarized below.

1. The assumption of a linear isotherm is realistic. The distribution coefficient, K_d, does not appear to be a function of concentration and operational factors affecting concentration.
2. K_d varies inversely as the pH over the range 2.85–3.25. A change of 0.01 pH unit shifts the position of a peak approximately one fraction volume (1 ml).
3. The variance (spread of peak) bore an inverse relationship to sodium ion concentration. Under the particular experimental conditions a change of 0.002 in the normality of sodium concentration changed the position of the glycine peak about one fraction.

References p. 107

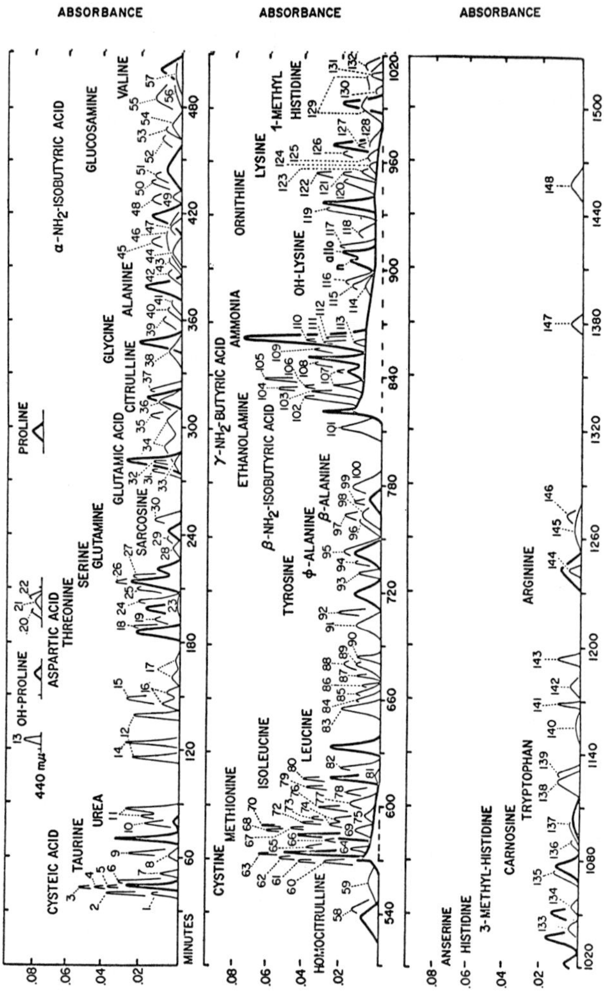

Fig. 42A. Position of amino acids, peptides, amino sugars and amino alcohols on the chromatogram according to Hamilton[43].

Fig. 42B. Ninhydrin reactive compounds corresponding to number peaks in *A*.

1. Penicillaminic acid
2. *O*-Phospho-4-hydroxyproline
3. *O*-Phosphoserine
4. *O*-Phosphothreonine
5. Homocysteic acid
6. Cysteine sulfinic acid
7. Glycerylphosphorylaminoethanol
8. Glucosamine-6-phosphate
9. *O*-Phosphoethanolamine
10. Levulinic acid

11. *threo*-β-Hydroxyaspartic acid
12. *erythro*-β-Hydroxyaspartic acid
13. 3-Hydroxyproline
14. Glucosaminic acid
15. *O*-Phosphohydroxylysine
16. Methionine sulfoxides
17. Ethionine sulfoxides
18. Methionine sulfone
19. β-Methylaspartic acid
20. *allo*-3-Hydroxyproline
21. *allo*-4-Hydroxyproline
22. Thioproline
23. 2-Thiolhistidine
24. *allo*-Threonine
25. Ethionine sulfone
26. Asparagine
27. 2-Amino-3-ureidopropionic acid
28. α-Methylserine
29. Muramic acid (3-*O*-Carboxyethyl-D-glucosamine)
30. Homoserine
31. Glutamic acid-γ-methyl ester
32. Glutamic acid-γ-ethyl ester
33. α-Methylglutamic acid
34. Isovalthines
35. *S*-Methylcysteine
36. Felinine
37. Felininic acid
38. Penicillamine
39. β-2-Thienylserine
40. 2-Aminoadipic acid
41. Alanylasparagine
42. *meso*-Lanthionine (peak 1)
43. Formiminoglycine (peak 1)
44. Glycylasparagine
45. *S*-Ethylcysteine
46. Glutathione (oxidized)
47. *meso*-Lanthionine (peak 2)
48. Glycylserine
49. Isoserine
50. Allylglycine
51. α-Amino-*n*-butyric acid
52. *threo*-β-Phenylserine
53. Glutamylglutamic acid
54. Isovaline
55. Manosamine
56. Formiminoglycine (peak 2)
57. Galactosamine
58. Pipecolic acid
59. Glycylglycylglycylglycine
60. *meso*-Cystine
61. Glutamic acid-γ-hydrazide
62. α-Aminopimelic acid
63. Glycylsarcosine
64. Alanylglycylglycine
65. Alanylalanine
66. 3,5-Dinitrotyrosine
67. α Methylmethionine
68. Phenylglycine
69. Alanine anhydride
70. Djenkolic acid
71. (Deleted)
72. Norvaline
73. Glycylglycine
74. Cystathionine
75. α-Phenyl-α-alanine
76. *allo*-Isoleucine
77. Glycylglycylglycine
78. Glycylalanine
79. Methionine sulfoximine
80. α-ε Diaminopimelic acid
81. 2-Methylleucine
82. Ethionine
83. Norleucine
84. 3,4-Dihydroxyphenylalanine (DOPA)
85. β-2-Thienylalanine
86. α-Hydroxy-δ-aminovaleric acid
87. 1-Aminocyclopentane-1-carboxylic acid
88. Alanylvaline
89. *meta*-Tyrosine
90. γ-Amino-β-hydroxybutyric acid
91. Glycylvaline
92. *ortho*-Fluorophenylalanine
93. *meta*-Fluorophenylalanine
94. Isoglutamine
95. Alanylmethionine
96. *ortho*-Tyrosine
97. Leucylglycine
98. Leucylglycylglycine
99. *para*-Fluorophenylalanine
100. Alanylnorvaline
101. Glycylnorvaline
102. Alanylleucine
103. Glycylethionine
104. δ-Aminolevulinic acid, methylester
105. Homocystine
106. Glycylleucine
107. Argininosuccinic acid
108. Alanylnorleucine
109. Glycylnorleucine
110. Alaninol (2-Amino-1-propanol)
111. Glycine amide
112. Glycylmethionine
113. β-Amino-*n*-butyric acid
114. Glycyltyrosine
115. Glycamine (methylamine)
116. α-Aminocapyryllic acid
117. Glycylphenylalanine
118. Alanylphenylalanine
119. 2,4-Diaminobutyric acid
120. Alamine (ethylamine)
121. δ-Amino-*n*-valeric acid
122. 2,3-Diaminopropionic acid
123. α-Amino-γ-phenylbutyric acid
124. 2-Amino-1-butanol
125. 5-Hydroxytryptophan
126. Kyneurenine
127. Phenylalanylglycine
128. β-Phenyl-β-alanine
129. 3-Aminotyrosine
130. 6-Hydroxytryptophan
131. δ-Aminolevulinic acid
132. Leucyltyrosine
133. Glycyldehydrophenylalanine
134. 4-Aminophenylglycine
135. Homocarnosine
136. 4-Hydroxytryptophan
137. 3-Iodotyrosine
138. Canavanine
139. 2-Amino-3-guanidinopropionic acid
140. ε-Amino-*n*-caproic acid
141. 2-Amino-4-guanidinobutyric acid
142. 3,5-Dibromotyrosine
143. Leucine amide
144. Homocysteine thiolactone
145. 3,5-Diiodotyrosine
146. 5-Methyltryptophan
147. Glycyltryptophan
148. Homoarginine

References p. 107

4. Greater sensitivity is obtained using columns of small diameter. $W = C_1 (Z)^{\frac{1}{2}} = C_2 (A)$ where W is the peak width; Z, the column length (cm); A, the area of cross-section (cm^2); and C_1, C_2 are constants.
5. Resolution varies with column length in proportion to the $\frac{1}{2}$ power. Increased resolution accompanies increased cross-linking of the resin and decreased particle size.

The system finally devised by Hamilton uses a single 0.636 × 125 cm column packed with 17.5×10^{-4} cm spherical particles. One pump circulates ninhydrin solution and a second pump delivers buffer to the top of the column. The various buffers are selected by solenoid operated valves. The time for a complete amino acid analysis is 21 hours and an accuracy of ± 5% is claimed for 0.01 μM quantities. The operation schedule for this system is shown in Table 16 and the positions of 185 ninhydrin reactive substances on the chromatogram are illustrated in Figs. 42 A and B.

TABLE 16

OPERATION SCHEDULE FOR METHOD OF HAMILTON[43]

Time (min)	*Temperature changes*
0–140	Column at 45° C
141	Change to 60° C
781	Change to 45° C
981	Change to 60° C
	Buffer and operational changes
0–460[a]	pH 2.875 buffer
461	Change to pH 3.80 buffer[b]
731	Change to pH 4.30 buffer[c]
1211	Change to column wash[d]
1241	Change to pH 2.875 (regeneration)
1265	Ninhydrin pump off
1270	Recorder off
1420	Circulating bath motor off
1425	Buffer pump off
1430	Program timer plus V_4 solenoid off

[a] 1 min corresponds to effluent volume from column of 0.5 ml.
[b] Emerges from column at min. 570.
[c] Emerges from column at min. 810.
[d] 0.2*N* NaOH +0.3% BRIJ 35 + 0.1% Versene.

Accelerated procedure of Spackman

An improved system for the amino acid analyzer capable of providing two complete analyses per day was reported by Spackman[44] in 1963. The essential details are as follows:

Column 1: dimensions, 0.6 × 10 cm; resin particle distribution 19–25 μ; analysis time 1.5 h; for basic amino acids.

Column 2: dimensions, 0.9 × 60 cm; resin particle distribution 19–31 μ; analysis time 5 h; for acidic and neutral amino acids.

Buffers and ninhydrin are the same as for the 22 hour method but the flow rates are increased to 40 ml/h and 20 ml/h respectively. A longer reaction coil is necessary and the columns are operated at 52°. A roughly two-fold increase in sensitivity is obtained and accuracy and precision are within ± 3% for sample loads down to 0.05 μM. A further modification[78] which makes use of a newly available 8% cross-linked spherical particle resin reduces the total analysis time to 4 h. 5 min.

Internal standards

In order to monitor the daily reliability of determinations an internal standard may be incorporated into samples before application to the column. Norleucine[45], β-thienylalanine[46,47] and β-guanidine-α-aminopropionic acid[45,47] have been used for this purpose and the first of these is probably the most stable to acid hydrolysis[48]. The value of using an internal standard to assess errors due to mechanical losses in preparing protein hydrolysates was pointed out by Walsh and Brown[45]. A known quantity of the standard *e.g.* norleucine, is added to a measured quantity of a protein solution before hydrolysis. The percentage recovery of the internal standard in the hydrolysate provides a direct measure of the percentage recovery of the amino acids of the protein solution.

Analysis of *replicate samples* simultaneously is feasible under certain circumstances. The limit to the number of samples that may be applied to a column will obviously depend on the complexity of the sample. Quadruplicate application of a mixture of glutamic acid, aspartic acid, glycine and valine, on a standard 150 cm column at 50° with *no* buffer change is possible. Samples are applied after 15–20 min pumping periods[47].

References p. 107

Polymorphism

Some proteins occur in two or more genetically different forms and the question arises as to whether automated amino acid analysis is sufficiently accurate to detect small differences in composition. The studies of Piez *et al.*[49] on the β-lactoglobulins illustrate very well one approach to the problem.

TABLE 17

AMINO ACID COMPOSITION OF β-LACTOGLOBULINS A AND B[49]

Data converted to μmoles in 1.002 mg β-lactoglobulin A and 0.997 mg β-lactoglobulin B

Amino acid	*Corrected values (µmoles)*			*Residues protein equivalent µmoles/0.055*		*nearest integer*	
	β-lacto-globulin A	*β-lacto-globulin B*	*difference*	*β-lacto-globulin A*	*β-lacto-globulin B*	*β-lacto-globulin A*	*β-lacto-globulin B*
Aspartic acid[a]	0.8811	0.8268	0.0543	16.0	15.0	16	15
Threonine	0.444	0.444		8.1	8.1	8	8
Serine	0.381	0.380		6.9	6.9	7	7
Glutamic acid	1.373	1.371		25.0	25.0	25	25
Proline	0.449	0.453		8.2	8.2	8	8
Glycine	0.1670	0.2207	0.0537	3.0	4.0	3	4
Alanine	0.7626	0.8184	0.0558	13.9	14.9	14	15
Half-cystine	0.212	0.226		3.9	4.1	4	4
Valine	0.546	0.490	0.056	9.9	8.9	10	9
Methionine	0.206	0.205		3.7	3.7	4	4
Isoleucine	0.533	0.533		9.7	9.7	10	10
Leucine	1.205	1.201		21.9	21.9	22	22
Tyrosine	0.196	0.196		3.6	3.6	4	4
Phenylalanine	0.213	0.208		3.9	3.8	4	4
Lysine	0.821	0.815		14.9	14.8	15	15
Histidine	0.103	0.101		1.9	1.8	2	2
Arginine	0.156	0. 157		2.8	2.9	3	3
Amide N	(0.716)	(0.703)		(13.0)	(12.8)	(13)	(13)
Cysteine[b]	0.058	0.059		1.1	1.1	1	1
Tryptophane[c]	0.138	0.138		2.5	2.5	2	2
Total	8.845	8.843				162	162
Minimal molecular weight						18,364	18,278

[a] Those amino acids the concentration of which did not change on hydrolysis are italicized.

[b] Determined on separate samples by reaction with N-ethylmaleimide.

[c] Determined on separate samples by ultraviolet spectroscopy.

Bovine β-lactoglobulin exists in two forms, A and B, the sub-units of which have a molecular weight of 18,300. Samples of A and B (7 mg) were hydrolyzed under nitrogen at 106° in 4 ml of constant boiling HCl during 24, 48 and 96 h. The manipulation of the analytical data will be described briefly with reference to Table 17.

(1) The total number of μmoles of those amino acids that did not change with time of hydrolysis was determined.

(2) Each of the partial totals was made equal to the average by multiplying all the values in each hydrolysate by an appropriate factor.

(3) Corrections for losses or incomplete hydrolysis were then applied and the three hydrolysates averaged.

(4) Assuming the two species of protein contain the same total number of amino acid residues per protein equivalent, the values for β-lactoglobulin B were multiplied by a factor to equalize the total number of micromoles in both proteins.

These manipulations are designed to obviate errors due to sampling. The data were finally converted to μmoles in approximately 1 mg of protein (Table 17). Column 3 shows the differences between the two proteins. Assuming single residue substitutions the differences represent microequivalents of a single residue per mg, and dividing each of the values by the averaged difference gives the number of residues per protein equivalent (columns 4–7). Piez *et al.*[49] provide cogent argument for the validity of the mathematical manipulation and conclude that β-lactoglobulin A has one more residue each of aspartic acid and valine, and one less residue each of glycine and alanine, than β-lactoglobulin B in a sub-unit of 18,300 molecular weight. The purity of the proteins analyzed is always a major consideration but in this instance the investigation was aided by the neat separation of the two species on a DEAE column. Nonetheless, small percentages of impurities would be undetectable.

Statistical analysis

The two egg yolk lipoproteins α- and β-lipovitellin are very similar and early work[50] had indicated that any difference in their amino acid composition was small and comparable with differences between preparations of the same protein. To avoid error arising from the presence of non-protein contaminants Cook *et al.*[51] used the ratio of one amino acid to another as the criterion of similarity and a statistical approach was adopted. This consisted of an analysis of variance of the logarithm of the amount of the more stable amino acids. Fig. 43 shows a plot of the amount in log m$M \times 10$ of each amino acid in decreasing order. The overall similarity in composition is obvious from the slopes of the curves but the starred points

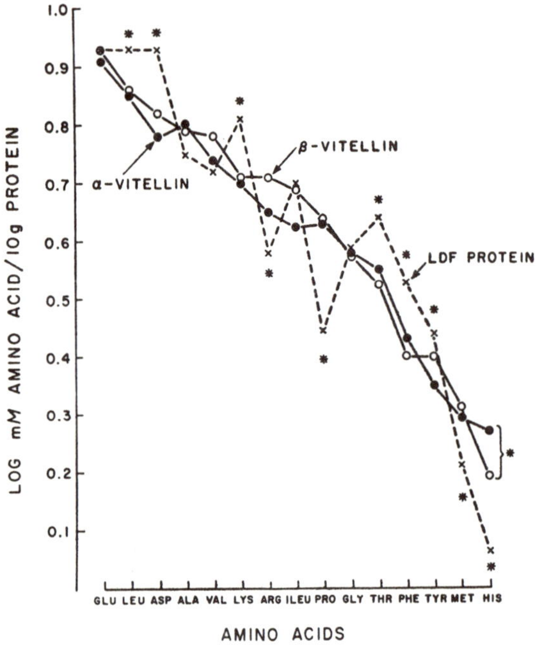

Fig. 43. Content of fifteen amino acids in the three egg-yolk lipoproteins. Starred points indicate amino acids present in significantly higher or lower proportions than in the other proteins. Cook *et al.*[51].

indicate amino acids present in significantly higher or lower proportions than in the other proteins. The low density fraction of egg yolk proteins was also included in the comparison.

The variancy ratio was defined as

$$\frac{\text{mean square (log acid} \times \text{log protein)}}{\text{mean square (log acid} \times \text{log preparation same protein)}}$$

or

$$\frac{\text{disproportionality in the amino acid}}{\text{disproportionality in amino acid in different preparations of same protein}}$$

and a comparison was made between the disproportionality of each amino acid and its standard error. The only significant difference between the vitellins is the lower histidine content of β-vitellin, whereas the low density fraction differs substantially from the other two proteins. These differences, however, were not detected in peptide maps of the three proteins. The sub-units of the vitellins and the protein moiety of LDF protein all have a molecular weight of approximately 2×10^5.

Absolute concentration of protein

The weight concentration of a protein in a solution is readily calculated from the weights of the individual amino acids determined by amino acid analysis, not forgetting to include a value for tryptophan (see p. 349). From the sum of the individual amino acids must be deducted the weight of one molecule of water per peptide bond or $\frac{(n-1) \times 18}{1000}$ mg for n μM total amino acids. Walsh and Brown[45] have discussed the advantages of using the concentration of a specific amino acid residue to define specific parameters *e.g.* enzymic activity, ultraviolet absorbancy, nitrogen content, partial specific volume, etc. If the molecular weight and amino acid composition are known the values for the particular amino acid may be converted to molar quantities of the protein.

ESTIMATION OF CYSTEIC ACID AND RELATED SUBSTANCES

Cysteic acid, S-sulphocysteine and other strongly acidic substances pass through cation exchangers unretarded and although it is possible to estimate these substances in the eluates under ideal conditions a more satisfactory approach is to use an anionic exchanger. Schram *et al.*[52] have described conditions for the oxidation of proteins containing cystine with performic acid, and subsequent chromatography of the cysteic acid on Dowex 2—X10, which permit an overall recovery of 90 $\pm$ 2%. Cysteic acid and methionine are

References p. 107

both sensitive to a large excess of performic acid and Moore[53] subsequently recommended the removal of the excess reagent with hydrogen bromide. The method to be described cannot be used for the complete analysis of a protein owing to the destruction of tryptophan, tyrosine and histidine. With care the recoveries of cysteic acid and methionine sulphone are respectively 94 ± 2% and 100 ± 2%.

Performic acid is prepared by the addition of 1 vol. 30% hydrogen perioxide to 9 vol. 88% formic acid. The solution is allowed to stand for 1 h at room temperature and then cooled at 0°. Two ml of the reagent is added to the protein sample which should contain approximately 0.1 mg of cystine (usually 2–5 mg of protein). The mixture is allowed to stand at 0° for 4 h for soluble proteins, or overnight for insoluble materials. HBr 48% (0.30 ml) is added with swirling of the vessel in the ice-bath. The solution is then concentrated by rotary evaporation at 40°, the bromine which distills over being absorbed in 1 *N* NaOH added to the condenser. The residue is hydrolyzed with 6 *N* HCl (3 ml) under reduced pressure at 110° for 18 h.

Dowex 1-X8 (minus 400 mesh) in the hydrochloride form is washed on a Büchner funnel with *N* HCl. *N* sodium chloroacetate is then passed through until the effluent is free from chloride and a slurry of the resin in pH 3.4 sodium chloroacetate (0.20 *N* in NaOH, 0.25 *N* in monochloroacetic acid, containing 3 ml of BRIJ-35 solution/l) is used to prepare a 0.9 × 60 cm column. After equilibration of the column the sample or an aliquot is applied to the column in the same buffer and the flow-rate of eluent maintained at 4 ml/h. Cysteic acid emerges at an effluent volume of approximately 60 ml; the colour yield for cysteic acid in the system is 0.99 relative to leucine but the exact value should be determined under the user's experimental conditions. For the determination of half-cystine as carboxymethylcysteine see p. 88. Other agents for the oxidation of cysteine and cystine, *e.g.* ozone and periodate, are discussed in Ch. 11.

Conversion to the S-sulpho derivatives and chromatography on a basic resin may prove useful in determining the cysteine and glutathione content of tissue extracts. In an actual case a sample containing 3 mg of glutathione and 0.5 mg of cysteine was treated with 2 ml of 0.2 *M* phosphate buffer at pH 7, 2 ml of 0.1 *M* sodium sulphite at pH 7, and 0.4 ml of 0.5 *M* potassium-*o*-iodosobenzoate (see p. 124). After 30 min at room temperature, one-tenth of the mixture was chromatographed on a 0.9 × 5.0 cm column of Dowex 1-X8 in the chloroacetate form, with 0.1 *M* sodium chloroacetate at pH 3.1 as eluent. The chromatogram obtained is shown in Fig. 44. The ninhydrin colour yield of S-sulpho cysteine relative to leucine in buffer at pH 3.1 is 0.83.[54]

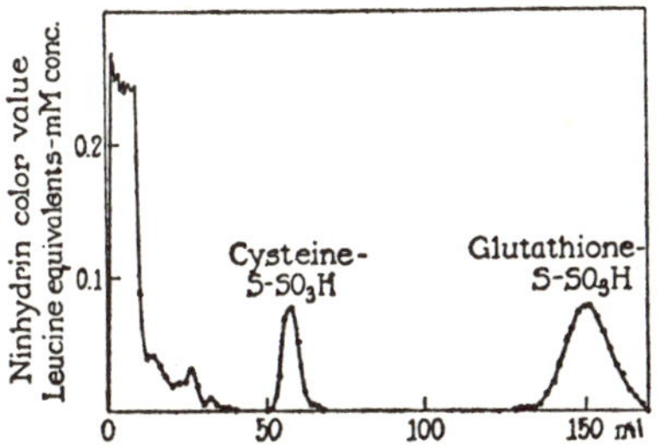

Fig. 44. Chromatography of an extract of cat liver to which cysteine had been added and which had then been treated with sulphite and *o*-iodosobenzoate. Column, Dowex 1-X8, 0.9 × 5.0 cm. (From Bailey and Cole[54].)

SEPARATION OF PEPTIDES

The conditions for the separation of amino acids by ion-exchange chromatography are clearly defined. When peptide mixtures are to be dealt with no hard and fast rules of procedure can be applied as each unknown mixture will present its own difficulties. The problem can only be approached on the basis of experience.

Attempts by Schroeder *et al.*[55], and Kay and Schroeder[56] to fractionate small peptides arising from the acid hydrolysates of gelatin and silk fibroin on Dowex 50-X8 were only partially successful. For satisfactory chromatography of peptides a resin was required which would permit greater penetration of large molecules into the resin matrix. The work of Dowmont and Fruton[57] in 1952 suggested that a low percentage cross-linked resin might be suitable for this purpose and, in fact, useful separations of peptide mixtures derived from lysozyme using Dowex 50-X4 were later reported by Thompson[58]. The properties of Dowex 50-X2 were studied by Hirs *et al.*[59] who were able to chromatograph peptides of up to 20–25 residues obtained from a tryptic hydrolysate of oxidized ribonuclease (Fig. 45).

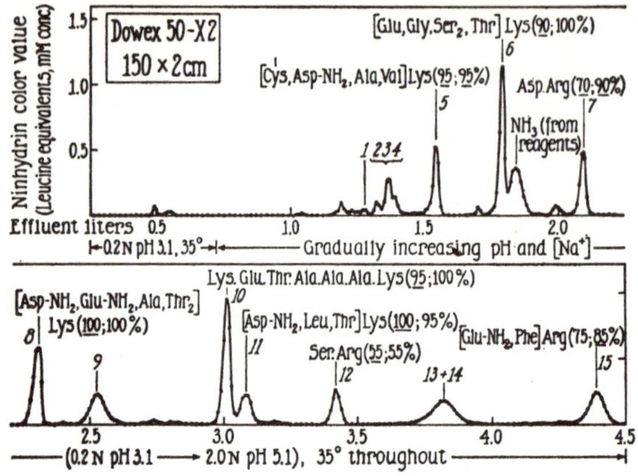

Fig. 45. The peptides in a 20 h tryptic hydrolysate of oxidized ribonuclease. Load, 200 mg of protein hydrolysate; mixing chamber for the influent buffer had a volume of 2450 ml. The figures in parentheses give the yield of each peptide after 3 and 20 h of hydrolysis by trypsin (from Hirs *et al.*[59]).

Peptides obtained by treating this protein with chymotrypsin and pepsin were fractionated similarly by the Rockefeller Institute group[60,61] (Fig. 46).

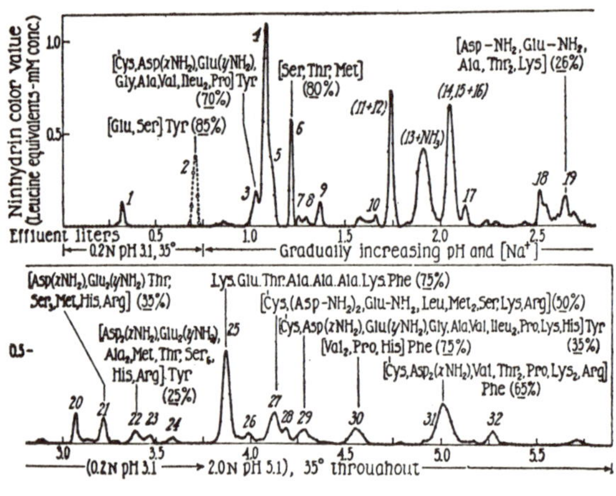

Fig. 46. The peptides in a 24 h chymotryptic hydrolysate of oxidized ribonuclease. Column, Dowex 50-X2, 1.8 × 150 cm; load, 200 mg of protein hydrolysate; 2450 ml mixing chamber from Hirs *et al.*[60].

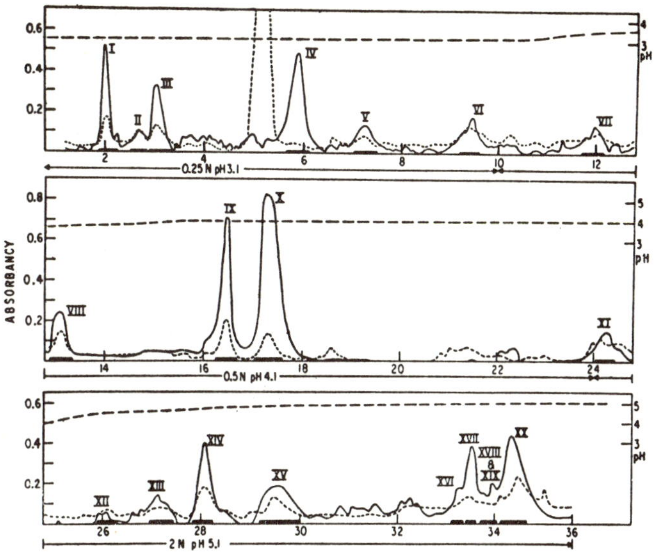

Fig. 47. Separation of the soluble peptides from the chymotryptic digest of 48 μmoles of oxidized papain. Column, Dowex 50-X2, 3 × 150 cm. Analysis by ninhydrin before (----) and after alkaline hydrolysis (——). Light and Smith[62].

Low cross-linked cation exchangers are being used with varying degrees of success in structural studies on other proteins. Fig. 47 shows an elution curve for the soluble peptides derived from oxidized papain by digestion with chymotrypsin (Light and Smith[62]). Of the 187 amino acid residues found in papain, 117 residues were accounted for in the purified peptides. In a similar way Jollès *et al.*[63] purified 17 peptides derived from heat-denatured, hen's-egg lysozyme. The laborious de-salting operations required in much of the early work are now avoided by using volatile buffers composed of ammonia, pyridine or trimethylamine, and formic or acetic acids[64–68]. The fractions obtained from the column are then available for examination by paper chromatography or by paper electrophoresis (Margoliash[69]). Alternatively, when an amino acid analyzer is available, peptide mixtures may be resolved further on a 15 cm column of Amberlite IR-120.

The use of anionic exchangers for peptide fractionation is a comparatively recent development. Rudloff and Braunitzer[70] have

References p. 107

described a procedure utilizing Dowex 1-X2 in the acetate form and consider a net charge effect to be responsible for the separations. Tryptophan and cysteine peptides are not retained by the resin. Dowex 1 has also been used by Anderer *et al.*[71] and Funatsu[72] in their work on the tobacco mosaic virus protein and by Konigsberg and Hill[73] in studies on human haemoglobin.

Fractionation of peptides on Dowex 50-X2 and Dowex 50-X4

The resin is prepared in the manner described for Amberlite IR-120 on p. 72, except that the Dowex resin is stirred with 2 *N* NaOH at room temperature instead of on the steam-bath. Flow rates through the lower cross-linked resin are slower than those obtained with resins of high cross-linking and in order to insure a satisfactory flow rate using the former resin equal volumes of 120–200 mesh and through-200 mesh are mixed. For analytical and exploratory work 1 × 150 cm columns are employed; a preparative column which will take loads of 200 mg or more of hydrolysate should have a diameter of 2–3 cm. The contraction of these columns is quite alarming, but useful fractionations are possible from runs lasting as long as two weeks.

Eluents used for the fractionation of peptides from oxidized ribonuclease[59–61] were the same as those used for the chromatography of amino acids and contained BRIJ35 but no thiodiglycol. The effluents from the 1 cm and 2 cm diameter columns are collected in 2-ml and 10-ml fractions respectively and suitable aliquots taken for analysis by the ninhydrin method. Analysis after alkaline hydrolysis provides a rough guide as to the size of the peptides eluted and also insures that peptides that give a low ninhydrin colour are not overlooked. Alkaline hydrolysis may be carried out with 2.5 *N* NaOH at 90° for 2½ h followed by neutralization of the contents of the tubes (alkali resistant glass) with aqueous acetic acid (1:1). Figs. 45 and 46 show the separation of peptides from a 20 h tryptic hydrolysate and from a 24 h chymotryptic hydrolysate of oxidized ribonuclease. The degree of purity of the eluted peptides is judged from the amino acid composition. Integral molar ratios are expected for a pure peptide. When a mixture of peptides is indicated further fractionation is sometimes possible under different cor.dinations of pH or temperature.

Canfield's work on the structure of egg white lysozyme illustrates the use of Dowex 50-X4 with pyridine buffer eluents. The resin was first washed with 1 *N* NaOH followed by 1 *N* HCl and then treated with 2 *N* pyridine-acetate, pH 5, to establish pyridine as the cation. An aliquot of tryptic digest (350 mg), adjusted to pH 2.8 with formic acid, was transferred to a column of dimensions 2.0 × 150 cm. Gradient elution of increasing pH and salt concentration was used, the eluent being delivered from an eight-chambered Varigrad at 35°. After reaching pH 5.18 the temperature was raised to 50° and the final eluent was 0.5 *N* ammonium acetate. The compositions of

TABLE 18

COMPOSITION OF PYRIDINE–ACETIC ACID BUFFERS USED TO SEPARATE TRYPTIC PEPTIDES OF REDUCED, CARBOXYMETHYLATED LYSOZYME ON DOWEX 50-X4 COLUMN[67]

Chamber No.	*Concentration of acetic acid (M)*	*Acetic acid*	*Pyridine (ml)*	*H_2O to make:*	*pH*
1	0.10	23	4	4000	3.80
2	0.20	23	5	2000	3.90
3	0.30	34	15	2000	4.25
4	0.40	46	30	2000	4.47
5	0.50	58	50	2000	4.65
6	0.75	86	100	2000	4.87
7	1.0	115	150	2000	4.96
8	2.0	230	400	2000	5.18

the buffers used are listed in Table 18. In three instances more than one peptide occurred in one peak and these composite zones were resolved further by electrophoresis at pH 3.6. It is important to equilibrate the columns properly when pyridine buffers are used; several days' equilibration may be required for the removal of ninhydrin positive substances from the resin. In general peptides (particularly those containing aromatic residues) are less retarded in systems containing organic cations[68].

Fractionation of peptides on Dowex 1-X2

In the method of Braunitzer and collaborators[70,74,75] Dowex 1-X2 (200–400 mesh) available in the Cl^- form was converted to the free

References p. 107

base with 1 *N* ammonium hydroxide and then into the acetate form with glacial acetic acid. The resin was washed well with deionized water and the 'fines' removed by decantation. For the fractionation of peptides derived from haemoglobin A columns of dimensions 2 × 150 cm and 3 × 15 cm were used. They were fitted with a heating mantle kept at 35 ± 0.5°. The volatile buffers were prepared from aqueous solutions of collidine and pyridine adjusted to the correct pH with glacial acetic acid and made up with de-gassed water to give the required percentage composition. The sample of peptide mixture was adjusted to 0.5 to 0.7 of a pH unit higher than that of the initial eluent. To regenerate the columns the resin was removed and allowed to stand in several portions of glacial acetic acid. This treatment was followed by conversion to the free base and then to the acetate form again.

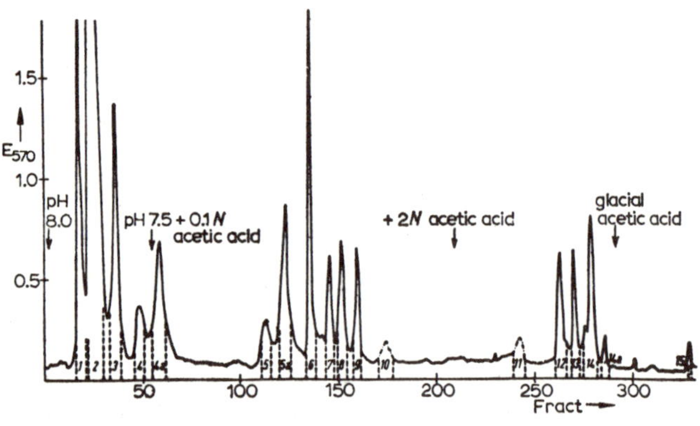

Fig. 48. Separation of the peptides from the tryptic hydrolysate of globin. Column, Dowex 1-X2, 2 × 150 cm. Start: 1% collidine acetate, pH 8.0; gradient, (1% collidine + 1% pyridine acetate, pH 7.5) + 0.1 *N* acetic acid; gradient, +2 *N* acetic acid; final, glacial acetic acid. from Rudloff and Braunitzer[70].

A chromatogram of peptides obtained from a tryptic hydrolysate of globin is shown in Fig. 48. After removal of the volatile buffer by concentration *in vacuo* a sample of each pooled peak was examined by paper chromatography.

Peptides which were strongly basic and also those containing the three tryptophan residues of haemoglobin A were eluted first;

the neutral peptides and many histidyl peptides emerged during gradient elution with 0.1 N acetic acid; gradient elution with 2 N acetic acid yielded the acidic group of peptides.

In similar studies on human haemoglobin by Konigsberg and Hill[73] the first eluent was a buffer at pH 8.4 containing 111 ml of N-ethylmorpholine and 11 ml acetic acid/l. This was followed by more acidic pyridine buffers.

REFERENCES

1 S. M. Partridge and R. C. Brimley, *Biochem. J.*, 51 (1952) 628.
2 S. M. Partridge and R. C. Brimley, *Biochem. J.*, 48 (1951) 313.
3 S. M. Partridge and R. C. Brimley, *Biochem. J.*, 49 (1951) 153.
4 W. H. Stein and S. Moore, *Cold Spring Harbour Symposia Quant. Biol.*, 14 (1950) 179.
5 S. Moore and W. H. Stein, *J. Biol. Chem.*, 192 (1951) 663.
6 S. Moore and W. H. Stein, *J. Biol. Chem.*, 211 (1954) 893.
7 S. Moore, D. H. Spackman and W. H. Stein, *Anal. Chem.*, 30 (1958) 1185.
8 D. H. Spackman, W. H. Stein and S. Moore, *Anal. Chem.*, 30 (1958) 1190.
9 P. N. Campbell, S. Jacobs and T. S. Work, *Chem. Ind. (London)*, (1955) 117.
10 P. B. Hamilton, *Anal. Chem.*, 30 (1958) 914.
11 F. S. Steven and G. R. Tristram, *Biochem. J.*, 83 (1962) 245.
12 G. King, *Drug Cosmetic Ind.*, 90 (1962) 24.
13 D. R. Whitaker and G. A. Strasdine, *Can. J. Biochem. Physiol.*, 41 (1963) 1105.
14 S. Moore and W. H. Stein, *J. Biol. Chem.*, 176 (1948) 367.
15 S. Moore and W. H. Stein, *J. Biol. Chem.*, 211 (1954) 907.
16 E. C. Cocking and F. W. Yemm, *Analyst*, 80 (1955) 209.
17 W. Troll and R. K. Cannan, *J. Biol. Chem.*, 200 (1953) 803.
18 P. B. Hamilton and R. A. Anderson, *J. Biol. Chem.*, 211 (1954) 95.
19 D. R. Kominz, *J. Chromatog.*, 9 (1962) 253.
20 B. Kassell and M. Laskowski Sr., *J. Biol. Chem.*, 236 (1961) 1996.
21 M. Ikawa and E. E. Snell, *J. Biol. Chem.*, 236 (1961) 1955.
22 P. E. Wilcox, E. Cohen and W. Tan, *J. Biol. Chem.*, 228 (1957) 999.
23 G. R. Tristram and R. H. Smith, *Advan. Protein Chem.*, 18 (1963) 227.
24 S. Moore, *J. Biol. Chem.*, 238 (1963) 235.
25 J. C. Fletcher and A. Robson, *Biochem. J.*, 87 (1963) 553.
26 B. Lewis, A. Robson and E. M. Tiler, *J. Textile Inst.*, 51 (1960) T653.
27 F. Sanger and E. O. P. Thompson, *Biochim. Biophys. Acta*, 71 (1963) 468.
28 P. D. Bragg and L. Hough, *Biochem. J.*, 78 (1961) 11.
29 P. A. Anastassiadis and R. H. Common, *Can. J. Biochem. Physiol.*, 36 (1958) 413.
30 E. F. Walborg, Jr. and D. N. Ward, *Biochim. Biophys. Acta*, 78 (1963) 304.
31 A. M. Crestfield, W. H. Stein and S. Moore, *J. Biol. Chem.*, 238 (1963) 2413.

[32] K. A. PIEZ, E. B. TOOPER AND L. S. FOSDICK, *J. Biol. Chem.*, 194 (1952) 669.
[33] C. ROLLINS, L. JENSEN AND A. N. SCHWARTZ, *Anal. Chem.*, 34 (1962) 711.
[34] C. H. W. HIRS, S. MOORE AND W. H. STEIN, *J. Biol. Chem.*, 195 (1952) 669.
[35] C. H. W. HIRS, S. MOORE AND W. H. STEIN, *J. Am. Chem. Soc.*, 76 (1954) 6063.
[36] D. H. SPACKMAN, W. H. STEIN AND S. MOORE, *Federation Proc.*, 15 (1956) 358.
[37] H. ROSEN, C. W. BERARD AND S. M. LEVENSON, *Anal. Biochem.*, 4 (1962) 213.
[38] H. ROSEN, *Arch. Biochem. Biophys.*, 67 (1957) 10.
[39] K. HANNIG, *Clin. Chim. Acta*, 4 (1959) 51.
[40] P. B. HAMILTON, *Anal. Chem.*, 32 (1960) 1779.
[41] P. B. HAMILTON, D. C. BOGUE AND R. A. ANDERSON, *Anal. Chem.*, 32 (1960) 1782.
[42] P. B. HAMILTON, *Ann. N.Y. Acad. Sci.*, 102, 1 (1962) 55.
[43] P. B. HAMILTON, *Anal. Chem.* 35 (1963) 2055.
[44] D. H. SPACKMAN, *Federation Proc.*, 22 (1963) 499.
[45] K. A. WALSH AND J. R. BROWN, *Biochim. Biophys. Acta*, 58 (1962) 596.
[46] F. L. SIEGEL AND M. K. ROACH, *Anal. Chem.*, 33 (1961) 1628.
[47] D. STEFANYE AND L. SPERO, *Anal. Biochem.*, 8 (1964) 265.
[48] M. J. CRUMPTON AND J. M. WILKINSON, *Biochem, J.*, 88 (1963) 228.
[49] K. A. PIEZ, E. W. DAVIE, J. E. FOLK AND J. A. GLADNER, *J. Biol. Chem.*, 236 (1961) 2912.
[50] G. BERNARDI AND A. H. COOK, *Biochim. Biophys. Acta*, 44 (1960) 96.
[51] W. H. COOK, R. W. BURLEY, W. G. MARTIN AND J. W. HOPKINS, *Biochim. Biophys. Acta*, 60 (1962) 98; see also R. J. BLOCK, *Ann. N.Y. Acad. Sci.*, 94 (1961) 31.
[52] E. SCHRAM, S. MOORE AND E. J. BIGWOOD, *Biochem. J.*, 57 (1954) 33.
[53] S. MOORE, *J. Biol. Chem.*, 238 (1963) 235.
[54] J. L. BAILEY AND R. D. COLE, *J. Biol. Chem.*, 234 (1959) 1733.
[55] W. A. SCHROEDER, L. M. KAY, J. LEGETTE, L. HONNEN AND F. C. GREEN, *J. Am. Chem. Soc.*, 76 (1954) 3556.
[56] L. M. KAY AND W. A. SCHROEDER, *J. Am. Chem. Soc.*, 76 (1954) 3564.
[57] Y. P. DOWMONT AND J. S. FRUTON, *J. Biol. Chem.*, 197 (1952) 27.
[58] A. R. THOMPSON, *Biochem. J.*, 61 (1955) 253.
[59] C. H. W. HIRS, S. MOORE AND W. H. STEIN, *J. Biol .Chem.*, 219 (1956) 623.
[60] C. H. W. HIRS, W. H. STEIN AND S. MOORE, *J. Biol. Chem.*, 221 (1956) 151.
[61] J. L. BAILEY, S. MOORE AND W. H. STEIN, *J. Biol. Chem.*, 221 (1956) 143.
[62] A. LIGHT AND E. L. SMITH, *J. Biol. Chem.*, 235 (1960) 3144.
[63] P. JOLLÈS, J. JOLLÈS AND J. JAUREGUI, *Biochim. Biophys. Acta*, 31 (1959) 96.
[64] J. VANĚČEK, B. MELOUN, V. KOSTKA, B. KEIL AND F. ŠORM, *Biochim. Biophys. Acta*, 37 (1960) 169.
[65] V. TOMÁŠEK, V. HOLEYŠOVSKY, O. MIKEŠ AND F. ŠORM. *Biochim. Biophys. Acta*, 38 (1960) 570.
[66] G. KREIL, *Z. Physiol. Chem.*, 334 (1963) 154.
[67] R. E. CANFIELD, *J. Biol. Chem.*, 238 (1963) 2691.
[68] J. R. KIMMEL, G. K. KATO, A. C. M. PAIVA AND E. L. SMITH, *J. Biol. Chem.*, 237 (1962) 2525.
[69] E. MARGOLIASH, Discussion to F. SANGER, *J. Polymer Sci.*, 49 (1961) 3.
[70] V. RUDLOFF AND G. BRAUNITZER, *Z. Physiol. Chem.*, 323 (1961) 129.

[71] F. A. ANDERER, H. UHLIG. E. WEBER AND G. SCHRAMM, *Nature*, 186 (1960) 922.
[72] G. FUNATSU, *Biochemistry (Tokyo)*, 3 (1964) 1351, 1356.
[73] W. KONIGSBERG AND R. J. HILL, *J. Biol. Chem.*, 237 (1962) 2547.
[74] G. HOBOM AND G. BRAUNITZER, *Z. Physiol. Chem.*, 335 (1963) 1.
[75] N. HILLSCHMANN AND G. BRAUNITZER, *Z. Physiol. Chem.*, 335 (1963) 21.
[76] E. J. COHN AND J. T. EDSALL, *Proteins, Amino Acids and Peptides*, Reinhold, New York, 1943.
[77] J. T. EDSALL AND J. WYMAN, *Biophysical Chemistry*, Vol. 1, Academic Press Inc., New York, 1958.
[78] J. V. BENSON, JR. AND J. A. PATTERSON, *Anal. Chem.*, 37 (1965) 1108.

CHAPTER 4

DISULPHIDE BONDS

One of the most popular pursuits in biochemistry today is an attempt to understand the unique role of sulphur in biological processes. Some idea of the wide range of topics covered by investigators in this field may be gained from an interesting review by Boyer[1] and from the recent *Symposium on Sulphur in Proteins* (Benesch[2]). The biochemical problems are closely linked with an appreciation of the properties of SH and SS groups in proteins and peptides and these fundamental aspects are the subject of extensive studies in protein laboratories throughout the world. One major problem is how to deal with these reactive groups during structural studies on proteins and this chapter is an account of the methods which have been developed to meet this particular problem.

DISULPHIDE INTERCHANGE REACTIONS

Ryle and Sanger[3] observed that after hydrolysis of insulin with cold concentrated hydrochloric acid many more cystine peptides are obtained than can be accounted for by a unique structure for the protein. It was concluded that an interchange of the various peptides was taking place through fission of the disulphide bond — a situation that was complicating further the task of determining which half-cystine residues were joined together in the original insulin. The phenomenon can be readily demonstrated by dissolving cystine and N,N′-bis-dinitrophenyl-cystine in a mixture of concentrated hydrochloric acid and acetic acid (1:2 by vol.) and leaving the mixture at room temperature for 3 to 4 days. After dilution of the mixture with water and extraction of the remaining bis-dinitrophenyl-cystine with ethyl acetate, the coloured aqueous layer can be shown to contain mono-dinitrophenyl-cystine.

It has been suggested that interchange in strongly acid solution proceeds through sulphenium ions[4] according to the scheme

$$R_1SSR_1 + H^+ \rightleftharpoons \begin{bmatrix} R_1SSR_1 \\ | \\ H \end{bmatrix}^+ \rightleftharpoons R_1SH + S^+$$

$$R_1S^+ + R_2SSR_2 \rightleftharpoons R_1SSR_2 + R_2S^+$$
$$R_2S^+ + R_1SSR_1 \rightleftharpoons R_1SSR_2 + R_1S^+$$

Disulphide interchange also takes place in neutral and alkaline solutions and is considered to be initiated by trace amounts of a thiol

$$R_1SSR_1 + R_2S^- \rightleftharpoons R_1SSR_2 + R_1S^-$$
$$R_2SSR_2 + R_1S^- \rightleftharpoons R_2SSR_1 + R_1S^-$$

SH reagents such as N-ethylmaleimide inhibit the interchange in neutral solution; in strongly acid solution, however, rearrangement is largely prevented in the presence of a thiol.

Ryle and Sanger[3] were able to work out satisfactory conditions for the acid hydrolysis of insulin; minimum rearrangement of cystine peptides occurred in the presence of thioglycollic acid in 10 N H_2SO_4 at 37° during 10 days or in the same solvent at 100° during 2 hours. Under the milder conditions of hydrolysis by enzymes rearrangement is avoided by adding N-ethylmaleimide to the digest. These ideas were put to good purpose by Ryle *et al.*[5] in their studies on the isolation of cystine peptides from both acid and enzymic hydrolysates of insulin. The addition of N-ethylmaleimide to buffers during chromatography or electrophoresis of cystine peptides was a useful precaution. With the protein ribonuclease, Spackman *et al.*[6] found that there was no advantage in adding N-ethylmaleimide to tryptic digests and resort was made to hydrolysis by pepsin at pH 2.

An interesting application of the rearrangement of disulphides to protein analysis has been described by Glazer and Smith[7]. Protein disulphide is allowed to exchange over some days with bis-dinitrophenylcystine in 9.6 N HCl until the mono-dinitrophenyl-cystine concentration of the aqueous solution after extraction with ethyl acetate is constant. The rate of exchange varies widely with different proteins and the equilibrium values of the reaction can be used in an accurate estimation of the half-cystine + cystine content.

CLEAVAGE OF DISULPHIDE BONDS

Cystine residues may form the links between two different peptide chains as in insulin (Fig. 49), or they may hold together different

References p. 131

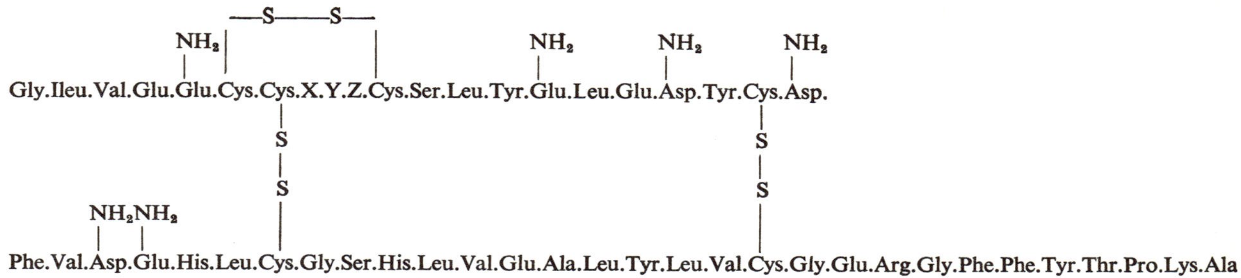

	Horse	Ox	Sheep	Pig	Whale
X	Thr	Ala	Ala	Thr	Thr
Y	Gly	Ser	Gly	Ser	Ser
Z	Ileu	Val	Val	Ileu	Ileu

Fig. 49. The structure of insulin.

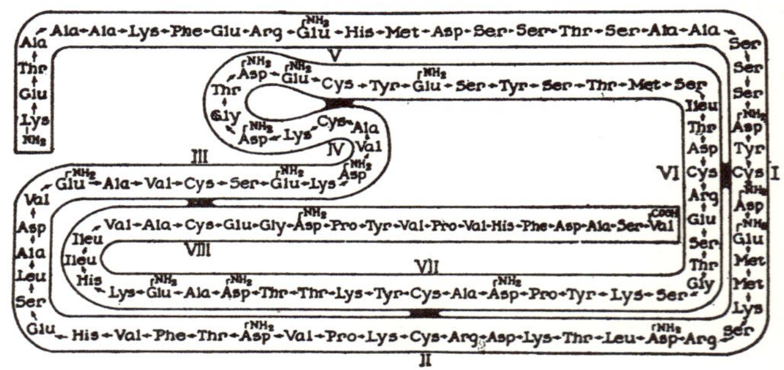

Fig. 50. A two dimensional diagram of the structure of bovine ribonuclease showing the arrangement of the disulphide bonds (from Spackman *et al.*[6] and incorporating the revision of Smyth *et al.*[69] and Potts *et al.*[70]).

parts of a single chain as in ribonuclease (Fig. 50). Sequence studies may be profitably carried out provided the cystine residues are first converted to more stable derivatives. The advantages of this preliminary step are three-fold. First, the problem of rearrangement reactions is solved; second, proteins in which the cystine residues are split are usually more susceptible to attack by proteolytic enzymes. Finally, for a protein containing more than one chain the step releases the component chains for separation and individual study. Three procedures are available for splitting disulphide bonds and the choice of method depends on the particular set of circumstances.

CLEAVAGE BY OXIDATION

The method used by Sanger for cleavage of the disulphide bonds in insulin was oxidation by performic acid. This procedure, which is based on earlier work of Toennies and Homiller[8], converts cystine to cysteic acid.

$$R_1S{-}SR_2 \xrightarrow[\text{acid}]{\text{Performic}} R_1SO_3H + R_2SO_3H$$

Performic acid is a powerful oxidizing agent and its action is not limited to the cystine residues. Tryptophan is converted to kynurenine and other degradation products, methionine is converted almost quantitatively to the sulphone and tyrosine may also be modified. The history of the use of the reagent in sequence studies brings out clearly the extra difficulties involved as larger and more complicated proteins are chosen for investigation. Insulin contains neither methionine nor tryptophan but a modified tyrosine (probably a chlorinated derivative) arises from the action of the reagent in the presence of chloride ions[9]. Ribonuclease contains the additional amino acid methionine but no difficulty was experienced by the Rockefeller Institute group in recognizing the corresponding sulphone in partial hydrolysates. In fact the fully oxidized derivative is a much better material to deal with than the sulphoxide or intact methionine. The yield of cysteic acid residues from a protein is probably not quantitative as can be seen from the recovery quoted on p. 99. Participation of serine and threonine residues in acyl shifts or formylation reactions under the influence of the reagent cannot be excluded (see p. 153). Troubles arise when proteins contain tryptophan and this might explain the gradual drift away from oxidative splitting of disulphide bridges in recent years. In illustration of the oxidative procedure examples from the studies on insulin and ribonuclease are given.

Preparation of A and B chains from oxidized insulin[10]

Insulin (250 mg) is dissolved in 9 ml formic acid and treated with 1 ml of 30% hydrogen peroxide for 15 min at room temperature. The mixture is diluted with 10 ml of water, concentrated *in vacuo* to a volume of 1–2 ml, and the oxidized protein precipitated with acetone. The precipitate is washed free from formic acid by repeated centrifugation of acetone suspensions and dried in the air. The yield is 250 mg.

The B chain (phenylalanyl) is less soluble than the A chain (glycyl) and can be precipitated at pH 6.5. The freshly oxidized insulin (1 g) is dissolved in 12 ml of a solution 0.1 *M* with respect to ammonia and acetic acid and the pH brought to 6.5. The precipitate is washed with 10 ml of 0.01 *M* acetic acid, dissolved in 5 ml of 0.1 *N* hydrochloric acid and fractionally precipitated with 40 ml of ethanol. The small precipitate is washed well with 80% ethanol and the combined supernatant and washings concentrated to near dryness *in vacuo* and treated with a large volume of acetone. The yield of phenylalanyl chain precipitated as a faintly buff-coloured solid is 140–210 mg.

The glycyl chain is prepared from the ammonium acetate solution left after precipitation at pH 6.5. The pH of the supernatant is adjusted back to 4.5 by the addition of 0.1 *M* acetic acid, and after removal of the small precipitate, is concentrated almost to dryness. The residue is taken up in 3 ml of water, mixed with 3 ml of 50% w/v ammonium acetate at pH 5.5 and the small precipitate discarded. The remaining solution is concentrated *in vacuo* and the ammonium acetate removed in a vacuum desiccator over H_2SO_4 and NaOH. The yield of glycyl chain as a white powder is 250–360 mg.

Preparation of oxidized ribonuclease

In the procedure of Hirs[11] performic acid is prepared by adding

0.50 ml of 30% hydrogen peroxide to 9.5 ml of 99% formic acid and the resulting solution allowed to stand at 25° for 2 h in a stoppered flask. In another flask, 200 mg of ribonuclease (8% moisture) is dissolved completely in 5.0 ml of 99% formic acid and 1.0 ml of anhydrous methanol added with stirring. Addition of the latter prevents freezing of the solution later on. The quantity of performic acid used is 12 times that required to transform all of the cystine in the protein to cysteic acid and at the same time convert all the methionine to the corresponding sulphone (90% yield). The protein solution and the performic acid are placed in the separate arms of a glass-stoppered test-tube fitted with a side-arm, cooled for 30 min in a bath at —10°, and then mixed by tipping the tube. After a reaction time of 2.5 h the contents of the tube are rinsed with 50 ml of ice water into a flask containing 350 ml of water at 0° and the mixture freeze-dried. Final traces of performic acid are removed by dissolving the material in 8.0 ml of water and freeze-drying a second time.

Hirs found that a performic acid oxidation carried out at 0° instead of —10° transformed 10% of the tyrosine to chlorotyrosine. At higher temperatures more extensive conversion took place. If chloride ion is removed from the protein solution before oxidation it is safe to use a higher temperature such as 0°. Chloride-free ribonuclease was prepared by passing a 10% solution of the protein in 0.01 *M* acetic acid through a short column of Dowex 2-X10 in the acetate form (1 ml of resin per ml of protein solution). The filtrate was then freeze-dried.

CLEAVAGE BY REDUCTION

The literature on the reduction of disulphides and on the properties of thiols is voluminous and of the many comprehensive reviews of the subject that by Cecil and McPhee[12] supplies most of the information likely to be of interest to protein chemists. In the field of protein chemistry investigators have been concerned mainly with the development of precise analytical methods for the estimation of SS and SH groups in intact proteins. For many years mild oxidizing agents have been used for the estimation of SH groups in simple substances, but these reagents, which include iodine, ferricyanide, iodosobenzoate, porphyrindin and phosphotungstic acid, are no longer considered suitable for proteins. Many have been shown to

References p. 131

be non-specific and interpretation of the results is not easy. An account of these methods will not be given here but practical details may be found in the review by Chinard and Hellerman[13]. More reliable methods available today make use of alkylating agents or mercaptide forming agents in conjunction with amperometric titration.

Amperometric titration of SH groups

The well-known technique of polarography is based on the current–voltage curves arising at an electrode when diffusion is the rate-determining stage in the discharge of ions. The name *amperometric titration* is given to the special case when the voltage applied across the reference electrode and indicator electrode is constant, and the current increments are plotted against the volume of titrating reagent added.

In the application to the titration of SH groups silver or mercuric ions are added to the solution to form soluble mercaptides with the sample to be investigated. Under well-defined conditions only a small residual current flows through the cell until all the SH groups are blocked; the end-point of the titration is then marked by the appearance of the diffusion current due to excess titrating reagent. A plot is made of current *versus* volume of reagent added and the intersection of the extrapolated branches of the curve is the end-point. A very sharp inflexion is obtained with simple substances but with more complicated molecules such as proteins the removal of heavy metal ions from the solution is slower and the curve gradually reaches linearity towards the end of the titration.

One method used in many laboratories today in various modifications is essentially that of Benesch *et al.*[14]. In this method which is both sensitive and specific, protein SH groups are titrated with silver in the form of a complex with tris(hydroxymethyl)aminomethane (Tris), in neutral, buffered solutions. It has given useful results with a number of proteins and will now be described in detail.

A special reference electrode must first be prepared. The one used by Benesch was the Hg–HgO–saturated $Ba(OH)_2$ electrode of Samuelson and Brown[15] which has a potential of —0.10 V *versus* the saturated calomel electrode. It is made up as shown in Fig. 51. A layer of mercury is placed at the bottom of the flask, this is covered

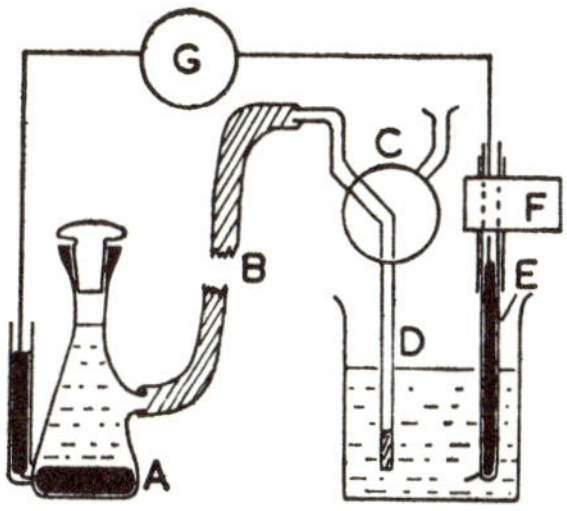

Fig. 51. Apparatus for amperometric titration. A. Reference electrode; B. Tygon tubing; C. three-way tap; D. bridge ending in paper plug; E. rotating platinum electrode; F. electric motor; G. galvanometer.

with a slurry of red mercuric oxide and solid barium hydroxide, and the flask is filled with water saturated with the two reagents. The electrode is stable for several months when protected from the light. The writer has found a cell filled with 0.2 *N* sodium hydroxide equally satisfactory and less trouble to make up. The bridge from the electrode consists of a short piece of glass tubing fitted with a rolled filter paper plug soaked in saturated potassium chloride solution. The glass tubing is attached to a convenient length of Tygon tubing (15–60 cm) leading to a 3-way stop-cock, which in turn, is connected to the reference cell by a short length of Tygon tubing. The bridge is filled with the filtered electrode solution as far as the 3-way tap and with saturated potassium chloride from the tap to the paper plug. The arrangement permits daily flushing of the KCl part of the bridge.

The indicator electrode is made from a short piece of platinum wire (No. 18 gauge) sealed into a piece of stout-walled glass tubing which is filled with mercury. This electrode has to be rotated at speeds of about 800 r.p.m. A convenient arrangement uses a rubber-tyred pulley fitted with a hollow shaft which is rotated by a frictional cone drive from a constant-speed motor. Electrical contact is maintained by a platinum lead dipping into the mercury well at the top of the electrode.

Titrations are carried out in 100-ml beakers containing a total of 20–30 ml of solution. The rotating platinum electrode and the bridge from the reference electrode are immersed in the solution and the wire leads from the electrodes are connected to a galvano-

References p. 131

meter of sensitivity 0.1 to 0.2 μA/mm. The titrating solution is 1 or 2×10^{-3} *M* silver nitrate which can be delivered from a micrometer syringe pipette or added in small portions from a 50 μl pipette. When not in use the silver nitrate solution must be stored in the dark. Since many proteins need to be denatured before the maximum number of binding sites become available the titrations are often carried out in concentrated urea solutions. Urea may be purified by passing an 8 *M* solution over a bed of Amberlite MB-1 cation- and anion-exchange resin; the purified solutions are stable for about two weeks at 4°. The sample for assay should contain 0.5 to 1.0 micromole of SH groups and is dissolved in 20–30 ml of a solution which is 0.2 *M* in Tris (adjusted to pH 7.2–7.4 with nitric acid), 0.01 *M* in potassium chloride, and when required 8 *M* in urea. The potassium chloride assists rapid return of the current to zero at the beginning of the titration. A titration curve obtained by Cole *et al.*[16] for human haemoglobin is shown in Fig. 52. The extrapolated value for the haemoglobin corresponds to 5.9 SH groups per molecule of molecular weight 66,000.

Other reagents have been used for assay of SH groups. Allison and Cecil[20] have made a detailed study of phenylmercuric hydroxide using a dropping-mercury indicator electrode and these authors consider the reagent to be more specific than the silver-Tris complex.

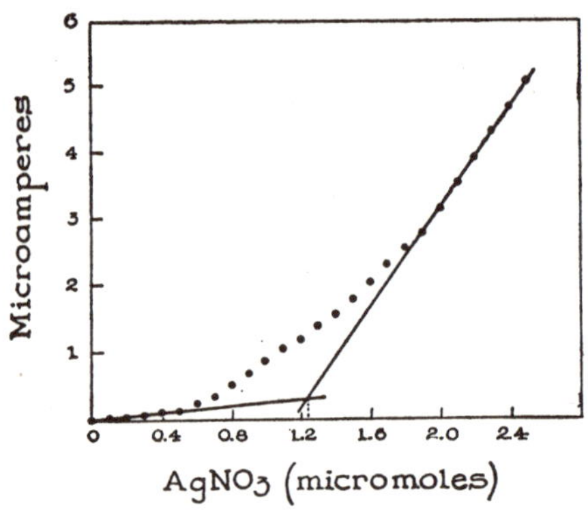

Fig. 52. Amperometric titration of human haemoglobin (13.85 mg) with 2×10^{-3} *M* $AgNO_3$. The titration was begun 2 min after the protein solution was added to the titration mixture and 50-μl aliquots of $AgNO_3$ were added at one minute intervals. The extrapolated value (vertical dashed line) corresponds to 5.9 Ag ions bound per molecule of haemoglobin. (From Cole *et al.*[16].)

This conclusion is based on the particular case of denatured haemoglobin which in the presence of excess silver ions will bind more than the six atoms per molecule required by theory; titration with phenylmercuric hydroxide has not indicated more than six SH groups per molecule. The reactions of mercuric chloride, phenyl mercuric oxide, *p*-chloromercuribenzenesulphonate and silver nitrate with the SH groups of normal adult human haemoglobin in its various forms is discussed by Cecil and Snow[17].

The spectrophotometric method of Boyer[18] offers a useful alternative to amperometric titration. It is based on the change in the differential absorption in the 250 mμ region accompanying mercaptide formation when *p*-chloromercuribenzoate (PCMB) is mixed with a thiol. Proteins absorb strongly in the same wavelength region and this is an obvious disadvantage to the method. Isles and Jocelyn[19] determined protein SH groups by measuring the loss of PCMB from solution when the protein is precipitated in the presence of the reagent and the following details are from the work of these authors.

p-Chloromercuribenzoate is purified by reprecipitation several times from a solution in dilute NaOH, and standardized by iodine titration[18]. Sodium *p*-chloromercuribenzoate (0.16 μmole) in water (0.2 ml) is added to a mixture of the protein (0.1 μmole) in water (1.0 ml) and 0.2 *M* phosphate buffer, pH 7.4 (0.5 ml). The volume is adjusted to 2 ml and 6% (w/v) metaphosphoric acid (1 ml) is added 10 min later. The mixture is centrifuged and the extinction of the protein–buffer–PCMB supernatant is measured at 247 mμ against that of a protein–buffer–water supernatant prepared in the absence of PCMB. This value is subtracted from the extinction of a solution of the PCMB (0.2 ml; 0.16 μmole) in 2% (w/v) metaphosphoric acid in 33 m*M* phosphate buffer, pH 7.4, read against a buffer–metaphosphoric acid blank. The loss of PCMB is calculated from the decrease in extinction and is taken as equivalent to the thiol content of the precipitated protein.

Alkylation of thiols

In general reagents that combine with SH groups possess an active halogen or a reactive double bond. Although they have been extensively used for many years as 'blocking' agents for SH enzymes

References p. 131

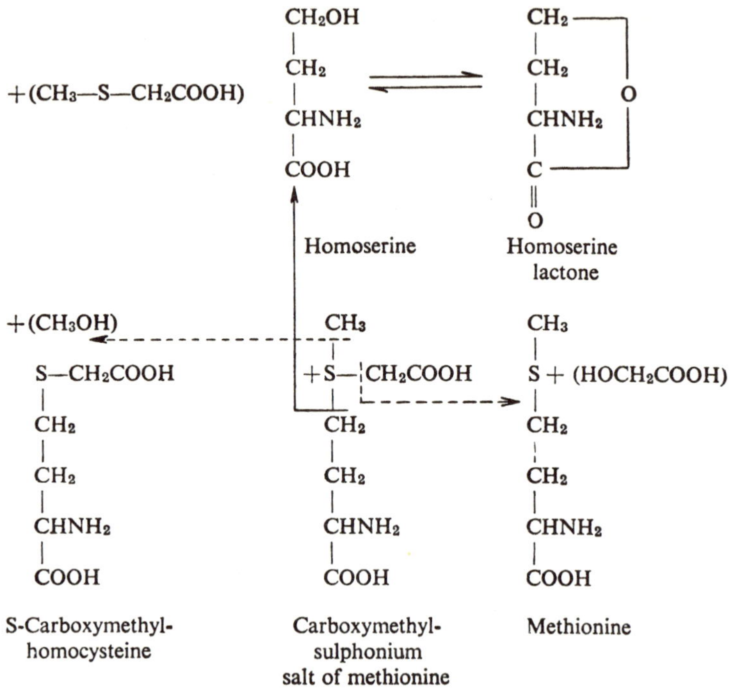

Fig. 53. Products of the decomposition of the carboxymethylsulphonium salt of methionine. Gundlach *et al.*[22].

and other proteins their use in quantitative work has been more limited.

Moore *et al.*[21] have shown that carboxymethylation of chymotrypsinogen with iodoacetic acid is practically quantitative in 20 min at pH 8.5 and at 25°. About 2% of ε-carboxymethyllysine is formed in the process and 10% of the methionine was unaccounted for. Compounds which may be expected to arise from the decomposition of the carboxymethyl sulphonium salt of methionine are shown in the scheme formulated by Gundlach *et al.*[22] (Fig. 53). Methionine is to some extent regenerated from the sulphonium salt under the normal conditions of acid hydrolysis of proteins. Only very slight decomposition of S-carboxymethyl-cystine occurs during acid hydrolysis when oxygen is excluded. Alkylation of histidine residues has been discussed by Crestfield *et al.*[23] in connection with a specific inactivation of ribonuclease involving the formation of a single residue of 1-carboxymethyl histidine at position 119.

Iodoacetic acid and *p*-nitrophenyl acetate are inhibitor and substrate respectively for the enzyme glyceraldehyde-3-phosphate dehydrogenase; experiments conducted at the μmole level using these compounds as the ^{14}C forms have indicated that the SH groups blocked in each case occur exclusively in the same unique octadecapeptide sequence of amino acids[24,25]. In a similar study[26] the essential SH groups were labelled with N-(4-dimethylamino-3,5-dinitrophenyl) maleimide and the peptides identified by their yellow colour.

SH groups react quantitatively with maleimide and N-ethylmaleimide (NEM) by addition across the double bond, but caution should be exercised in the use of NEM as a specific reagent for thiol groups of proteins. The reagent has been shown to react with the imidazole group of histidine[27] and with the α-amino group of certain amino acids[27,28]. S-(1,2-dicarboxyethyl)-L-cysteine[29] is formed in 86% yield by hydrolysis of NEM-modified proteins for 72 h in 6 *N* HCl at 110°. A high degree of specificity for thiol groups is exhibited when reaction mixtures are maintained below neutrality and excess reagent is avoided.

Acrylonitrile is a promising new alkylating agent. SH groups add across the double bond forming S-cyanoethyl derivatives which, in turn, are converted to S-carboxyethyl derivatives on acid hydrolysis[30]. According to Weil and Seibles[31] the reaction is specific with no complications.

Reduction procedures

Thiols

Reagents that have been used for the reductive cleavage of disulphide bonds are cysteine, reduced glutathione, thioglycollic acid and β-mercaptoethylamine. The result may be represented by the simple equation

$$R_1SSR_1 + 2R_2SH \rightleftharpoons 2R_1SH + R_2SSR_2$$

from which it is obvious that in order to drive the reaction in the desired direction a considerable excess of added thiol is required.

References p. 131

Over long periods an interchange of radicles will establish an equilibrium according to the equation

$$R_1SSR_1 + R_2SH \rightleftharpoons R_1SSR_2 + R_1SH$$

Equilibrium constants have been determined for these exchange reactions in a number of systems[32]. A few examples will illustrate the use of thiols in the cleavage of protein SS groups.

Sela *et al.*[33] treated ribonuclease (12 mg/ml) with thioglycollic acid (400 moles assuming a mol. wt. for the protein of 13,683 and a moisture content of 9.38%) in 8 *M* urea at pH 8.5 (aqueous trimethylamine) during 4.5 h at room temperature. The reduced protein was precipitated with acetone and 1 *N* HCl (39:1) at —5°. The product was washed three times with the solvent and finally twice with cold ether. Titration with PCMB indicated 8 SH groups per mole. Two disulphide bonds are apparently split quite readily, but the remaining two bonds require the strong urea medium. The authors claim to have regenerated 12–19% of the specific activity of native ribonuclease by bubbling air through a solution of the fully reduced protein in 0.01 *M* phosphate at pH 7–8 for 68 h. The liberated SH groups were blocked successfully by treatment with iodoacetic acid (700 moles) for 2 h at pH 8.5. Constant pH was maintained by the addition of a 5% solution of trimethylamine using an autotitrator.

There have been occasional reports in the literature that the sulphur content of proteins increases after treatment with thioglycollic acid. White[34] has demonstrated differences in the properties of ribonuclease treated with pure and impure samples of the reagent and concluded that polythioglycolides in the impure sample probably couple with the free amino groups in the protein. In this work the reagent was purified by vacuum distillation (b.p. 105° at 10 mm) and stored at 5°.

Lindley[35] reduced insulin with 0.1 *M* lithium thioglycolate at pH_5 in 8 *M* urea and separated the products by zone electrophoresis in paper moistened with the same solution. Reduction of insulin in the absence of urea followed by coupling with iodoacetamide gave a product in which only $\frac{1}{3}$ of the disulphide bonds had been split. Further reduction in 8 *M* urea then yielded the separate chains.

β-Mercaptoethylamine has been used in studies on serum proteins. Markus and Karush[36] employed the reagent in the concentration range 2.5×10^{-3} *M* to 2.0×10^{-1} *M* and at pH 7.0–7.4. Removal of the reagent after a reduction was done by passing the solution through a short column of Dowex 50 in the sodium form. Human serum albumin in the native form was found to contain only one susceptible bond; 17 disulphide bonds were found in the protein fully denatured with sodium decyl sulphate.

Mercaptoethanol has also been used a great deal in regeneration and structural studies. Anfinsen and Haber[37] reduced ribonuclease with the reagent at a level of 1 µl/mg protein. After treatment for 4.5 h at room temperature the mixture was acidified with acetic acid and the reduced protein separated from the reagents by gel filtration on Sephadex. By virtue of its sterically favourable cyclic disulphide, dithiothreitol (*threo* isomer of 2,3-dihydroxy-1,4-dithiolbutane) reduces disulphides quantitatively according to the following schemes[71]:

$$RSSR + HSCH_2(CHOH)_2CH_2SH \longrightarrow RSH + RSSCH_2(CHOH)_2CH_2SH$$

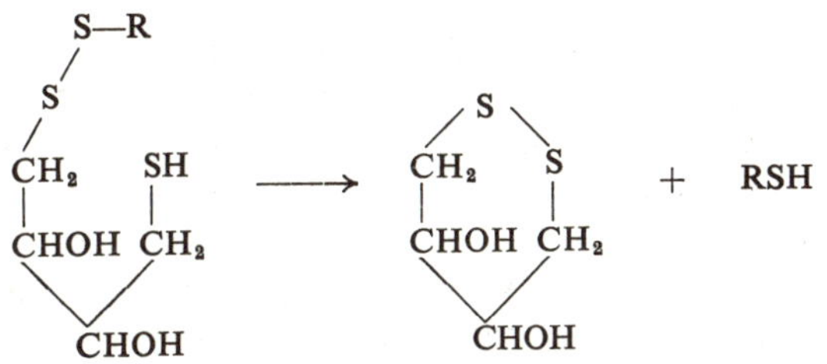

The reagent has the very low redox potential of –0.33 V at pH 7, is surprisingly stable to atmospheric oxidation and possesses little odour. Succinylation of proteins aids reductive cleavage [72,73].

Sodium borohydride

This metallo hydride is a milder reducing agent than lithium borohydride (see p. 231) and may be used in aqueous solutions at pH 9–10. In the protein field the reagent was first used by Edman and Diehl[38] who reported the reduction of two of the three disulphide bonds in insulin. Moore *et al.*[21] worked out the conditions for the reduction of ribonuclease, chymotrypsinogen, trypsinogen, lysozyme, and β-lactoglobulin. Complete reduction of ribonuclease in 100 min was obtained using 0.5% solutions of both the protein and sodium borohydride at 51°. In 8 *M* urea at the same temperature reduction was complete in 50 min. A similar rapid reduction was possible with chymotrypsinogen and β-lactoglobulin, but the rate of cleavage of lysozyme was slower. Other recent applications have been reported by Stauff and Duden[39], and Brown[40]. The conditions used by Brown were 0.10–2.00 μmoles protein + 1 ml of a freshly prepared solution of sodium borohydride (25 mg/ml) in a total volume of 5 ml.

Conditions for the carboxymethylation of proteins after reduction with sodium borohydride that bring about minimum side reactions are as follows[21]. After the reduction the excess sodium borohydride is not destroyed but the alkaline mixture is treated immediately with iodoacetic acid (50 moles). The pH is brought to 8.5 by slow addition of hydrochloric acid and the alkylation allowed to proceed at 25° for 20 min. The solution is then diluted with two volumes of water and dialyzed at 4° for 48 h against several changes of stirred distilled water containing in suspension a mixed-bed ion-exchange resin. Crestfield *et al.*[41] have provided evidence for the heterogeneity of the product of reduction of ribonuclease by sodium borohydride. The carboxymethylated protein contained material which was retarded on a Sephadex column and arose presumably from peptide bond cleavage. On the other hand, carboxymethylated ribonuclease prepared with mercaptoethanol gave a single sharp peak when examined by gel filtration. It was concluded, therefore, that sodium borohydride is a useful reagent if the objective is the titration of the SH groups produced; if structural studies are contemplated, the reagent is less satisfactory than mercaptoethanol.

CLEAVAGE BY SULPHITE

The action of sulphite on disulphides may be represented by the equilibrium shown in equation

$$R.S.S.R + SO_3^{2-} \rightleftharpoons RS^- + RS.SO_3^-$$

The reaction is at the same time a reduction and an oxidation and provides a highly specific means of fission. From kinetic studies with simple substances Cecil and McPhee[42] concluded that the reaction of sulphite above pH 9 is a simple reversible bimolecular reaction, but below pH 9 the mechanism is complex. Later studies of McPhee[43] showed that substances carrying a positive charge such as bis-(β-aminoethyl)disulphide (cystamine) and bis(γ-aminopropyl) disulphide (homocystamine) react much more rapidly than those with no net charge. It is reasonable to assume then that rates of reaction of sulphite with disulphide groups in proteins might be similarly influenced by the charge environment.

Since the early observation of Heffter[44] in 1907 many workers have studied the reaction with proteins but interest today is focussed on the possibility of converting all cysteine and cystine groups in proteins to S-sulphonate groups. For convenience cysteine sulphonic acid is termed S-sulphocysteine and proteins containing these groups are termed S- sulphoproteins.

The simplest way of converting a disulphide wholly to the S-sulpho derivative is by the process of autoxidation. If the reaction is too slow, as in the case of some proteins, a mild oxidizing agent is added to the system and an oxidation-reduction cycle maintained until SH groups can no longer be detected.

$$\mathrm{R.S.S.R.} + SO_3^{2-} \rightarrow \mathrm{R.S.SO_3^-} + RS^-$$

air

or mild oxidizing agent

This principle was used by Clarke[45] in 1932 for the preparation of S-sulphocysteine by aeration of an ammoniacal solution of cysteine containing excess sulphite. An improved procedure for making the crystalline sodium salt of S-sulphocysteine is that of Segel and Johnson[46]. The introduction of S-sulphocysteinyl groups into proteins by an oxidation–reduction cycle was described by Bailey[47] in 1957 and later by Bailey and Cole[48]. Details of the method are given below.

Kolthoff and Stricks[49] have described the amperometric titration of cysteine and cystine by cupric sulphate in alkaline solutions of sulphite. The overall reactions for a disulphide and a thiol may be represented by the equations

$$\mathrm{RSSR} + 2Cu^{2+} + 2SO_3^{2-} \rightleftharpoons 2RSSO_3^- + 2Cu^+$$

$$\mathrm{RSH} + 2Cu^{2+} + SO_3^{2-} \rightleftharpoons RSSO_3^- + 2Cu^+ + H^+$$

The sole product from both cysteine and cystine is S-sulphocysteine. These reactions in which cupric ions are reduced to cuprous ions have been made the basis of an alternative method for the introduction of S-sulphocysteinyl groups into proteins by Swan[50]. In this procedure the protein to be treated is dissolved in a solution containing a two to fourfold excess of sodium sulphite, and 10% excess of a freshly prepared cupric ammonium hydroxide solution at pH 10 is then added. The two methods of oxidative fission with sulphite have been compared in recent studies on S-sulphofibrinogen[51] and on S-sulpho derivatives of soluble proteins of feather keratin[52,57].

Preparation of S-sulphoproteins

(1) 100 mg of protein are dissolved at 38° in 5 ml of 8 *M* (or saturated) urea buffered at pH 7.4 with 0.2 *M* Tris buffer. Initial reduction of disulphide bonds is carried out by adding 0.62 ml of 1.0 *M* Na_2SO_3 (or 1.25 ml of 0.5 *M* Na_2SO_3 in 8 *M* urea) at pH 7.4 during 10 min. The SH groups released are then reoxidized with either 1.25 ml of 0.5 *M* iodosobenzoate in 8 *M* urea, or with 1.25 ml of 0.5 *M* sodium tetrathionate in 8 *M* urea. The latter reagent is prepared from sodium thiosulphate and iodine as described by Gilman *et al.*[53]; the 0.5 *M* solution is adjusted to pH 7 with *N* NaOH. After another interval of 10 min the treatment with sodium sulphite is repeated and further oxidation-reduction cycles completed. Finally, the salts

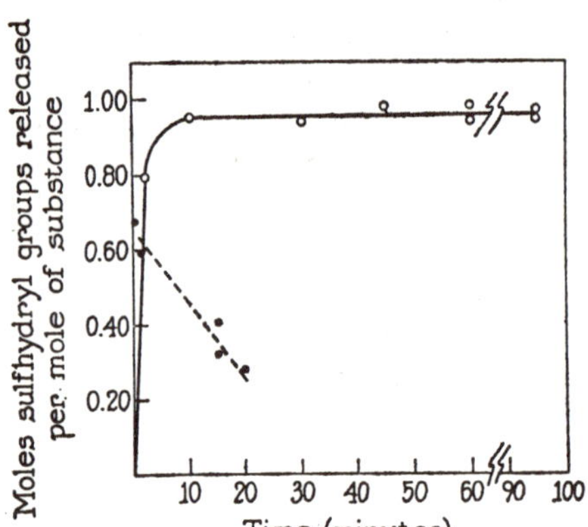

Fig. 54. The reaction of sulphite with cystine (●----●) and oxidized glutathione (○——○). Sulphite, 0.2 *M*; urea, 8 *M*. Bailey and Cole[48].

and urea are removed by dialysis against distilled water or 0.01 *M* sodium sulphite at pH 7. The S-sulphoprotein is recovered from salt-free solutions by freezedrying.

It is well established that different proteins show different reactivities with sulphite and the exact conditions for complete conversion to an S-sulphoprotein will vary slightly in each particular case. The reaction with chymotrypsinogen proceeds very rapidly at 38° without the aid of an oxidizing agent additional to atmospheric oxidation. On the other hand, ribonuclease requires an oxidizing agent and concentrated urea to bring about complete reaction. With most peptides and some proteins the rates of reaction of sulphite and tetrathionate with SS and SH respectively are considerably faster than the rate at which the two reagents neutralize each other. For each individual protein it is advisable to follow the course of reduction and oxidation by amperometric titration. This may be done after addition of 25–50 microequivalents of 1.0 *M* sodium sulphite to the sample dissolved in 8 *M* urea buffered at pH 7.4 with 0.2 *M* Tris–HNO_3. Figs. 54 and 55 show the formation of SH groups by the action of sulphite on cystine and oxidized glutathione and on ribonuclease and chymotrypsinogen.

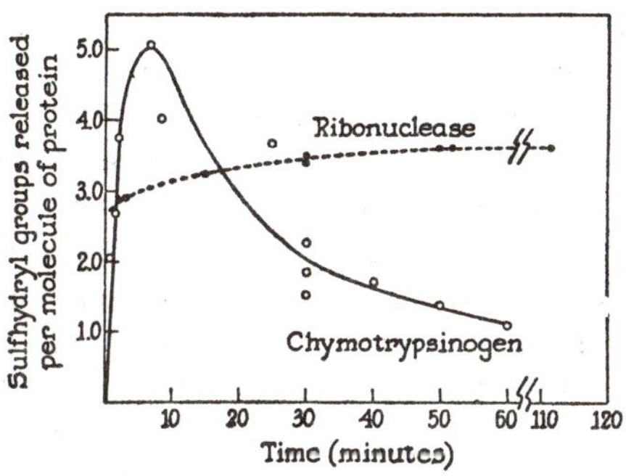

Fig. 55. Formation of SH groups by the action of sulphite on proteins. The protein concentration was 2%; sulphite, 0.3 *M*; urea, 8 *M*; temperature 38°. (From Bailey and Cole[48].)

The specificity of the reactions has been checked by hydrolysis of S-sulphochymotrypsinogen. Apart from a slight loss (less than 5%) of tryptophan all other amino acids excluding cystine are re-

References p. 131

covered quantitatively[48]. On hydrolysis of S-sulphocysteine with 6 N HCl at 110° for 22 h, cystine is recovered in a yield of 80% which is the same as the recovery of 81% obtained when cystine itself is hydrolyzed.

(2) The following is taken from Pechère *et al.*[54] and refers to the preparation of S-sulphochymotrypsinogen and S-sulphotrypsinogen. 500–700 mg of protein are dissolved in 15 ml of 0.005 N hydrochloric acid and diluted with 25 ml of 0.001 N hydrochloric acid. Urea (48 g) is added in portions with stirring, the pH being maintained at 3 by occasional addition of 1 N HCl. 10 ml of 1.5 M sodium sulphite is added and sufficient concentrated ammonium hydroxide to bring the pH to 10.2. Finally, 2 ml of 2 M $Cu(NO_3)_2$ is added in small amounts keeping the pH constant and the mixture is brought to a volume of 100 ml with distilled water. After a reaction period of 1 h at room temperature copper ions are removed by six dialyses of 12 h each according to the schedule (a) 0.02 N HCl brought to pH 9.0 with ammonia (b) 0.01 N acid, 0.01 M EDTA, pH 8.5, (c) and (d) 0.005 N acid, 0.005 M EDTA, pH 8.0, (e) and (f) 0.005 N acid or water at pH 8.0. In this work the course of disulphide splitting was estimated by a polarographic method and checked by the use of [^{35}S-]sulphite which converts the cystine residues to [^{35}S-]sulphocysteine residues.

Another example of the use of cuprammonium sulphite is from Weil and Seibles[55] who prepared S-sulpho derivatives of α- and β-lactoglobulin which contained respectively 8 and 10 $S–SO_3$ groups. For these proteins 8 M urea was unnecessary. The crystalline protein (1 g) was dissolved in 50 ml of water with the aid of a little 0.5 N ammonium hydroxide to give a solution of pH 9.0. Cupric ammonium solution was added (0.25 g of $CuSO_4.5H_2O$ in 40 ml water adjusted to pH 9.0 with conc. ammonium hydroxide) followed by sodium sulphite solution (0.83 g in 10 ml water) and the mixture kept at room temperature for 2 h. To remove cupric ions the solution was dialyzed against 2 litre of 0.1 M sodium citrate, pH 7, at 4° during 3 days. The protein was precipitated by bringing the pH to 4.0 for the α-protein and 5.0 for the β-protein. After a further dialysis against distilled water the solution was freeze-dried.

Comments

The physical properties of the sulphoproteins seem to be similar to those of other derivatives of the same proteins in which disulphide bonds have been split and converted to anionic groups. Some sulphoproteins have a strong tendency to aggregate and may be held in solution only under narrowly defined conditions of pH and ionic strength. The facile elimination of the sulphonate group is discussed below in connexion with the regeneration of insulin activity from the two separated insulin chains.

Preparation of insulin chains

The A and B chains of oxidized insulin have been of value in protein chemistry as models in exploratory work on chemical degradation methods and as model substrates in enzyme specificity studies. The S-sulpho derivatives of the A and B chains are obtained in high yield by the sulphite reaction and the following preparative details will be of interest to those requiring large peptides as model substances.

0.1 *M* Copper sulphate (2 ml 200 μmoles) is added to insulin (200 mg; 30 μmoles) and sodium sulphite (60 mg anhydrous; 500 μmoles) dissolved in 0.16 *M* ammonium acetate buffer, pH 10.2 (20 ml) at room temperature. After a reaction period of 1 h an aliquot (5 ml) is transferred to a 2 $\times$ 50 cm column of Sephadex G-50 and eluted with 0.05 *M* ammonium carbonate buffer, pH 9. The deep violet copper complex of the B chain S-sulpho derivative emerges first from the column, followed by the deep blue derivative of the A chain. Copper is removed by passing each derivative through a short column of Dowex 50-X8 in the $NH^{+}{}_{4}$ form[52]. The eluates are made 0.05% with respect to EDTA (dissodium salt) and then exhaustively dialyzed against distilled water.

In the presence of a limited excess of sulphite, and with air as the oxidant, a dimer of the B chain is formed in about 27% yield[56]. High yields of both chains are obtained by the tetrathionate method[48], and by reaction with sulphite in the presence of phenylmercuric hydroxide[58] or mercuric chloride[59]. Cecil and Wake[59] have shown that the majority of intra-chain disulphide bonds do not react with sulphite alone and the following preparation of insulin A chain 6,11-disulphide-7,20-bis-S-sulphonate has been given by Meienhofer and Brinkhof[60]. Insulin (1 g) is treated at pH 7.2 with a 250-fold

References p. 131

excess of sodium sulphite and the mixture stirred for 5 h at 35° under nitrogen. After exhaustive dialysis against water the A and B chains are precipitated at pH 4.8 with dilute HCl and separated by chromatography on a 2.2 × 14 cm column of Dowex 50-X2 (200–400 mesh) using a pH 1.9 buffer containing 5% acetic and 5% formic acid as eluent. Yields of the A and B derivatives are 82 and 90% respectively.

Regeneration of insulin from the two chains

In 1960 Dixon and Wardlaw[61] reported the regeneration of hormonal activity from the completely inactive S-sulpho glycyl and S-sulpho phenylalanyl chains after reduction with mercaptoethanol and reoxidation. The yields of resynthesized insulin in this work and later studies[62] were 0.5–5%. Independently, the group at the *Institute of Biochemistry in Shanghai* reported[63,64] the isolation of a crystalline substance which was identical with insulin in crystalline form and possessed an activity of 18.4 International Units/mg. The fascinating problem of the resynthesis of insulin from the separate reduced chains has attracted many workers for many years. From a mathematical analysis made by Kauzman[65] of the relative probabilities of the 12 isomers containing one each of the two chains it appears that formation of the native insulin structure has a probability of about 5%. The specific activity of the regenerated products obtained by the Chinese workers was usually 5–10% of that of crystalline insulin and occasionally as high as 20%. This suggests that of the 12 possible isomers the insulin structure is one of the most stable.

In regeneration studies the prior isolation of the two chains free from cross-contamination is of paramount importance. The glycyl chain has been purified by zone electrophoresis in cellulose powder. The latter was first washed thoroughly with 0.1 *M* citric acid–8 *M* urea and then while moist, was packed into a slab of dimensions 21.5 × 5.5 × 0.35 cm. The sample was applied 4 cm from the cathode end and submitted to electrophoresis for 18 h at 120 V and 17mA. The progress of migration was followed with the Pauly reagent and at the end of the run the zone was cut out and eluted with a small volume of the citric acid–urea solution. The powder obtained after dialysis of the solution and freeze-drying was slightly contaminated with impurities from the cellulose powder but no phenylalanyl chain was detectable.

Insulin activity was regenerated from the pure chains in the following way. The S-sulpho glycyl chain (40 mg) was dissolved in 8.4 ml of 8 *M* urea and the pH adjusted to 5. This solution was incubated for 10 h at 29° under nitrogen with 1.6 ml of a solution which was 0.5 *N* in thioglycolate adjusted to pH 5 (800 μmoles), and 8 *M* in urea. The extent of reduction was 60%. The S-sulpho phenylalanyl chain (60 mg) was similarly reduced with 400 μmoles of thioglycolate in a total volume of 10 ml. The extent of reduction was about 40%. The reduced chains were precipitated from their respective reaction mixtures by the addition of 10–20 volumes of acetone, and regeneration of active material was carried out at pH 8.5 at room temperature (29°) in aqueous solution at a total concentration of 5 mg/ml. The mixture was examined for biological activity when SH groups were no longer detectable. Both insulin chains have now been synthesized[66,67] and the remarkable achievement of a completely synthetic insulin of 0.6–1.0% activity was reported by Meienhofer and collaborators[68].

REFERENCES

1 P. D. Boyer in: *The Enzymes* (P. D. Boyer, H. Lardy and K. Myrbäck, eds.) Vol. I, Part II, Academic Press, New York, 1959.
2 R. Benesch *et al.*, (Eds.), *Symposium on Sulfur in Proteins*, Academic Press, New York, 1959.
3 A. P. Ryle and F. Sanger, *Biochem. J.*, 60 (1955) 535.
4 R. E. Benesch and R. Benesch, *J. Am. Chem. Soc.*, 80 (1958) 1666.
5 A. P. Ryle, F. Sanger, L. F. Smith and R. Kitai, *Biochem. J.*, 60 (1955) 541.
6 D. H. Spackman, W. H. Stein and S. Moore, *J. Biol. Chem.*, 235 (1960) 648.
7 A. N. Glazer and E. L. Smith, *J. Biol. Chem.*, 236 (1961) 416.
8 G. Toennies and R. P. Homiller, *J. Am. Chem. Soc.*, 64 (1942) 3054.
9 E. O. P. Thompson, *Biochim. Biophys. Acta*, 15 (1954) 440.
10 F. Sanger, *Biochem. J.*, 44 (1949) 126.
11 C. H. W. Hirs, *J. Biol. Chem.*, 219 (1956) 611.
12 R. Cecil and J. R. McPhee, *Advan. Protein Chem.*, 14 (1959) 225.
13 F. P. Chinard and L. Hellerman, *Methods Biochem. Anal.*, 1 (1954) 1.
14 R. E. Benesch, H. A. Lardy and R. Benesch, *J. Biol. Chem.*, 216 (1955) 663.
15 G. J. Samuelson and D. J. Brown, *J. Am. Chem. Soc.*, 57 (1935) 2711.
16 R. D. Cole, W. H. Stein and S. Moore, *J. Biol. Chem.*, 233 (1958) 1359.
17 R. Cecil and N. S. Snow, *Biochem. J.* 82 (1962) 247.

18 P. D. Boyer, *J. Am. Chem. Soc.*, 76 (1954) 4331.
19 T. E. Isles and P. C. Jocelyn, *Biochem. J.* 88 (1963) 84.
20 A. C. Allison and R. Cecil, *Biochem. J.*, 69 (1958) 27.
21 S. Moore, R. D. Cole, H. G. Gundlach and W. H. Stein, *Symposium on Proteins*, 4th International Congress of Biochemistry, Vienna, 1958.
22 H. G. Gundlach, S. Moore and W. H. Stein, *J. Biol. Chem.*, 234 (1959) 1761.
23 A. M. Crestfield, S. Moore and W. H. Stein, *J. Biol. Chem.*, 238 (1963) 2413, 2421.
24 J. I. Harris, B. P. Meriwether and J. H. Park, *Nature*, 197 (1963) 154.
25 R. N. Perham and J. I. Harris, *J. Mol. Biol.*, 7 (1963) 321.
26 H. L. Segal and A. H. Gold, *J. Biol. Chem.*, 238 (1963) PC 2589.
27 D. G. Smyth, A. Nagamatsu and J. S. Fruton, *J. Am. Chem. Soc.*, 82 (1960) 4600.
28 A. Riggs, *J. Biol. Chem.*, 236 (1961) 1948.
29 D. G. Smyth, O. O. Blumenfeld and W. Konigsberg, *Biochem. J.*, 91 (1964) 589.
30 Y. Tomimatsu, J. J. Bartulovich and W. H. Ward, *Textile Res. J.*, 29 (1959) 293.
31 L. Weil and T. S. Seibles, *Arch. Biochem. Biophys.*, 95 (1961) 470.
32 I. M. Kolthoff, W. Stricks and R. C. Kapoor, *J. Am. Chem. Soc.*, 77 (1955) 4733.
33 M. Sela, F. H. White and C. B. Anfinsen, *Science*, 125 (1957) 691.
34 F. H. White, Jr., *J. Biol. Chem.*, 235 (1960) 383.
35 H. Lindley, *J. Am. Chem. Soc.*, 77 (1955) 4927.
36 G. Markus and F. Karush, *J. Am. Chem. Soc.*, 79 (1957) 134.
37 C. B. Anfinsen and E. Haber, *J. Biol. Chem.*, 236 (1961) 1361.
38 P. Edman and K. Diehl, *2nd International Congress of Biochemistry Paris, Résumés*, 1952, p. 51.
39 J. Stauff and R. Duden, *Biochem. Z.*, 331 (1958) 10.
40 W. D. Brown, *Biochim. Biophys. Acta*, 44 (1960) 365.
41 A. M. Crestfield, S. Moore and W. H. Stein, *J. Biol. Chem.*, 238 (1963) 622.
42 R. Cecil and J. R. McPhee, *Biochem. J.*, 60 (1955) 496.
43 J. R. McPhee, *Biochem. J.*, 64 (1956) 22.
44 A. Heffter, *Med. Naturw. Arch.*, 1 (1907) 81; *Chem. Zbl.*, (1907) 822.
45 H. T. Clarke, *J. Biol. Chem.*, 97 (1952) 235.
46 I. H. Segel and M. J. Johnson, *Anal. Biochem.*, 5 (1963) 330.
47 J. L. Bailey, *Biochem. J.*, 67 (1957) 21P.
48 J. L. Bailey and R. D. Cole, *J. Biol. Chem.*, 234 (1959) 1733.
49 I. M. Kolthoff and W. Stricks, *J. Am. Chem. Soc.*, 73 (1951) 1728.
50 J. M. Swan, *Nature*, 180 (1957) 643.
51 A. Henschen, *Arkiv Kemi*, 22 (1964) 1.
52 B. S. Harrap and E. F. Woods, *Biochem. J.* 92 (1964) 8.
53 A. Gillman, F. S. Philips, E. S. Koelle, R. P. Allen and E. St. John, *Am. J. Physiol.*, 147 (1946) 115.
54 J. F. Pechére, G. H. Dixon, R. H. Maybury and H. Neurath, *J. Biol. Chem.*, 233 (1958) 1364.

[55] L. WEIL AND T. S. SEIBLES, *Arch. Biochem. Biophys.*, 84 (1959) 244.
[56] S. J. LEACH, J. M. SWAN AND L. A. HOLT, *Biochim. Biophys. Acta.*, 78 (1963) 196.
[57] B. S. HARRAP AND E. F. WOODS, *Biochem. J.*, 92 (1964) 8.
[58] R. CECIL AND U. E. LOENING, *Biochem. J.*, 76 (1960) 146.
[59] R. CECIL AND R. G. WAKE, *Biochem. J.*, 82 (1962) 401.
[60] J. MEIENHOFER AND O. BRINKHOF, *Nature*, 199 (1963) 1095.
[61] G. H. DIXON AND A. C. WARDLAW, *Nature*, 188 (1960) 721.
[62] S. WILSON, G. H. DIXON AND A. C. WARDLAW, *Biochim. Biophys. Acta*, 62 (1962) 483.
[63] YU-CANG DU, YU-SHANG ZHANG, ZI-XIAN LU AND CHEN-LU TSOU, *Sc. Sinica (Peking)*, X (1) (1961) 84.
[64] CHEN-LU TSOU, YU-CANG DU AND GENG-JUN XÜ, *Sci. Sinica (Peking)*, X (3) (1961) 332.
[65] W. KAUZMAN, in: *Symposium on Sulfur in Proteins* (Edited by R. BENESCH *et al.*), Academic Press, New York, 1959.
[66] P. G. KATSOYANNIS AND K. SUZUKI, *J. Am. Chem. Soc.*, 85 (1963) 2659.
[67] Y. WANG AND COLLABORATORS, *Acta Chim. Sinica*, 29 (1963) 190.
[68] J. MEIENHOFER AND COLLABORATORS, *Z. Naturforsch.*, 18b (1963) 1120.
[69] D. G. SMYTH, W. H. STEIN AND S. MOORE, *J. Biol. Chem.*, 237 (1962) 1845.
[70] J. T. POTTS, A. BERGER, J. COOKE AND C. B. ANFINSEN, *J. Biol. Chem.*, 237 (1962) 1851.
[71] W. W. CLELAND, *Biochemistry*, 3 (1964) 480.
[72] A. F. S. A. HABEEB, *Fed. Proc.*, 24 (1965) 224.
[73] J. LENARD AND S. J. SINGER, *Nature*, 210 (1966) 536.

Chapter 5

SELECTIVE CLEAVAGE OF PEPTIDE CHAINS

Following cleavage of the disulphide bonds, a protein will either fall apart releasing the component chains (like insulin) or it will unfold into one long peptide chain (like ribonuclease). Not all proteins contain cystine but other possibilities of cross-links exist, *e.g.* phosphate ester links. Also the three-dimensional aspect must not be overlooked for interaction of amino acid side chains with each other or with points along the peptide chain undoubtedly occurs. Completely alien substances like metals, pigments and sugars often play a vital role in maintaining the unique structure required for biological function. The molecule of human haemoglobin consists of four peptide chains (two α- and two β-chains) associated with four haem groups which are the oxygen carriers. The structures of the two different haemoglobin chains as reported by Braunitzer *et al.*[1] and that of myoglobin[2–4] are illustrated in Fig. 56. The recently reported structure for the tobacco mosaic virus protein subunit (Fraenkel-Conrat[5]) is remarkable in that no cross-linking of the chain of 158 amino acid residues occurs (Fig. 57). An interesting homology is apparent in the linear structures of trypsinogen (Walsh *et al.*[6]) and chymotrypsinogen (Hartley[7]). If the N-terminal isoleucine residue formed upon activation of the zymogens is made the common starting point for the comparison (Walsh and Neurath[8]), about 40% of the amino acid residues occur in identical positions in the two chains (Fig. 58).

The labour involved in sequence analysis would be reduced considerably if ways could be found to fragment long peptide chains into smaller peptides containing ten to fifteen amino acid residues and then into a second series of larger peptides to establish overlaps. This ideal situation rarely arises. In practice the problem has been tackled in a number of ways and the studies on insulin, ribonuclease and haemoglobin reflect three different approaches. With insulin in the early stages, partial acid hydrolysates of dinitrophenylated peptides were examined (see Ch. 6), enzymes being used later to provide larger peptides. The rapid elucidation of the ribonuclease structure was made possible by refinements in amino acid analysis.

Peptides obtained by the action of enzymes were analyzed after repeated subtraction of the N-terminal residues by the Edman technique (see Ch. 6). An ingenious feature of the studies on haemoglobin (Braunitzer) was the use of alternating enzymic hydrolysis. In this procedure the enzymes trypsin and pepsin were used alternately and the specificity of the latter was enhanced considerably by limiting the reaction times. A similar approach was adopted by Konigsberg and collaborators[11,12] in parallel studies on haemoglobin. The technical problems associated with the fractionation of complex hydrolysates and the determination of a complete structural formula for a protein have been solved by the monumental labours of a few dedicated groups of workers. We know now the complete amino acid sequence for insulin, glucagon, ribonuclease, haemoglobin, tobacco mosaic virus protein, lysozyme, chymotrypsinogen, cytochrome-*c*, and bacterial ferredoxin, and for corticotrophin and other peptide hormones. Studies are nearing completion for papain, myoglobin and trypsinogen, and are well advanced for one or two other proteins. The sequence studies have been carried out on partial acid hydrolysates or on hydrolysates obtained by the action of various proteolytic enzymes. Occasionally, confirmation of particular sequences has been possible using purely chemical methods of selective cleavage of peptide bonds.

PARTIAL ACID HYDROLYSIS

Hydrolysis of proteins by acids under milder conditions than those required for a complete breakdown into their constituent amino acids is firmly established among the techniques in protein chemistry as a valuable method for fragmenting peptide chains. The main problem in using this method is to ascertain the conditions necessary to produce the minimum degree of complexity in the resulting mixture of peptides. Unfortunately, the mathematical approach to the problem is too complicated to be of value and as has been pointed out by Sanger[13] there is no phase during the course of acid hydrolysis when the composition of the hydrolysate is not complex except towards the end.

The relative rate of hydrolysis of a given peptide bond must depend on various factors both steric and electrostatic which influence the approach of hydrogen ions. Synge[14] made a kinetic study

References p. 159 *Text continued p. 140*

	A																				B										
α	Val		Leu	Ser	Pro	Ala	Asp	Lys	Thr	Asp	Val	Lys	Ala	Ala	Try	Gly	Lys	Val	Gly	Ala	His	Ala	Gly	Glu	Tyr	Gly	Ala	Glu	Ala	Leu	Glu
	1		2	3	4	5	6	7	8	9	10	11	12	13	14	15	16	17	18	19	20	21	22	23	24	25	26	27	28	29	30
β	Val	His	Leu	Thr	Pro	Glu	Glu	Lys	Ser	Ala	Val	Thr	Ala	Leu	Try	Gly	Lys	Val	Asp			Val	Asp	Glu	Val	Gly	Gly	Glu	Ala	Leu	Gly
	1	2	3	4	5	6	7	8	9	10	11	12	13	14	15	16	17	18	19			20	21	22	23	24	25	26	27	28	29
M	Val		Leu	Ser	Glu	Gly	Glu	Try	(Glu,	Leu,	Val,	Leu,	His,	Val,	Try)	Ala	Lys	Val	Glu	Ala	Asp	Val	Ala	Gly	(His,	Gly,	Glu)	Asp	Ileu	Leu	Ileu
	1	2	3	4	5	6	7	8	9	10	11	12	13	14	15	16	17	18	19	20	21	22	23	24	25	26	27	28	29	30	31

						C							CD						D								E				
α	Arg	Met	Phe	Leu	Ser	Phe	Pro	Thr	Thr	Lys	Thr	Tyr	Phe	Pro	His	Phe		Asp	Leu	Ser	His						Gly	Ser	Ala	Glu	Val
	31	32	33	34	35	36	37	38	39	40	41	42	43	44	45	46		47	48	49	50						51	52	53	54	55
β	Arg	Leu	Leu	Val	Val	Tyr	Pro	Try	Thr	Glu	Arg	Phe	Phe	Glu	Ser	Phe	Gly	Asp	Leu	Ser	Thr	Pro	Asp	Ala	Val	Met	Gly	Asp	Pro	Lys	Val
	30	31	32	33	34	35	36	37	38	39	40	41	42	43	44	45	46	47	48	49	50	51	52	53	54	55	56	57	58	59	60
M	Arg	Leu	Phe	Lys	Ser	His	Pro	Glu	Thr	Leu	Glu	Lys	Phe	Asp	Arg	Phe	Lys	His	Leu	Lys	Thr	Glu	Ala	Glu	Met	Lys	Ala	Ser	Glu	Asp	Leu
	32	33	34	35	36	37	38	39	40	41	42	43	44	45	46	47	48	49	50	51	52	53	54	55	56	57	58	59	60	61	62

																	EF										F				
α	Lys	Gly	His	Gly	Lys	Lys	Val	Ala	Asp	Ala	Leu	Thr	Asp	Ala	Val	Ala	His	Val	Asp	Asp	Met	Pro	Asp	Ala	Leu	Ser	Ala	Leu	Ser	Asp	Leu
	56	57	58	59	60	61	62	63	64	65	66	67	68	69	70	71	72	73	74	75	76	77	78	79	80	81	82	83	84	85	86
β	Lys	Ala	His	Gly	Lys	Lys	Val	Leu	Gly	Ala	Phe	Ser	Asp	Gly	Leu	Ala	His	Leu	Asp	Asp	Leu	Lys	Gly	Thr	Phe	Ala	Thr	Leu	Ser	Glu	Leu
	61	62	63	64	65	66	67	68	69	70	71	72	73	74	75	76	77	78	79	80	81	82	83	84	85	86	87	88	89	90	91
M	Lys	Lys	His	Gly	Val	Thr	Val	Leu	Thr	Ala	Leu	Gly	Ala	Ileu	Leu	Lys	Lys	Lys	Gly	His	His	Glu	Ala	Glu	Leu	Lys	Pro	Leu	Ala	Glu	Ser
	63	64	65	66	67	68	69	70	71	72	73	74	75	76	77	78	79	80	81	82	83	84	85	86	87	88	89	90	91	92	93

			FG							G																			GH		
α	His	Ala	His	Lys	Leu	Arg	Val	Asp	Pro	Val	Asp	Phe	Lys	Leu	Leu	Ser	His	Cys	Leu	Leu	Val	Thr	Leu	Ala	Ala	His	Leu	Pro		Ala	Glu
	87	88	89	90	91	92	93	94	95	96	97	98	99	100	101	102	103	104	105	106	107	108	109	110	111	112	113	114		115	116
β	His	Cys	Asp	Lys	Leu	His	Val	Asp	Pro	Glu	Asp	Phe	Arg	Leu	Leu	Gly	Asp	Val	Leu	Val	Cys	Val	Leu	Ala	His	His	Phe	Gly		Lys	Glu
	92	93	94	95	96	97	98	99	100	101	102	103	104	105	106	107	108	109	110	111	112	113	114	115	116	117	118	119		120	121
M	His	Ala		Thr	Lys	His	Lys	Ileu	Pro	Ileu	Lys	Tyr	Leu	Glu	Phe	Ileu	Ser	Glu	Ala	Ileu	Ileu	His	Val	Leu	His	Ser	Arg	His	Pro	Gly	Asp
	94	95	96	97	98	99	100	101	102	103	104	105	106	107	108	109	110	111	112	113	114	115	116	117	118	119	120	121	122	123	124

			H																									HC			
α	Phe	Thr	Pro	Ala	Val	His	Ala	Ser	Leu	Asp	Lys	Phe	Leu	Ala	Ser	Val	Ser	Thr	Val	Leu	Thr	Ser	Lys	Tyr	Arg						
	117	118	119	120	121	122	123	124	125	126	127	128	129	130	131	132	133	134	135	136	137	138	139	140	141						
β	Phe	Thr	Pro	Pro	Val	Glu	Ala	Ala	Tyr	Glu	Lys	Val	Val	Ala	Gly	Val	Ala	Asp	Ala	Leu	Ala	His	Lys	Tyr	His						
	122	123	124	125	126	127	128	129	130	131	132	133	134	135	136	137	138	139	140	141	142	143	144	145	146						
M	Phe	Gly	Ala	Asp	Ala	Glu	Gly	Ala	Met	Asp	Lys	Ala	Leu	Glu	Leu	Phe	Arg	Lys	Asp	Ileu	Ala	Ala	Lys	Tyr	Lys	Glu	Leu	Gly	Tyr	(Glu-NH_2,	Gly)
	125	126	127	128	129	130	131	132	133	134	135	136	137	138	139	140	141	142	143	144	145	146	147	148	149	150	151	152	153	154	155

Fig. 56. The structure of the α- and β-chains of the molecule of normal, adult haemoglobin, compared with the sequence of the peptide chain of whale myoglobin (M). Braunitzer *et al.*[1], Edmundson[3] and Smyth[4]. A- - - - - - etc., denotes helical structure: CD, EF, etc., non helical sections.

1 5 NH_2 10 15
Acetyl N-Ser → Tyr → Ser → Ileu → Thr → Pro → Thr → Ser → Glu → Phe → Val → Phe → Leu → Ser → Ser

30 NH_2 25 NH_2 20
Ala ← Asp ← Thr ← CySH ← Asp ← Leu ← Ileu ← Leu ← Glu ← Ileu ← Pro ← Asp ← Ala ← Try ← Ala

NH_2 NH_2 35 NH_2 NH_2 NH_2 40 NH_2 45
Leu → Gly → Asp → Glu → Phe → Glu → Thr → Glu → Glu → Ala → Arg → Thr → Val → Glu → Val

60 NH_2 55 50 NH_2 NH_2
Val ← Thr ← Val ← Glu ← Pro ← Ser ← Pro ← Lys ← Try ← Val ← Glu ← Ser ← Phe ← Glu ← Arg

65 70 NH_2 75
Arg → Phe → Pro → Asp → Ser → Asp → Phe → Lys → Val → Tyr → Arg → Tyr → Asp → Ala → Val

90 85 80
Arg ← Thr ← Asp ← Phe ← Ala ← Gly ← Leu ← Leu ← Ala ← Thr ← Val ← Leu ← Pro ← Asp ← Leu

NH_2 95 NH_2 NH_2 NH_2 100 NH_2 105
Asp → Arg → Ileu → Ileu → Glu → Val → Glu → Asp → Glu → Ala → Asp → Pro → Thr → Thr → Ala

120 115 110
Ala ← Val ← Thr ← Ala ← Asp ← Asp ← Val ← Arg ← Arg ← Thr ← Ala ← Asp ← Leu ← Thr ← Glu

125 NH_2 130 135
Ileu → Arg → Ser → Ala → Asp → Ileu → Asp → Leu → Ileu → Val → Glu → Leu → Ileu → Arg → Gly

150 145 140 NH_2
Leu ← Gly ← Ser ← Ser ← Ser ← Glu ← Phe ← Ser ← Ser ← Arg ← Asp ← Tyr ← Ser ← Gly ← Thr

155 158
Val → Try → Thr → Ser → Gly → Pro → Ala → Thr

Fig. 57. Sequence of the 158 amino acid residues in the protein subunit of tobacco mosaic virus. Data of Anderer *et al.*[9] and of Fraenkel-Conrat *et al.*[5]. Reproduced by permission of the National Academy of Sciences, Washington, and incorporating recent work[10].

References p. 159

```
                     1   2   3   4   5   6   7   8   9   10  11  12  13  14  15  16  17  18
Chymotrypsinogen     cys-gly-val-pro-ala-ile-gln-pro-val-leu-ser-gly-leu-ser-arg-ILE-VAL-GLY-
Trypsinogen                                              val-asp-asp-asp-asp-lys-ILE-VAL-GLY-
                                                         1   2   3   4   5   6   7   8   9

19  20  21  22  23  24  25  26  27  28  29  30  31  32  33  34  35  36  37  38  39  40  41
asp-glu-glu-ala-val-pro-gly-ser-trp-PRO-trp-GLN-VAL-SER-LEU-gln-asp-lys-thr-GLY-phe-HIS-PHE-
gly-tyr-thr-cys-gly-ala-asn-thr-val-PRO-tyr-GLN-VAL-SER-LEU-asn-        -ser-GLY-tyr-HIS-PHE-
10  11  12  13  14  15  16  17  18  19  20  21  22  23  24  25              26  27  28  29  30

42, 43  44  45  46  47  48  49  50  51  52  53  54  55  56  57  58  59  60  61  62  63  64
CYS-GLY-GLY-SER-LEU-ILE-ASN-glu-asn-TRP-VAL-VAL-thr-ALA-ALA-HIS-CYS-gly-val-thr-thr-ser-asp-
CYS-GLY-GLY-SER-LEU-ILE-ASN-ser-gln-TRP-VAL-VAL-ser-ALA-ALA-HIS-CYS-tyr-lys-ser-gly-ile-gln-
31  32  33  34  35  36  37  38  39  40  41  42  43  44  45  46  47  48  49  50  51  52  53

65  66  67  68  69  70  71  72  73  74  75  76  77  78  79  80  81  82  83  84  85  86  87
VAL-val-val-ala-gly-glu-phe-asp-gln-gly-ser-ser-ser-glu-lys-ile-gln-lys-leu-lys-ile-ala-lys-
VAL-arg-leu-gly-glu-asp-asn-ile-asn-val-val-glu-gly-asp-glu-gln-phe-ile-ser-ala-ser-lys-ser-
54  55  56  57  58  59  60  61  62  63  64  65  66  67  68  69  70  71  72  73  74  75  76

88  89  90  91  92  93  94  95  96  97  98  99  100 101 102 103 104 105 106 107 108 109 110
val-phe-lys-asn-SER-lys-TYR-ASN-ser-leu-thr-ile-ASN-ASN-asn-ILE-thr-LEU-leu-LYS-LEU-ser-thr-
ile-val-his-pro-SER-    -TYR-ASN(pro,leu,thr,asn)ASN-ASN-asp-ILE-met-LEU-ile-LYS-LEU-lys-ser-
77  78  79  80  81       82  83  84  85  86  87  88  89  90  91  92  93  94  95  96  97  98

111 112 113 114 115 116 117 118 119 120 121 122 123 124 125 126 127 128 129 130 131 132 133
ALA-ALA-SER-phe-ser-gln-thr-VAL-ser-ala-val-cys-LEU-PRO-ser-ala-ser-asp-asp-phe-ala-ALA-GLY-
ALA-ALA-SER-leu-asn-ser-arg-VAL-ala-ser-ile-ser-LEU-PRO-thr-ser-cys-        -ala-ser-ALA-GLY-
```

```
THR-thr-CYS-val-thr-thr-GLY-TRP-GLY-leu-THR-arg-tyr-thr-asn-ala-asn-thr-PRO-ASP-arg-LEU-gln-
THR-gln-CYS-leu-ile-ser-GLY-TRP-GLY-asn-THR-lys-ser-ser-gly-thr-ser-tyr-PRO-ASP-val-LEU-lys-
120 121 122 123 124 125 126 127 128 129 130 131 132 133 134 135 136 137 138 139 140 141 142

157 158 159 160 161 162 163 164 165 166 167 168 169 170 171 172 173 174 175 176 177 178 179
gln-ala-ser-leu-PRO-leu-LEU-SER-asn-thr-asn-CYS-LYS-lys-tyr-trp-gly-thr-lys-ILE-lys-asp-ala-
cys-leu-lys-ala-PRO-ile-LEU-SER-asp-ser-ser-CYS-LYS-ser-ala-tyr-pro-gly-gln-ILE-thr-ser-asn-
143 144 145 146 147 148 149 150 151 152 153 154 155 156 157 158 159 160 161 162 163 164 165

180 181 182 183 134 185 186 187 188 189         190 191 192 193 194 195 196 197 198 199 200
MET-ile-CYS-ALA-GLY-ala-ser-gly-val-ser-       -SER-CYS-met-GLY-ASP-SER-GLY-GLY-PRO-leu-VAL-
MET-phe-CYS-ALA-GLY-tyr-leu-glu-gly-gly-lys-asn-SER-CYS-gln-GLY-ASP-SER-GLY-GLY-PRO-val-VAL-
166 167 168 169 170 171 172 173 174 175 176 177 178 179 180 181 182 183 184 185 186 187 188

201 202 203 204 205 206 207 208 209 210 211 212 213 214 215 216 217 218 219 220 221 222 223
CYS-lys-lys-asn-gly-ala-trp-thr-leu-val-GLY-ILE-VAL-ser-SER-TRP-GLY-SER-ser-thr-cys-ser-thr-
CYS-ser-gly-lys-leu-gln-               -GLY-ILE-VAL-   -SER-TRP-GLY-SER-gly-cys-ala-gln-lys-
189 190 191 192 193 194                 195 196 197     198 199 200 201 202 203 204 205 206

224 225 226 227 228 229 230 231 232 233 234 235 236 237 238 239 240 241 242 243 244 245 246
ser-thr-PRO-GLY-VAL-TYR-ala-arg-VAL-thr-ala-leu-VAL-asn-TRP-val-gln-GLN-THR-leu-ALA-ala-ASN
asn-lys-PRO-GLY-VAL-TYR-thr-lys-VAL-cys-asn-tyr-VAL-ser-TRP-ile-lys-GLN-THR-ile-ALA-ser-ASN
207 208 209 210 211 212 213 214 215 216 217 218 219 220 221 222 223 224 225 226 227 228 229
```

Fig. 58. A scheme of the linear structures of chymotrypsinogen and trypsinogen (Walsh and Neurath[8]).

TABLE 19

HYDROLYSIS OF DIPEPTIDES IN STRONG ACID[14]

Peptide	*Relative velocity of hydrolysis (Gly.Gly = 1)*
Gly.Gly	1
Gly.Ala	0.62
Ala.Gly	0.62
Gly.Leu	0.40
Gly.Try	0.35
Gly.Val	0.31
Leu.Gly	0.23
Leu.Leu.	0.048
Leu.Try	0.041
Val.Gly	0.015

of the hydrolysis of a number of dipeptides by a mixture of 10 *N* hydrochloric acid and glacial acetic acid (1:1 by vol) at 37°. Valyl peptides, and to a lesser extent leucyl peptides, were found to be the most stable. This stabilizing effect was not apparent when the bulky side chain was attached to a C-terminal residue as in Gly.Val. Synge's results are shown in Table 19. The positive basic groups might be expected to oppose the approach of hydrogen ions and this argument has been used to explain the stability of dipeptides which accumulate towards the end of a hydrolysis. Thus Stein *et al.*[15] found that a 43 h hydrolysis of silk fibroin in concentrated hydrochloric acid at 40° yielded a mixture containing 25% of its nitrogen as free amino acids and 75% in the form of peptides of average length 2.05 residues.

The conditions used by Sanger and co-workers in their studies on insulin were concentrated hydrochloric acid at 37° for 3 to 8 days. In the particular case of the diisopropoxyphosphinyl derivatives of chymotrypsin and trypsin Naughton *et al.*[16] found that partial hydrolysis carried out with 5.7 *N* HCl at 100° for 10 min was equivalent to 12 *N* HCl at 37° for 24 h.

In dilute acid solution an inversion of the order of amino acids in a dipeptide might take place via the intermediary diketopiperazine. This was demonstrated for glycylvaline in 0.1 *N* hydrochloric acid by Sanger and Thompson[17].

Peptide bonds involving serine and threonine are very labile in acid media and the mechanism of this specific hydrolysis is discussed on p. 153. Tryptophan is destroyed by acids. Another specific hydrolysis studied by Partridge and Davis[18], Blackburn and Lee[19], and Grannis[20] is the preferential release of free aspartic acid when a protein is boiled with acetic acid or oxalic acid. Under these conditions free glutamic acid is also released but at a much slower rate. It has been suggested that in sufficiently dilute acid the acidic groups still retain a negative charge and by attracting hydrogen ions, labilize neighbouring peptide bonds. By repeated extraction with 0.25 *M* oxalic acid at 100° for 1 h. Partridge *et al.*[21] were able to bring the insoluble protein elastase completely into solution as two derived proteins. This result was predicted from the facts that elastin is low in aspartic acid and also low in the hydroxyamino acids. An interesting extension of this type of specific degradation has been described by Schultz[22,23]. Ribonuclease was treated with 0.03 *N* hydrochloric acid (1–2 mg protein/ml) at 105° for 48 h and the partial hydrolysate submitted to high voltage electrophoresis in 0.1 *M* formic acid followed by chromatography in the solvent butanol–acetic acid–water (3:1:1, 90 min at 60°). In this way a peptide map was prepared which revealed 12 ninhydrin positive peptide spots and a single aspartic acid spot. Ribonuclease contains 15 aspartic acid (or asparagine) residues per molecule.

A reaction whereby the normal α form of aspartyl peptides is converted into the $\alpha\beta$ and β forms during acid hydrolysis was demonstrated by Swallow and Abraham[24] and later confirmed by Naughton *et al.*[16]. This interconversion is illustrated in the following scheme.

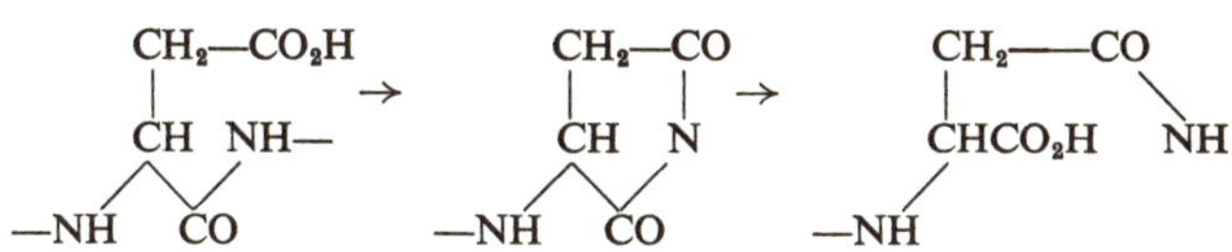

α-Aspartyl $\alpha\beta$-Aspartyl β-Aspartyl

HYDROLYSIS BY ENZYMES

Proteases are available from animal, plant, fungal and bacterial sources. Of the many enzymes (endopeptidases) that split internal peptide bonds, only trypsin, chymotrypsin and pepsin are sufficiently specific in their action for a controlled degradation of proteins.

References p. 159

The other enzymes like subtilisin, mould and bacterial proteases, are more useful for further degradation of the larger fragments. Leucine aminopeptidase and carboxypeptidase (exopeptidases) are useful for the removal of N-terminal and C-terminal amino acids respectively (see pp. 216 and 222).

A number of factors control the rate at which these enzymes attack certain peptide bonds. Apart from the spatial arrangement inherent in amino acid sequence the folding of the chain imposes further structural relationships. Many native proteins are resistant to attack by enzymes but are readily digested after denaturation or oxidation. The splitting of certain bonds may modify the behaviour of adjacent susceptible bonds. The split products themselves tend to influence the specificity of an enzyme and lead to unexpected cleavage reactions.

Although enzymes are more stable at low temperatures digests at 37–40° allow faster reaction rates. Difficult bonds may be split using the forcing conditions of a high enzyme-substrate ratio. The optimum conditions for the controlled degradation of a protein are obtained by construction of hydrolysis rate curves. Typical curves for the hydrolysis of oxidized ribonuclease by trypsin and chymotrypsin[25], and pepsin[26] are illustrated in Figs. 59 and 60. The rate of hydrolysis by trypsin is very slow after 6 hours and the final increase in ninhydrin colour corresponds to the splitting of twelve

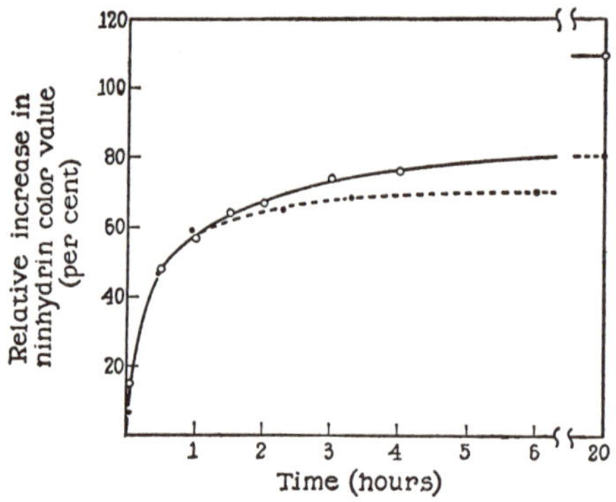

Fig. 59. The rate of hydrolysis of oxidized ribonuclease (1% solution) by chymotrypsin (○) and trypsin (●) at pH 7.0 and 25° (enzyme concentration, 0.005%). The curves are corrected for the small amount of autodigestion occurring in a control solution of the enzyme (from Hirs *et al.*[25]).

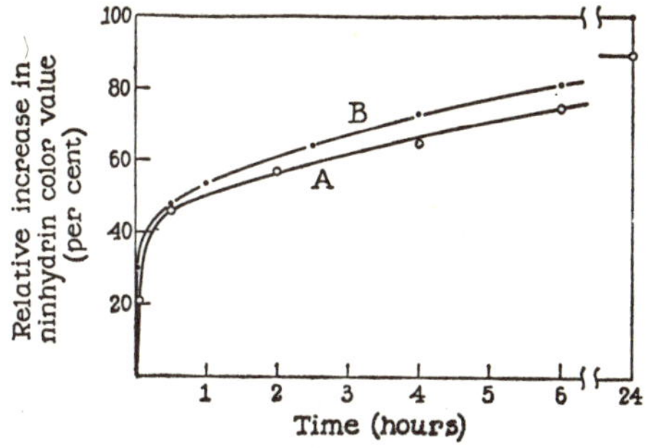

Fig. 60. The rate of hydrolysis of oxidized ribonuclease by pepsin at 25°. A, pepsin concentration, 0.02%; protein concentration, 4%; pH 2.2. B, pepsin concentration 0.02%; protein concentration, 1%; pH 1.9. (from Bailey *et al.*[26].)

bonds. Digestion by chymotrypsin proceeds at a steady rate throughout the first 20 hours with the release of 18 main peptides. Non-specific cleavages might be expected to be minimal when the rate of increase in amino groups becomes steady. The action of pepsin on the protein at two different enzyme-substrate ratios gives increments in ninhydrin colour corresponding to a theoretical cleavage of seven and eleven bonds. Actually, the increase in colour is due to incomplete hydrolysis of a larger number of bonds.

A more precise evaluation of the number of peptide bonds broken can be obtained using a recording autotitrator[27] and assuming complete titration of the free amino groups at pH 9.0. Corrections must be applied which take into account the amounts of alkali necessary to bring the intact protein and the reaction mixture separately from the operating pH to pH 9.0. At temperatures above 25° the pK of the peptides decreases by 0.0257 per degree[28].

The sites of attack of various endopeptidases with reference to the peptide chain of oxidized ribonuclease and the oxidized A and B chains of insulin are shown in Figs. 61 and 62.

Trypsin

There apears to be only one trypsinogen in bovine pancreatic juice and a number of enzymes bring about activation to form the protease, trypsin. The enzyme is used in the pH range 7–9 but owing to the strong tendency to autolize it is not possible to keep the enzyme active for long periods at high pH. The autolysis is retarded considerably

References p. 159 *Text continued p. 146*

```
                                           NH2
Lys.Glu.Thr.Ala.Ala.Ala.Lys.Phe.Glu.Arg.Glu.His.Met.Asp.Ser.Ser.Thr.Ser.Ala.Ala.
 1   2   3   4   5 | 6   7 | 8 | 9  10 | 11  12  13  14  15  16  17  18| 19  20
                   P       T  CP      T                               C

        NH2     I  NH2NH2                 NH2
         |      |   |  |                   |
Ser.Ser.Ser.Asp.Tyr.Cys.Asp.Glu.Met.Met.Lys.Ser.Arg.Asp.Leu.Thr.Lys.Asp.Arg.——
21  22  23  24  25 | 26  27  28  29  30  31 | 32  33 | 34  35 | 36  37 | 38  39 |
                   C                        T       T        C       T        T

II          NH2                                    NH2         III  NH2
|            |                                      |           |    |
Cys.Lys.Pro.Val.Asp.Thr.Phe.Val.His.Glu.Ser.Leu.Ala.Asp.Val.Glu.Ala.Val.Cys.Ser.Glu.——
40  41  42  43  44  45 | 46 | 47  48  49  50  51 | 52  53  54  55 | 56  57  58  59  60
                       P   CP                   P               P

    NH2         IV      NH2         NH2NH2 V     NH2
     |           |       |           |   |  |     |
Lys.Asp.Val.Ala.Cys.Lys.Asp.Gly.Thr.Asp.Glu.Cys.Tyr.Glu.Ser.Tyr.Ser.Tyr.Met.Ser.Ileu.Thr.——
61 | 62 | 63  64  65  66 | 67  68  69  70  71  72  73 | 74  75  76 | 77  78  79 | 80  81  82
   T    C               T                            C           C           C

    VI                                  NH2     VII             NH2     NH2
     |                                   |       |               |       |
Asp.Cys.Arg.Glu.Ser.Thr.Gly.Ser.Lys.Tyr.Pro.Asp.Ala.Cys.Tyr.Lys.Thr.Thr.Asp.Ala.Glu.Lys.——
83  84  85 | 86  87  88  89  90  91 | 92  93  94  95  96  97 | 98 | 99  100 101 102 103| 104 |
```

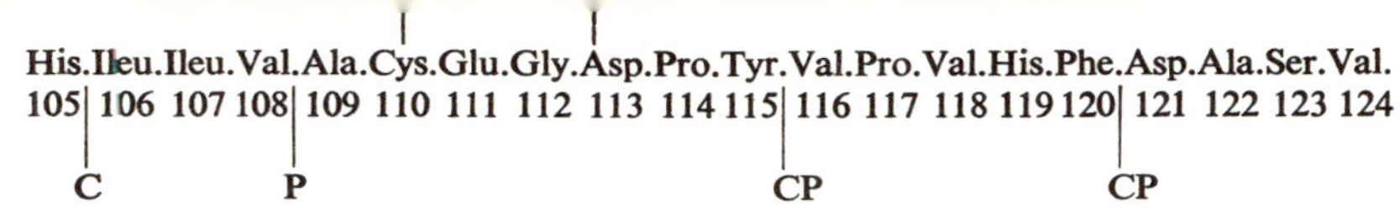

Fig. 61. Action of enzymes on the peptide chain of oxidized bovine ribonuclease. T. trypsin, C. chymotrypsin, P. pepsin. (From Hirs *et al.*[29] and incorporating the revision of Smyth *et al.*[30] and Potts *et al.*[31].)

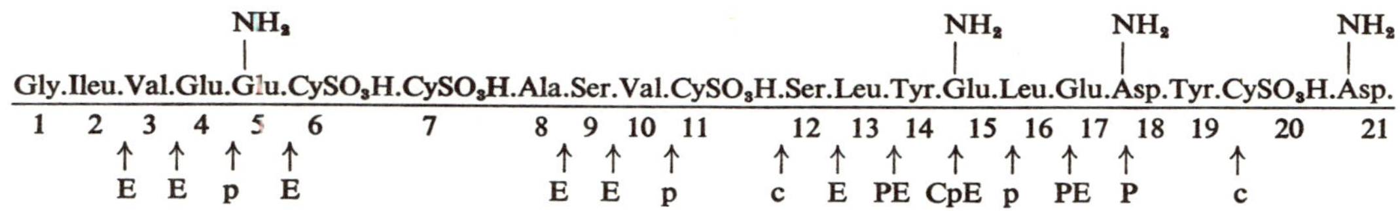

A chain of oxidized bovine insulin

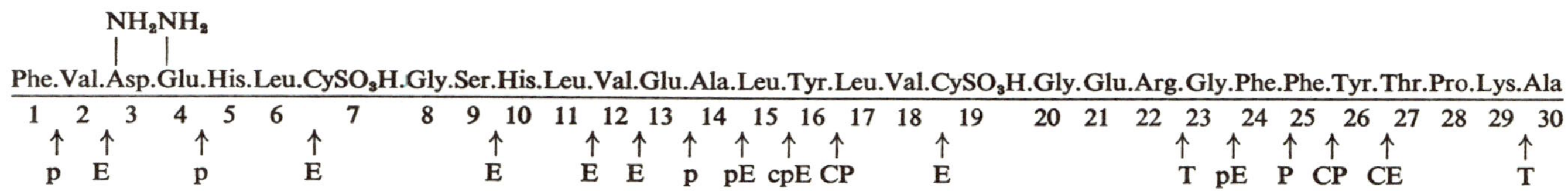

B chain of oxidized insulin

Fig. 62. Action of enzymes on the A and B chains of oxidized insulin. Major sites of attack indicated by capital initial of each enzume; sites of slow attack, small initial. T. trypsin, C. chymotrypsin, P. pepsin, E. elastase.

in the presence of calcium ions and it has been suggested that the metal ion prevents association between two different molecular species[32]. In the absence of Ca^{2+} the enzymes are completely inactivated in 6 h at pH 7.9 and 26°; in the presence of 0.001 *M* Ca^{2+} under the same conditions less than 5% of the activity is lost. The stabilizing effect of Ca^{2+} and of other ions on a number of proteins is a well-studied phenomenon[33]. Maximum stability is exhibited at pH 3; in 1 m*M* HCl trypsin at a concentration of 0.640 mg/ml has $E_{280\ m\mu}^{1\ cm}$ 1.0 (ref.[34]). The enzyme appears to be stable to low concentrations of urea and undergoes reversible denaturation in strong urea solutions. A series of natural inhibitors[35,36] are known and in common with chymotrypsin the enzyme is inactivated by diisopropylphosphofluoridate.

The most valuable property of trypsin is its narrowly restricted specificity which limits action exclusively to bonds linking the carboxyl group of a basic amino acid to the amino group of another amino acid or to the hydroxyl group of an alcohol. It was discovered many years ago that certain native proteins are resistant to the action of trypsin but that proteolysis takes place readily after denaturation of the proteins. On the basis of amino acid composition it is possible to predict accurately the number of peptides that might be expected in a tryptic digest of a denatured protein.

By the masking of specific basic groups or by the introduction of new ones the action of the enzyme may be limited or extended at will. For example, when the ε-amino groups of the lysine residues in ribonuclease are blocked by dinitrophenylation only arginyl bonds are split (Redfield and Anfinsen[37], Anfinsen *et al.*[38]). Formation of dithiocarbamyl[39]- or trifluoroacetyl[40]-derivatives will similarly limit the action of trypsin and both groups are easily removed for subsequent analytical work. Trifluoroacetyl groups (formed by reaction of ε-amino groups with ethyl thioltrifluoroacetate) are released by exposure to 1.0 *M* piperidine at 0°. Further sites for enzymic degradation are created by conversion of the cysteine residues of a reduced protein to S-(β-aminoethyl) residues by the action of β-bromoethylamine (Lindley[41]; Tietze *et al.*[42]), or better still with ethylenimine[43] which reacts quantitatively. In this way a method for the selective cleavage of peptide chains at cysteinyl residues may be realised. Arginine can be blocked by coupling with benzil ($C_6H_5COCOC_6H_5$)

which does not react with lysine[44]. Homoarginyl residues [—$CH(CH_2)_4NHC{=}NH(NH_2)$] have been formed in the α-lactoglobulin molecule by conversion of the ε-amino groups into guanidyl groups, but in this case the modified protein is resistant to the action of trypsin (Weil and Telka[45]). Bonds involving histidine or ornithine[46], and the sequences Arg.Pro, and Lys.Pro., are resistant. Acetyl insulin which is available commercially owes its enhanced stability to the fact that most of the ε-amino groups are blocked (Nord *et al.*[47]). The proximity of strongly acidic chains or unprotected α-amino or α-carboxyl groups tends to depress trypsin activity.

From time to time there have been reports of an apparent inherent chymotryptic activity of trypsin and there seems to be sufficient evidence now to have established a true cross reactivity. Recent studies of the non-specific reactions of both enzymes by Inagami and Sturtevant[48] have provided a useful means of detecting small amounts of cross contamination.

Transpeptidation reactions are well established for trypsin[49] and this undesirable complication has been expected in sequence studies. It is generally considered however that transpeptidation does not play a significant role in view of the high yields of peptides obtained in digests (50–100% of theory) and the fact that many peptides containing more than one basic group are not usually obtained.

Chymotrypsin

Unlike trypsinogen activation the formation of chymotrypsin from its zymogen is quite complicated. The processes that lead to the formation of the α-chymotrypsin family from chymotrypsinogen A make a fascinating story the details of which are to be found in an excellent review by Desnuelle[51].

Again, like trypsin the enzyme is most stable at pH 3 but its action on proteins is optimal in the pH range 7 to 9. In concentrated solutions at pH 7.8 the enzyme autolyzes into two different active forms, β- and γ-chymotrypsin. Chymotrypsin exhibits a broader specificity than trypsin but links involving the carboxyl group of the aromatic residues tyrosine, tryptophan and phenylalanine as amides or esters are very susceptible. Leucyl, methionyl, glutaminyl and asparaginyl bonds are split at a slower rate. Aromatic residues adjacent to proline are resistant.

Phenylmethanesulphonyl fluoride[52] inactivates chymotrypsin, probably by reaction at the hydroxyl group of the serine residue at the active site. Gold and Fahrney[53] have reported that under acid conditions an O-aspartyl linkage is formed at the active site via an intermediate oxazoline (see p. 153). At pH 7 the protein reverts to an active enzyme.

Pepsin

Commercial crystalline pepsin is obtained from gastric juice or gastric mucosa and is a mixture of enzymes. It is fairly stable at pH 5 to 6, is denatured above pH 6 and is optimally active on synthetic substrates at pH 4 and on proteins at pH values between 1 and 2. The enzyme autolyzes at low pH and it may be necessary to remove peptide contaminants that may have arisen in this way by passage through a short column of Dowex 50 at pH 2.9 at low temperature (Heirwegh and Edman[54]). During the formation of pepsin from pepsinogen a large fragment of the molecule is split off which contains 29 amino acids (Van Vunakis and Herriott[55]). This polypeptide is a natural inhibitor acting between pH 5 and 6.

The enzyme exhibits a very wide specificity. Although experiments with synthetic substrates have shown that a preferential attack is made at bonds involving the amino groups of aromatic amino acids there are nevertheless many instances where such bonds are resistant in proteins. Only peptide bonds are attacked; amides and esters are resistant. The specificity of pepsin is enhanced at low enzyme-substrate ratio by limiting the reaction time to about thirty minutes[56].

An interesting discovery is that transpeptidations are catalyzed by the enzyme (Neuman *et al.*[57]). For example, N-acetyl-L-tyrosyl-L-tyrosine when incubated with enzyme yields tyrosyl-tyrosine as well as tyrosine. Since pepsin is incapable of removing the N-acetyl group it is necessary to invoke transpeptidation in explanation of the results. The following scheme then is possible.

$$\begin{aligned} 2\ \text{Ac Tyr.Tyr} &\rightarrow \text{Ac Tyr} + \text{Ac Tyr.Tyr.Tyr} \\ \text{Ac Tyr.Tyr.Tyr} &\rightarrow \text{Ac Tyr} + \text{Tyr.Tyr} \\ \text{Ac Tyr.Tyr.Tyr} &\rightarrow \text{Ac Tyr.Tyr} + \text{Tyr} \end{aligned}$$

The milk-clotting enzyme, rennin, has a specificity similar to that of pepsin. The purification and partial characterization of prorennin has been described by Foltmann[58] and Bundy *et al.*[59].

Papain

The enzyme is prepared from papaya latex obtained from the green fruit of the tropical papaw. Kimmel and Smith[60] have described a preparation from the dried latex and the enzyme, which is a homogeneous protein, is the subject of an intensive structural investigation (Smith and Kimmel *et al.*[61,62]).

Papain is a sulphydryl enzyme and requires activation by the usual reducing agents such as hydrogen sulphide, cysteine, or KCN; conversely, rapid inactivation is brought about by the thiol reagents Hg^{2+}, *p*-mercuribenzoate and iodoacetate. Guanidine salts and urea at concentrations above 3 *M* cause irreversible inactivation. A remarkable property of the enzyme is the high temperature stability.

Maximum enzymic activity is in the pH range 5 to 7.5; below pH 4 irreversible inactivation is rapid. It is impossible to generalize about its action in large polypeptides or proteins as a broad hydrolytic action similar to that of pepsin is usually exhibited. On the other hand some points of interest have emerged from the extensive classical studies using synthetic substrates (Bergmann and Fruton[63], Kimmel and Smith[64]). The most sensitive substrates are those containing the basic amino acids as amides and this trypsin-like activity is much stronger than the other hydrolytic actions of the enzyme. Hydrolysis rates of peptide bonds involving glutamic acid are low at high pH and increase as the pH is lowered to about 4. Decreasing susceptibility with progressive ionization of a neighbouring α- or γ-carboxyl group appears to be a general phenomenon.

Papain possesses esterase activity on certain N-acetylated substrates an effect which apparently extends to thiol esters[65]. Transpeptidation reactions have been established by Fruton and co-workers and this work has been reviewed[37]. Other plant proteases are known, for example, ficin from the fig family, bromelin from the pineapple and asclepain from the milkweed. In general these enzymes exhibit a specificity similar to that of papain[66], but with the exception of ficin (Hammond and Gutfreund[67]) little is known about their purity.

References p. 159

Bacterial and mould proteases

Subtilisin was crystallized by Güntelberg and Ottesen[68] from the culture medium of *B. subtilis*. A remarkable specificity was found for the enzyme when it was shown that limited hydrolysis of ovalbumin produced a new crystalline protein, plakalbumin with simultaneous release of the hexapeptide Ala.Gly.Val.Asp.Ala.Ala (Ottesen and Wollenberger[69]). Another example of this limited action is the cleavage of the 20 N-terminal residues of native ribonuclease as a single peptide (Richards and Vithayathil[70]). The S-peptide and residual protein designated RNase S are enzymically inactive. The two components show great mutual affinity even in very dilute solution and the reconstituted enzyme regains most of the original activity.

When applied to other proteins the specificity of subtilisin is very wide; in the case of the B chain of oxidized insulin, for example, 18 of the total 29 bonds are split[71]. The pH optimum for the enzyme is in the region 10–11.

A similar bacterial protease has been described by Hagihara *et al.*[72] and named BPN^1 (*Bacillus protease*, strain N^1). This enzyme and subtilisin are both stabilized by Ca^{2+} and are inactivated by diisopropylphosphofluoridate. Proteases from *Clostridia*, *Pseudomonas*, *Streptococci* and *Streptomyces* are in various stages of characterization and all show a very broad specificity. One from *St. griseus*, Pronase, is reported to hydrolyze 87% of the total bonds in the egg albumin molecule and is the nearest approach on record to a universal proteolytic enzyme[73–75].

Iminopeptidases

Proline iminopeptidase has been prepared from *E. coli* by Katchalski and collaborators[76]. It acts as a specific exopeptidase cleaving exclusively N-terminal L-proline residues from both high and low molecular weight substrates; N-terminal hydroxy-L-proline residues are not split.

Iminodipeptidase (prolinase) is prepared from swine kidney cortex[77,78] although it is also present in *E. coli* and other micro organisms. The dipeptidase acts similarly to iminopeptidase on polyglycine, both enzymes are activated by manganic ions, and both show a similar pH-activity curve. Iminodipeptidase is strictly a dipeptidase acting on hydroxyprolyl as well as prolyl peptides.

Elastase

This enzyme was reported to be present in the pancreas by Balo and Banga[50] in 1950. The proteolytic action of the enzyme on elastin was recognized by Partridge and Davis[79] in 1955 and later workers studied the specificity. The action of a purified preparation obtained by chromatography on carboxymethyl-cellulose on fractions A and B of oxidized insulin has been described by Naughton and Sanger[80].

The method of estimation of elastase activity illustrates a simple and reasonably accurate technique that is often used to explore proteolytic activity in biological preparations. Powdered elastin (2 g) is suspended in a saturated aqueous solution (20–30 ml) of Congo red overnight, and the dyed protein filtered off. The material is well washed with water and dried with acetone and ether. A sample of the solution to be assayed is added to 3.5 ml of an elastin suspension (4–6 mg solid) in 0.05 M Na_2CO_3–HCl buffer (pH 8.8) in a 15-ml centrifuge tube. The amount of digestion is estimated by centrifuging down the undissolved elastin and reading the extinction at 495 mμ of the supernatant solution. E_{495} for a 0.1% solution of digested dyed elastin was found to be about 1.0.

Elastin attacks bonds involving neutral amino acids having aliphatic side chains and appears to have a chymotryptic activity of 1–2%. Activation of proelastase by trypsin has been discussed by Lamy and Tauber[81].

SELECTIVE CLEAVAGE BY CHEMICAL METHODS

Fission of peptide chains by chemical means exploits the unique reactivity of the side chains of particular amino acids in the labilization of an adjacent peptide bond. The residues investigated so far for this purpose have been the hydroxyamino acids, cysteine, tryptophan, methionine, the aromatics and the dicarboxylic acids.

Acyl migration at serine and threonine residues

Under the influence of acids and acid chlorides, β-hydroxyamides of amino acids and peptides undergo a rearrangement whereby the

References p. 159

acyl group migrates from nitrogen to oxygen. In the presence of alkali the ester reverts to the original amide (see below and p. 233).

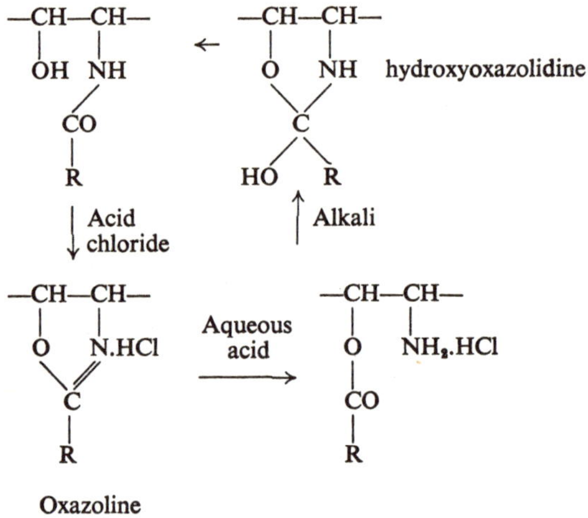

The first observation of an N,O acyl shift in an aliphatic system was made by Bergman *et al.*[82] in 1921 and shortly after, Bettzieche[83] observed a similar transformation when N-benzoylamino alcohols were treated with strong hydrochloric or sulphuric acid. In recent years ethanolic hydrogen chloride hass been used by Phillips and Baltzly[84] for the rearrangement of the β-hydroxyethylamides of a number of cyclic carboxylic acids. To explain the reversible nature of the process, these authors postulated the formation of a cyclic hydroxyoxazolidine. This cyclic structure which was originally proposed by Bergmann to account for the reverse transformation under the influence of alkali has had further support from the work of Welsh[85] on the inversion and retention of configuration during acyl migrations in derivatives of ephedrine. The mechanism put forward for retention of configuration is shown below and is analogous to the one used to represent the acid-catalyzed hydrolysis of esters. In the rearrangement, the alcoholic hydroxyl rather than a water molecule, acts as an electron donor to the carboxyl atom. During transformation of O–(1–^{14}C)-acetyl-DL-serine in the presence of CH_3COO^- only a small loss of activity to the medium occurs[86], as would be expected if a cyclic intermediate were involved.

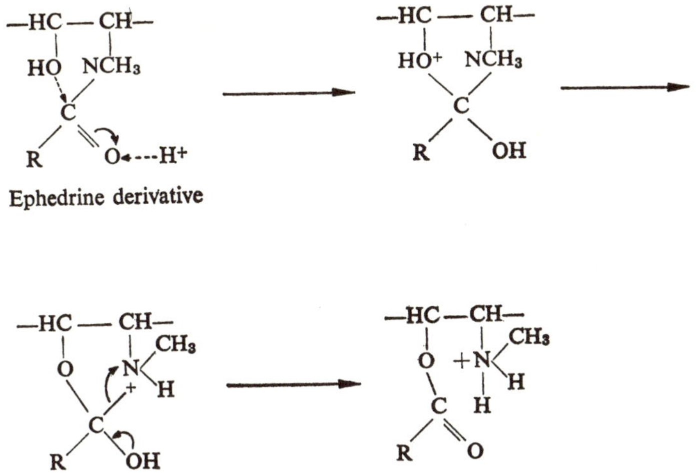

In the protein field the first observations bearing on the subject were made by Desnuelle and Casal[87] who showed that the peptide bonds in which serine and threonine are linked through their nitrogen atoms are more labile to acid hydrolysis than other peptide bonds. More definite information on acyl migrations in proteins was provided by Elliott[88] who treated silk fibroin with concentrated sulphuric acid for 72 h at 21° and obtained a product in which about 60% of the possible transfers at the serine residues had taken place. The estimated extent of rearrangement of the silk fibroin was based on

(1) a determination of the increase of amino nitrogen by the Van Slyke technique,

(2) recovery of DNP-serine after dinitrophenylation of the product at pH 5, and

(3) uptake of alkali during hydrolysis of ester groups in the rearranged product after protection of amino groups by acetylation.

Little non-specific splitting of bonds appears to take place in sulphuric acid and indeed it has been found that insulin retains its biological activity after sulphuric acid treatment[89]. It should be mentioned, however, that the treatment may cause sulphonation of tyrosine residues and it is uncertain how tryptophan residues behave.

References p. 159

Josefsson and Edman[90,91] have described the reversible inactivation of lysozyme and ribonuclease by anhydrous formic acid and have ascribed this phenomenon to N,O-peptidyl shifts. Lysozyme hydrochloride, for example, was dissolved in anhydrous formic acid (prepared by drying 98% formic acid over boric anhydride) and kept for 16 h at 25° when all the enzyme activity was lost. Activity was restored in aqueous solution at pH 7.5 in 5 h. The theoretical number of bonds that might be involved are those at the 10 serine and 7 threonine residues; alkali consumption at pH 7.5 corresponded to 11 groups. The most recent results of Josefsson[92] suggest that the released amino groups are formylated, but in contrast to this interpretation Rabinowitz[93], Smillie and Neurath[94], and Narita[95] seem to have established that the reagent simply formylates the hydroxyl groups,and the slow uptake of alkali by the modified protein is due to hydrolysis of the formyl esters. Formyl groups are undoubtedly attached to the modified protein as has been shown by treatment with radioactive formic acid – [^{14}C].

It has been the experience of the writer that increase of amino nitrogen as determined by the Van Slyke method is not an entirely reliable criterion for acyl migration. This was manifest in studies[96] on insulin modified by solution in various acid chlorides when little agreement was found between the amino nitrogen results, which approached the value required by theory, and the yields of amino alcohols following reduction of the modified insulin with lithium borohydride. More precise kinetic data are obviously required for a complete understanding of the properties of modified proteins formed in anhydrous acid media. The rates reported for the reverse O,N acyl migration in proteins all seem too slow; with model substances it is clear that whereas the N,O acyl migration is often a slow reaction rarely proceeding to completion, the reverse migration is fast and practically quantitative. Some observations of Tanford[97] make it doubtful whether a small percentage of hydroxyamino acids might be involved in acyl shifts in aqueous solutions of proteins at low pH. During titration to pH 2 little evidence for acyl shifts was found with ribonuclease, lysozyme, β-lactoglobulin, ovalbumin and human serum albumin.

Cystine residues

In 1957 Swan[98] suggested that conversion of cysteine and cystine residues to dehydroalanyl residues might lead to a potential method for specific cleavage at those sites in a peptide chain. The following scheme utilizing the reactivity of S-sulphocysteine was envisaged.

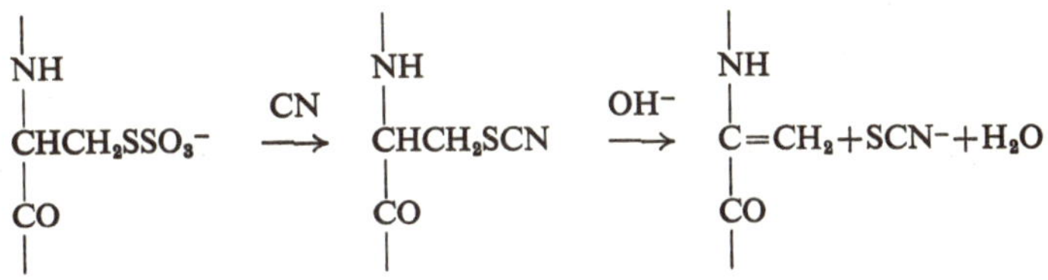

In a study of the action of cyanide on bovine serum albumin at pH 7, Wood and Catsimpoolas[99] demonstrated disulphide bond cleavage but failed to detect thiocyanate ion in the solution when the product was made alkaline. Moreover, the product obtained after acid hydrolysis was 2-iminothiazolidine-4-carboxylic acid. Ring formation involving a thiocyano group on a carbon atom adjacent to one carrying a free amino group[100] has been known in organic chemistry for many years and a similar mechanism operating in proteins might well provide a means for selective cleavage of the peptide bond at the cystine amino group. An enzymic method for splitting bonds at cysteinyl residues is discussed on p. 146.

N,S-acyl migration at cysteine residues has not been seriously considered as a means of specific cleavage. Thiazolines and thiazolidines, the sulphur analogues of the oxazolines and oxazolidines, discussed above, are known as simple compounds and there is spectral evidence to support the view that in strongly acid conditions reduced glutathione assumes a thiazoline structure[101]. Although the idea that such rings might exist in proteins was put forward many years ago by Linderstrøm–Lang and Jacobsen[102] there has been little evidence to support the suggestion. Thiazoline rings have been found in certain antibiotics, *e.g.* bacitracin-A, micrococcin-P and botromycin, and a thiazolidine structure has been shown for penicillin.

Tryptophan residues

The lability of tryptophan under the normal conditions of acid hydrolysis is a major obstacle in sequence studies and so far no satisfactory way has been found to convert this residue into a more stable derivative. The reagent N-bromosuccinimide (NBS) introduced by Patchornik *et al.*[103] has shown promise in experiments designed to split peptide chains at tryptophyl residues. The reaction which has been proposed to explain fission by NBS involves a cyclization to a γ-lactone. The following scheme illustrates this with a model substance ethyl N-carbobenzyloxytryptophylglycinate.

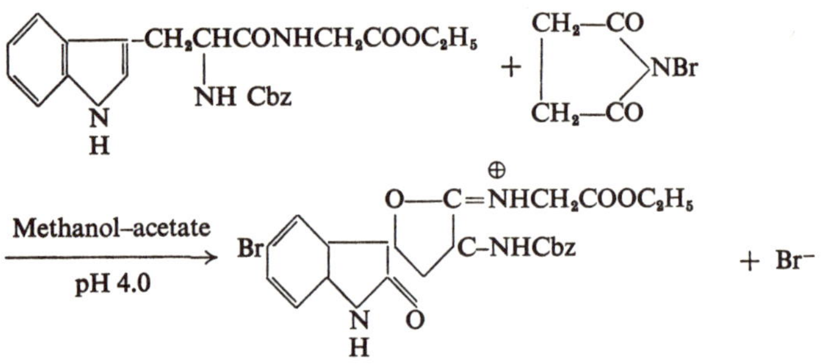

The reaction can be followed with a UV spectrophotometer[104]: as the indole absorption at 280 mμ disappears a new band at 240–250 mμ appears together with a low intensity band at 307 mμ. In aqueous acetate buffer at pH 4 the effect on the spectrum is linear up to a consumption of 1.5 moles/mole of tryptophan. Multiplication of the decrease in extinction at 280 mμ by an empirical factor (1.31) gives the extinction due to tryptophan in the peptide or protein.

The method was first tried out on glucagon when a good yield was obtained of the terminal tetrapeptide derived from Try.Leu.-Met.Asp.Thr.[104]. In later work[105] it was discovered that the amount of NBS taken up to obtain minimum extinction at 280 mμ varies with the nature of the protein and the medium used. Human serum albumin (Hg-dimer) consumed 20 moles of NBS per mole of tryptophan present, in 8 *M* urea; bovine serum albumin 10 moles in 10 *M* urea; lysozyme 2.3 moles in aqueous acetate–formate buffer, pH 4.15, 3.0 moles in lithium acetate, 2.7 moles in 66% acetic acid.

Addition of 6 moles NBS per protein subunit of TMV protein led to the release of 0.2–0.3 mole each of alanine and lysine as new N-terminal groups. Application of the method to specific cleavage of the α- and β-chains of bovine haemoglobin[106] was a little more successful. Bond cleavages generally average about 20–70%.

Tyrosine undergoes a similar reaction with NBS[107–109].

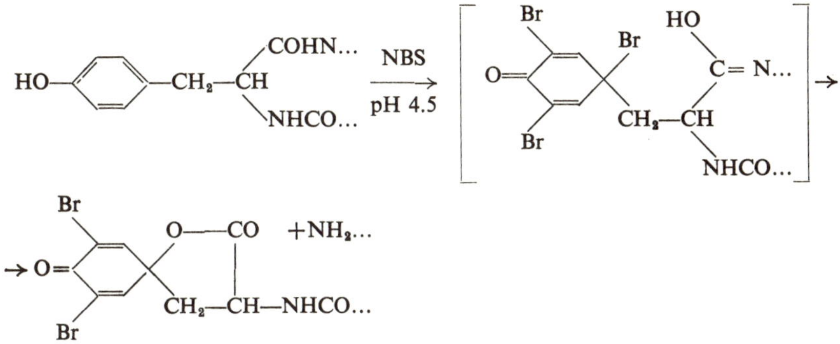

Although it is difficult to suppress tyrosine oxidation completely when NBS is used by itself, differential oxidation of tryptophan and tyrosine is possible, however, if the reaction is performed in an 8 *M* solution of urea in water[110]. The active agent under these circumstances is probably N-bromourea which reacts with tyrosine at a much slower rate.

Methionine residues

The appearance of homoserine and its lactone in hydrolysates of carboxymethyl proteins is due to the decomposition of the carboxymethyl sulphonium salt of methionine (see p. 120). Further work on the alkylation of methionine in peptide linkage[111] indicated that a satisfactory procedure for cleavage at the methionine carbonyl group could be developed. Cyanogen bromide is capable of forming a methionine sulphonium derivative sufficiently labile to undergo spontaneous intramolecular displacement[112]. Under acidic conditions the cyanosulphonium bromide derivative splits out methyl thiocyanate leading to the formation of the hydrobromide of the iminolactone of homoserine. The latter decomposes spontaneously

References p. 159

to homoserine lactone liberating a new N-terminus. The reaction mechanism proposed by Gross and Witkop[112] for cleavage with cyanogen bromide is shown below.

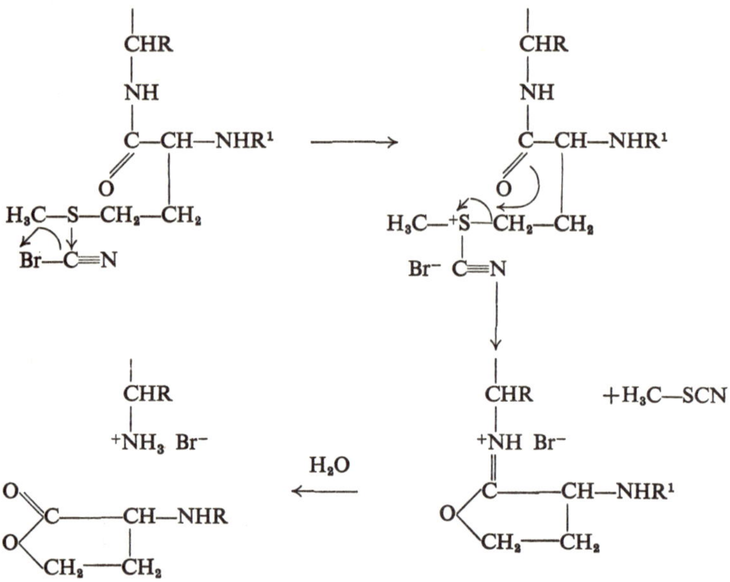

No side reactions occur at acid pH, except for a slow attack on the SH group of cysteine which could easily be blocked, and the operation is carried out at room temperature. Cleavage yields exceeding 90% for model peptides has been demonstrated.

Some interesting facts emerge from a recent report by Ambler[115] on the behaviour of peptides formed by cleavage with cyanogen bromide. The conversion to homoserine is slower with peptides containing C-terminal homoserine lactone than is the case with the free lactone. The lactone form of a peptide suffers little change during paper electrophoresis and the size of the peptide can be estimated from the difference in the electromobility of the lactone and open chain forms. C-terminal homoserine is readily removed from a peptide by the action of carboxypeptidase.

Application of the method in structural studies on ribonuclease[112] was most rewarding in that the results helped to rationalize the conflicting data regarding positions 11 and 18. The releasė of homoserine (lactone) from the Met29–Met30 sequence was also confirmed. In this work a 30-fold excess of cyanogen bromide (127 mg) was used to treat purified ribonuclease (137 mg) dissolved in 0.1 *N* HCl (13.7 ml). After a reaction time of 24 h the mixture was freeze-dried and the products fractionated by gel filtration. Edmundson[3] applied the method to sperm whale myoglobin and more recently Hofmann[113] reported successful cleavage of trypsinogen at its two methionine residues. Under the acid conditions employed, hydrolytic breakage of a particularly labile peptide bond involving leucine or isoleucine also occurred. A most interesting reaction takes place between cyanogen bromide and free amino acids at pH 7.6; the reagent activates one molecule of the amino acid, presumably by condensation and ring closure to a 5-imino-oxazolidone, and this highly reactive intermediate undergoes further rapid condensation with a second molecule of amino acid. The end product is an N-carbamido dipeptide[114].

REFERENCES

1 G. Braunitzer, R. Gehring–Müller, N. Hilschmann, K. Hilse, G. Hobom, V. Rudloff and B. Whittman–Liebold, *Z. Physiol. Chem.*, 325 (1961) 283.

2 A. B. Edmundson and C. H. W. Hirs, *J. Mol. Biol.*, 5 (1962) 663, 683, 706.

3 A. B. Edmundson, *Nature*, 198 (1963) 354.

4 Reported in review by D. G. Smyth, *Ann. Rep. Chem. Soc. (London)*, 60 (1964) 468.

5 A. Tsugita, D. T. Gish, J. Young, H. Fraenkel–Conrat, C. A. Knight and W. M. Stanley, *Proc. Natl. Acad. Sci. U.S.*, 46 (1960) 1463.

6 K. A. Walsh, K. S. V. Sampath Kumar and D. L. Kauffman, *Abstracts, 6th International Congress of Biochemistry, New York*, 2 (1964) 186.

7 B. S. Hartley, *Nature*, 201 (1964) 1284.

8 K. A. Walsh and H. Neurath, *Proc. Natl. Acad. Sci. U.S.*, 52 (1964) 884.

9 F. A. Anderer, H. Uhlig, E. Weber and G. Schramm, *Nature*, 186 (1960) 922.

10 H. Fraenkel Conrat, personal communication, 1962.

11 W. Konigsberg and R. J. Hill, *J. Biol. Chem.*, 237 (1962) 2547.

12 W. Konigsberg, J. Goldstein and R. J. Hill, *J. Biol. Chem.*, 238 (1963) 2028.

13 F. Sanger, *Advan. Protein Chem.*, 7 (1952) 1.

14 R. L. M. Synge, *Biochem. J.*, 39 (1945) 351.

15 W. H. Stein, S. Moore and M. Bergmann, *J. Biol. Chem.*, 154 (1944) 191.

16 M. A. Naughton, F. Sanger, B. S. Hartley and D. C. Shaw, *Biochem. J.*, 77 (1960) 149.

[17] F. SANGER AND E. O. P. THOMPSON, *Biochim. Biophys. Acta*, 9 (1952) 225.
[18] S. M. PARTRIDGE AND H. F. DAVIS, *Nature*, 165 (1950) 62.
[19] S. BLACKBURN AND G. R. LEE, *Biochem. J.*, 58 (1954) 227.
[20] G. F. GRANNIS, *Arch. Biochem. Biophys.*, 91 (1960) 255.
[21] S. M. PARTRIDGE, H. F. DAVIS AND G. S. ADAIR, *Biochem. J.*, 61 (1955) 11.
[22] J. SCHULTZ, Discussion to F. SANGER, *J. Polymer Sci.*, 49 (1961) 3.
[23] J. SCHULTZ, H. ALLISON AND M. GRICE, *Biochemistry*, 1 (1962) 694.
[24] D. L. SWALLOW AND E. P. ABRAHAM, *Biochem. J.*, 70 (1958) 364.
[25] C. H. W. HIRS, W. H. STEIN AND S. MOORE, *J. Biol. Chem.*, 221 (1956) 151.
[26] J. L. BAILEY, S. MOORE AND W. H. STEIN, *J. Biol. Chem.*, 221 (1956) 143.
[27] N. HAUGAARD AND E. S. HAUGAARD, *Compt. Rend. Trav. Lab. Carlsberg, Ser. Chim.*, 29 (1955) 350.
[28] E. J. COHN AND J. T. EDSALL, *Proteins, Amino Acids and Peptides*, Reinhold, New York, 1943.
[29] C. H. W. HIRS, S. MOORE AND W. H. STEIN, *J. Biol. Chem.*, 235 (1960) 633.
[30] D. G. SMYTH, W. H. STEIN AND S. MOORE, *J. Biol. Chem.*, 237 (1962) 1845.
[31] J. T. POTTS, A. BERGER, J. COOKE AND C. B. ANFINSEN, *J. Biol. Chem.*, 237 (1962) 1851.
[32] F. F. NORD AND M. BIER, *Biochim. Biophys. Acta*, 12 (1953) 56.
[33] J. YON, *Biochim. Biophys. Acta*, 31 (1959) 75.
[34] N.M. GREEN, J. Biol Chem., 205 (1953) 535.
[35] M. LASKOWSKI AND M. LASKOWSKI JR., *Advan. Protein Chem.*, 9 (1954) 203.
[36] J. J. RACKIS, H. A. SASAME, R. K. MANN, R. L. ANDERSON AND A. K. SMITH, *Arch. Biochem. Biophys.*, 98 (1962) 471.
[37] R. R. REDFIELD AND C. B. ANFINSEN, *J. Biol. Chem.*, 221 (1956) 385.
[38] C. B. ANFINSEN, M. SELA AND H. TRITCH, *Arch. Biochem. Biophys.*, 65 (1956) 156.
[39] T. C. MERIGAN, W. J. DREYER AND A. BERGER, *Biochim. Biophys. Acta.*, 62 (1962) 122
[40] R. F. GOLDBERGER AND C. B. ANFINSEN, *Biochemistry*, 1 (1962) 401.
[41] H. LINDLEY, *Nature*, 178 (1956) 647.
[42] F. TIETZE, J. A. GLADNER AND J. E. FOLK, *Biochim. Biophys. Acta*, 26 (1953) 659.
[43] M. A. RAFTERY AND R. D. COLE, *Biochem. Biophys. Res. Commun.*, 10, (1963) 467.
[44] H. A. ITANO AND A. J. GOTTLIEB, *Biochem. Biophys. Res. Commun.*, 12 (1963) 405.
[45] L. WEIL AND M. TELKA, *Arch. Biochem. Biophys.*, 71 (1957) 473.
[46] E. O. P. THOMSON, *Advan. Org. Chem.*, 1 (1960) 149.
[47] F. F. NORD, M. BIER AND L. TERMINIELLO, *Arch. Biochem. Biophys.*, 65 (1956) 120.
[48] T. INAGAMI AND J. M. STURTEVANT, *J. Biol. Chem.*, 235 (1960) 1019.
[49] J. DURELL AND J. S. FRUTON, *J. Biol. Chem.*, 207 (1954) 487.
[50] J. BALO AND I. BANGA, *Biochem. J.*, 46 (1950) 384.
[51] P. DESNUELLE, in: P. D. BOYER, H. LARDY AND K. MYRBÄCK (Eds.), *The Enzymes*, 2nd Ed., Vol. 4., Academic Press, New York, 1960, p. 93.
[52] J. ALEXANDER, I. B. WILSON AND R. KATZ, *J. Biol. Chem.*, 238 (1963) 741.

[53] A. M. GOLD AND D. FAHRNEY, *Biochemistry*, 3 (1964) 783.
[54] K. HEIRWEGH AND P. EDMAN, *Biochim. Biophys. Acta*, 24 (1957) 219.
[55] H. VAN VUNAKIS AND R. M. HERRIOTT, *Biochim. Biophys. Acta*, 22 (1956) 537.
[56] G. BRAUNITZER, N. HILSCHMANN AND R. MÜLLER, *Z. Physiol. Chem.*, 318 (1960) 284.
[57] H. NEUMAN, Y. LEVIN, A. BERGER AND E. KATCHALSKI, *Biochem. J.*, 73 (1959) 33.
[58] B. FOLTMANN, *Compt. Rend. Trav. Lab. Carlsberg*, 32 (27) (1962) 425.
[59] H. F. BUNDY, N. J. WESTBERG, B. M. DUMMEL AND C. A. BECKER, *Biochemistry*, 3 (1964) 923.
[60] J. R. KIMMEL AND E. L. SMITH, *J. Biol. Chem.*, 207 (1954) 515.
[61] E. L. SMITH, J. R. KIMMEL AND A. LIGHT, *5th International Congress of Biochemistry, Moscow, Symp. IV*, 1961.
[62] J. R. KIMMEL, G. K. KATO, A. C. M. PAIVA AND E. L. SMITH, *J. Biol. Chem.*, 237 (1962) 2525, 2537.
[63] M. BERGMANN AND J. FRUTON, *Advan. Enzymol.*, 1 (1941) 63.
[64] J. R. KIMMEL AND E. L. SMITH, *Advan. Enzymol.*, 19 (1957) 267.
[65] R. B. JOHNSTON, *J. Biol. Chem.*, 221 (1956) 1037.
[66] F. W. PUTMAN, C. W. EASLEY AND L. T. LYNN, *Biochim. Biophys. Acta*, 58 (1962) 279.
[67] B. R. HAMMOND AND H. GUTFREUND, *Biochem. J.*, 72 (1959) 349.
[68] A. V. GÜNTELBERG AND M. OTTESEN, *Nature*, 170 (1952) 802.
[69] M. OTTESEN AND A. WOLLENBERGER, *Compt. Rend. Trav. Lab. Carlsberg, Ser. Chim.*, 28 (1953) 463.
[70] F. M. RICHARDS AND P. J. VITHAYATHIL, *J. Biol. Chem.*, 235 (1960) 1029.
[71] H. TUPPY, *Monatsh.*, 84 (1953) 996.
[72] B. HAGIHARA, H. MATSUBARA, M. NAKAI AND K. OKUNUKI, *J. Biochem. (Tokyo)*, 45 (1958) 185.
[73] M. NOMOTO, Y. NARAHASHI AND J. MURAKAMI, *Rept. Inst. Phys. Chem. Res. (Tokyo)*, 35 (1959) 154.
[74] M. NOMOTO, Y. NARAHASHI AND M. MURAKAMI, *J. Biochem. (Tokyo)*, 48 (1960) 593.
[75] A. HIRAMATSU AND T. OUCHI, *J. Biochem. (Tokyo)*, 54 (1963) 462.
[76] S. SARID, A. BERGER AND E. KATCHALSKI, *J. Biol. Chem.*, 234 (1959) 1740; 237 (1962) 2207.
[77] R. E. NEUMANN AND E. L. SMITH, *J. Biol. Chem.*, 193 (1951) 97.
[78] N. C. DAVIS AND E. L. SMITH, *J. Biol. Chem.*, 200 (1953) 373.
[79] S. M. PARTRIDGE AND H. F. DAVIS, *Biochem. J.*, 61 (1955) 21.
[80] M. A. NAUGHTON AND F. SANGER, *Biochem. J.*, 78 (1961) 156.
[81] F. LAMY AND S. TAUBER, *J. Biol. Chem.*, 238 (1963) 939.
[82] M. BERGMAN, E. BRAND AND F. DREYER, *Ber.*, 54 (1921) 936.
[83] F. BETTZIECHE, *Z. Physiol. Chem.*, 146 (1925) 227.
[84] A. P. PHILLIPS AND M. E. BALTZLY, *J. Am. Chem. Soc.*, 69 (1947) 200.
[85] L. H. WELSH, *J. Am. Chem. Soc.*, 71 (1949) 3500.
[86] L. JOSEFSSON, *Biochim. Biophys. Acta*, 74 (1963) 774.
[87] P. DESNUELLE AND A. CASAL, *Biochim. Biophys. Acta*, 2 (1948) 64.
[88] D. F. ELLIOTT, *Biochem. J.*, 50 (1952) 542.

[89] M. B. GLENDENNING, D. M. GREENBERG AND H. FRAENKEL-CONRAT, *J. Biol. Chem.*, 167 (1947) 125.
[90] L. JOSEFSSON AND P. EDMAN, *Biochim. Biophys. Acta*, 25 (1957) 614.
[91] L. JOSEFSSON AND P. EDMAN, *Arkiv Kemi*, 12 (1958) 183.
[92] L. JOSEFSSON, *5th International Congress of Biochemistry, Moscow, Abstracts*, 1961, p. 33.
[93] J. L. RABINOWITZ, *Compt. Rend. Trav. Lab. Carlsberg, Ser. Physiol.*, 31 (1960) 483.
[94] L. B. SMILLIE AND H. NEURATH, *J. Biol. Chem.*, 234 (1959) 355.
[95] K. NARITA, *J. Am. Chem. Soc.*, 81 (1959) 1751.
[96] J. L. BAILEY, unpublished.
[97] C. TANFORD, *J. Am. Chem. Soc.*, 77 (1955) 1912.
[98] J. M. SWAN, *Nature*, 180 (1957) 643.
[99] J. L. WOOD AND N. CATSIMPOOLAS, *J. Biol. Chem.*, 238 (1963) PC2887.
[100] S. GABRIEL, *Ber. Deut. Chem. Ges.*, 22 (1889) 1139.
[101] M. CALVIN, in S. COLOWICK *et al.* (Eds.), *Glutathione: A Symposium*, Academic Press, New York, 1954, p. 3.
[102] K. LINDERSTRØM-LANG AND C. F. JACOBSEN, *J. Biol. Chem.*, 137 (1941), 443.
[103] A. PATCHORNIK, W. B. LAWSON AND B. WITKOP, *J. Am. Chem. Soc.*, 80 (1958) 4748.
[104] A. PATCHORNIK, W. B. LAWSON AND B. WITKOP, *J. Am. Chem. Soc.*, 80 (1958) 4747.
[105] L. K. RAMACHANDRAN AND B. WITKOP, *J. Am. Chem. Soc.*, 81 (1959) 4028.
[106] S. SASAKAWA, *J. Biochem. (Tokyo)*, 53 (1963) 188.
[107] B. WITKOP, *Advan. Protein Chem.* 16 (1961) 221.
[108] G. L. SCHMIR, L. A. COHEN AND B. WITKOP, *J. Am. Chem. Soc.*, 81 (1959) 2228.
[109] E. J. COREY AND L. F. HAEFELE, *J. Am. Chem. Soc.*, 81 (1959) 2225.
[110] M. FUNATSU, N. M. GREEN AND B. WITKOP, *J. Am. Chem. Soc.*, 86 (1964) 1846.
[111] W. B. LAWSON, E. GROSS, C. M. FOLTZ AND B. WITKOP. *J. Am. Chem. Soc.*, 83 (1961) 1509.
[112] E. GROSS AND B. WITKOP, *J. Biol. Chem.*, 237 (1962) 1856.
[113] T. HOFMANN, *Biochemistry*, 3 (1964) 356.
[114] J. SCHREIBER AND B. WITKOP, *J. Am. Chem. Soc.*, 86 (1964) 2441.
[115] R. P. AMBLER, *Biochem. J.*, 96 (1965) 32 P.

CHAPTER 6

N-TERMINAL SEQUENCE DETERMINATION

DINITROPHENYL METHOD OF SANGER

The condensation of 1-fluoro-2,4-dinitrobenzene (FDNB) with the free amino groups of insulin under mild conditions was described by Sanger[1] in 1945. When the dinitrophenyl (DNP)-protein was hydrolyzed the DNP derivatives of glycine and phenylalanine and ε-DNP-lysine were found in the hydrolysate. One mole each of DNP-glycine and DNP-phenylalanine were recovered per molecule of 6000 molecular weight. The inference was drawn that the insulin molecule was composed of two different chains, one containing N-terminal glycine and the other N-terminal phenylalanine. This was confirmed by later study of the N-terminal sequences[2].

For a preparation of 1-fluoro-2,4-dinitrobenzene see Cook and Saunders[3]. The pale yellow liquid boils at 130°/1 mm Hg and melts at 26° after crystallization from ethanol. In view of its vesicant properties the reagent must be handled with care. Original reaction conditions for the preparation of DNP-amino acids are as follows. Amino acid (100 mg) dissolved in the smallest volume of 10% aqueous sodium bicarbonate is shaken with a two-fold excess of 10% FDNB in ethanol for 2 h at room temperature. After removal of the ethanol *in vacuo* and extraction of the residue with peroxide-free ether the DNP-amino acids are precipitated by addition of hydrochloric acid. The purification of a number of the derivatives was described by Porter and Sanger[4]. An example of the modified procedure of Levy and Chung[5] is illustrated by the following preparation of DNP-L-leucine.

L-Leucine (1.30 g) and anhydrous sodium carbonate (2.0 g) are dissolved in 40 ml of water at 40°. One equivalent of FDNB (1.85 g) is added and the mixture stirred vigorously by means of a magnetic stirrer at 40°. After 30 min 3 ml of concentrated hydrochloric acid are added to give 2.89 g of crystalline DNP-L-leucine. Formation of the DNP-derivatives is complete within an hour with the majority of the amino acids, but the complete substitution of histidine requires a much longer time. Table 20 gives the melting point data for

References p. 218

TABLE 20 MELTING POINT

DNP-derivative	*Melting point °C*
α-N-DNP-L-arginine	252(d)
α-N-DNP-L-arginine hydrochloride	168
DNP-DL-alanine	172–173
DNP-L-alanine	178
DNP-DL-aspartic acid	196
Di-DNP-L-cystine	118–121(d), 161
Di-DNP-L-cysteine	155–169(d)
DNP-DL-glutamic acid	155–162
DNP-L-glutamic acid	134–136
DNP-glycine	192–193(d)
im-DNP-histidine (colourless)	
DNP-DL-isoleucine	168–172
DNP-DL-leucine	131.5–132.5
DNP-L-leucine	101
Di-DNP-L-lysine	173–174.5
ε-DNP-L-lysine hydrochloride (1 H_2O)	186
α-DNP-L-lysine	260(d)
DNP-DL-methionine	107–112, 117
DNP-DL-methionine sulphoxide	184.5(d)
DNP-DL-methionine sulphone	184.5(d)
δ-DNP-L-ornithine hydrochloride (1 H_2O)	223(d)
α-DNP-DL-ornithine	227(d)
DNP-DL-phenylalanine	204–206(d)
DNP-L-phenylalanine	186
DNP-L-proline	137–138.5
DNP-DL-serine	186–188(d), 199
DNP-DL-threonine	177–178
DNP-L-tryptophan	196–198(d)
Di-DNP-L-tyrosine	92–98(d)
O-DNP-L-tyrosine (1 H_2O) colourless	202(d)
DNP-DL-valine	182.5–183.5
DNP-L-valine (red)	132
DNP-L-valine (yellow tautomer)	131

DATA OF DNP-AMINO ACIDS

Solvent	References (p. 218)
water	4
conc. hydrochloric acid	4
acetone–ligroin	22
aqueous ethanol	11
water	4
aqueous acetic acid	22, 10
aqueous acetic acid	5
(crude product)	22
ethylacetate–chloroform	5
acetone–ligroin,	
acetone–ligroin, water	22, 1
acetic acid	6
acetone–ligroin	22
acetone–ligroin	22
aqueous acetic acid	5
aqueous methanol	
formic acid	22
dilute hydrochloric acid	1
water	1
ether–ligroin, ether	22, 4
ethanol	5
ethanol	5
N-hydrochloric acid	7
water	7
acetone–ligroin	22
aqueous methanol	1
ether–ligroin	22
acetone–ligroin, methanol	22, 4
acetone–ligroin, ether–ligroin	22
(crude product)	22
(crude product)	22
water	1
acetone–ligroin	22
ether–ligroin	8
ether–ligroin	9

a number of DNP-derivatives. In general the DNP-DL-amino acids crystallize more readily than the L-derivatives.

ε-DNP-L-lysine[4]

Copper carbonate is added to a boiling solution of 0.5 g L-lysine in 10 ml of water. The excess copper carbonate is filtered off and washed with 2–3 ml of water. Excess sodium bicarbonate is added followed by a solution of 1.5 g FDNB in 20 ml of ethanol. The mixture is shaken for 2 h at room temperature and the yellow-green precipitate filtered off and washed with water, ethanol and peroxide-free ether. It is then suspended in 5 ml water and sufficient N-hydrochloric acid added to produce a clear solution. The copper ion is removed by bubbling a stream of hydrogen sulphide through the cooled solution for 2 min. A trace of charcoal is added and the mixture filtered rapidly. The filtrate is taken to dryness *in vacuo* and the residue crystallized from 20% hydrochloric acid.

O-DNP-tyrosine[1]

1-Chloro-2,4-dinitrobenzene (2 g) in ethanol is added to N-acetyl tyrosine (0.55 g) dissolved in a small volume of 10% sodium bicarbonate solution. The mixture is heated on the water bath for 4 h, acidified, and the precipitated oil refluxed with 20% hydrochloric acid for 3 h to remove the acetyl group. The hydrolysate is concentrated to a small volume, the precipitate dissolved in warm nitric acid and the warm solution neutralized with pyridine. O-DNP-tyrosine crystallizes in colourless needles.

im-DNP-histidine[6]

α-Acetylhistidine is treated with FDNB by the standard procedure and the acetyl group removed by hydrolysis with 20% hydrochloric acid as above.

DNP-peptides

Derivatives of low molecular weight may be extracted from the reaction mixture with peroxide-free ether; derivatives of higher molecular weight may be extracted with ethyl acetate or *n*-butanol after removal of dinitrophenol with ether. Sanger and Thompson[12] have described a method for peptides which eliminates salts and allows the DNP-derivatives to be applied directly to a chromatogram. The peptide is eluted from strips of Whatman No. 3 paper and dissolved in 0.1 ml of 1% trimethylamine. Ethanol (0.2 ml) and FDNB (0.01 ml) are added and after a period of 2 h the excess FDNB is extracted three times with ether. The ether is removed and the residue hydrolyzed in 3 drops of 5.7 *N* hydrochloric acid (redistilled, metal free) at 105° for 8 h. This is conveniently carried out in a thick-walled capillary tube. The hydrolysate is diluted with 2 vol of water and the DNP-amino acids extracted three times with ether. Tyrosine and lysine when present in the body of a peptide chain will be converted to O-DNP-tyrosine and ε-DNP-lysine which are water soluble. Histidine is converted slowly to *im*-DNP-histidine which is also water soluble. Other derivatives not extracted from aqueous acid solutions are DNP-cysteic acid, Di-DNP-histidine, S-DNP-cysteine and α-DNP-arginine.

DNP-proteins are formed under the same conditions as described for the DNP-amino acids. Insoluble derivatives are washed with water, ethanol and ether; water-soluble DNP-proteins can be precipitated with ethanol and washed with ether. In a recent study of erythropoietin[13] the DNP-derivatives were prepared by reaction with excess FDNB at 43°, the pH being maintained at 8.0 by the addition of standard potassium hydroxide from the microburette of an autotitrator. The reaction was considered complete when a constant slope was reached for a plot of millilitres alkali against time. Šorm *et al.*[14] prepared water soluble DNP-proteins by slow reaction with FDNB in acetate buffer, pH 5.5, in the absence of alcohol.

End-group analysis

Reactions involving FDNB should be carried out in the dark. The photodecomposition of DNP-amino acids has been studied by Pollara and Von Korff[15]. Using DNP-[^{14}C]amino acids these authors were able to demonstrate very large and varied photodecomposition whereby volatile compounds were released. The results are illustrated in Fig. 63; samples kept in total darkness failed to lose activity over the same period. Although DNP-[2-^{14}C]tyrosine did not lose activity the presence of DNP-tyramine was detected after illumination. A further interesting observation is that N-methyl and N-ethyldinitroaniline are products formed from DNP-glycine

References p. 218

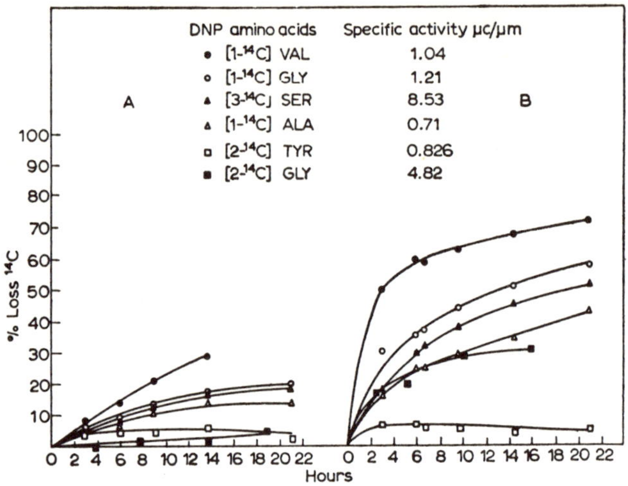

Fig. 63. Effect of light on the decomposition of DNP-(^{14}C)amino acids plated on stainless steel. A, compounds exposed to light 12 ft from window. B, compounds exposed to light 1 ft from window[15].

and DNP-alanine respectively. This very useful study indicates that decarboxylation is the main photolytic reaction in the solid state. Volatile acidic compounds of unknown structure are also produced to a smaller extent. In dilute aqueous solution above pH 7 photolysis follows a different course which results generally in oxidation and decarboxylation of the amino acid moiety, with simultaneous reduction of the aromatic residue (Russell[16]). The main product of photolysis between pH 2 and pH 5 is a 5-nitrobenzimidazole 3-oxide[101]. A new N-terminal degradation based on photodecomposition of DNP-peptides in organic solvents has been proposed by Akabori *et al.*[17].

Successful end-group analysis depends on an accurate knowledge of the stability of DNP-amino acids during the hydrolysis of the particular DNP protein or peptide under study. Approximate recoveries under various conditions were worked out by Porter and Sanger[4] and are given in Table 21. The optimum conditions of hydrolysis for a particular DNP-derivative are very much a matter of trial and error since in addition to the inherent instability of some amino acids the presence of large amounts of carbohydrates and other extraneous materials will affect recoveries.

In general hydrolysis is carried out under three sets of conditions; about 100 mg of DNP protein is usually sufficient for the estimation of 1 end group per mole unit of 100,000.

(1) 5.7 *N* hydrochloric acid at 105° for 16 h will release all DNP-amino acids in good yield with the exception of the glycine, proline and hydroxyproline derivatives. DNP-valyl peptides may require up to 40 h in 11 *N* hydrochloric acid at 105°.

(2) 11 *N* hydrochloric acid at 105° for 5 h will yield DNP-proline and DNP-hydroxyproline (see below).

(3) 5.7 *N* hydrochloric acid at 105° for 5 h will give DNP-glycine.

Among the decomposition products of DNP-proline are δ-chloro-α-DNP-amino valeric acid and α-chloro-δ-DNP-amino valeric acid[18]. In many cases recovery of DNP-proline is so low that the determination is based on the molecular extinction of the chloro-

TABLE 21

APPROXIMATE BREAKDOWN OF DNP-AMINO ACIDS[4]

Amino-acid derivative	*Hydrolysis in boiling 5.7 N HCl*		*Hydrolysis for 16 h in 12 N HCl at 105° (minimal figures) Amount unchanged (%)*
	Time of hydrolysis (h)	*Amount unchanged (%)*	
DNP-alanine	12	80	75
DNP-arginine	12	90	75
DNP-aspartic acid	24	60	75
Di-DNP-cystine	12	25	0
DNP-glutamic acid	12	75	75
DNP-glycine	8	40	50
DNP-hydroxyproline	4	40	50
DNP-isoleucine	12	80	75
DNP-leucine	12	80	75
Di-DNP-lysine	8	95	75
ε-DNP-lysine	12	95	75
DNP-methionine	12	75	75
DNP-phenylalanine	12	70	50
DNP-proline	2	10	50
DNP-serine	12	90	75
DNP-threonine	24	90	75
DNP-tryptophan	12	90	0
DNP-tyrosine	12	75	50
DNP-valine	12	80	75

derivatives. In a 4.5 h hydrolysis of DNP-prolylglycine the total recovery of DNP-proline and the chloro-DNP-amino acids is 33%[19]. It is claimed that hydrolysis of DNP-proteins by Dowex 50 in the H^+ form is more reliable than the strong acid hydrolysis technique[28].

Breakdown products of a more general nature are 2,4-dinitrophenol, 2,4-dinitroaniline and various artifacts which have been reported by Redfield and Anfinsen[20]. As DNP-cysteic acid is more stable to hydrolysis than Di-DNP-cystine a prior oxidation of the protein with performic acid before hydrolysis may be advantageous.

Moderate yields of free amino acids are obtainable from some of the corresponding DNP derivatives after hydrolysis with saturated baryta in a sealed tube at 100° for 1 h or with concentrated ammonium hydroxide at 100° for 2 h[21].

For end group determination insoluble DNP proteins are washed three times each with water, ethanol, and peroxide-free ether. Solutions of soluble DNP proteins are evaporated *in vacuo* at a low temperature to remove ethanol and excess FDNB extracted with ether. They may then be precipitated by acidification. It is usual to air dry samples for several days.

To determine the number of end groups per mole of protein it is necessary to know the protein content of the DNP-derivative. Air-dried DNP proteins generally contain about 70% protein but an accurate value must be obtained from a knowledge of the amino acid composition or the amide-N. The latter value for both the DNP protein and the original protein may be determined simply and rapidly by the method described on p. 348. Alternatively the amounts of three or four different amino acids should be determined.

COLUMN CHROMATOGRAPHY OF DNP-AMINO ACIDS

Chromatography on silica gel

Silica gel may be prepared in the following way[22]. Equal volumes of commercial waterglass, water and ice are stirred in a large beaker and concentrated hydrochloric acid is slowly poured in with stirring until the mixture is acid to thymol blue. During this operation the mixture will stiffen at first and then become more mobile as the correct pH is approached. It is allowed to stand for 3 h, a little hydro-

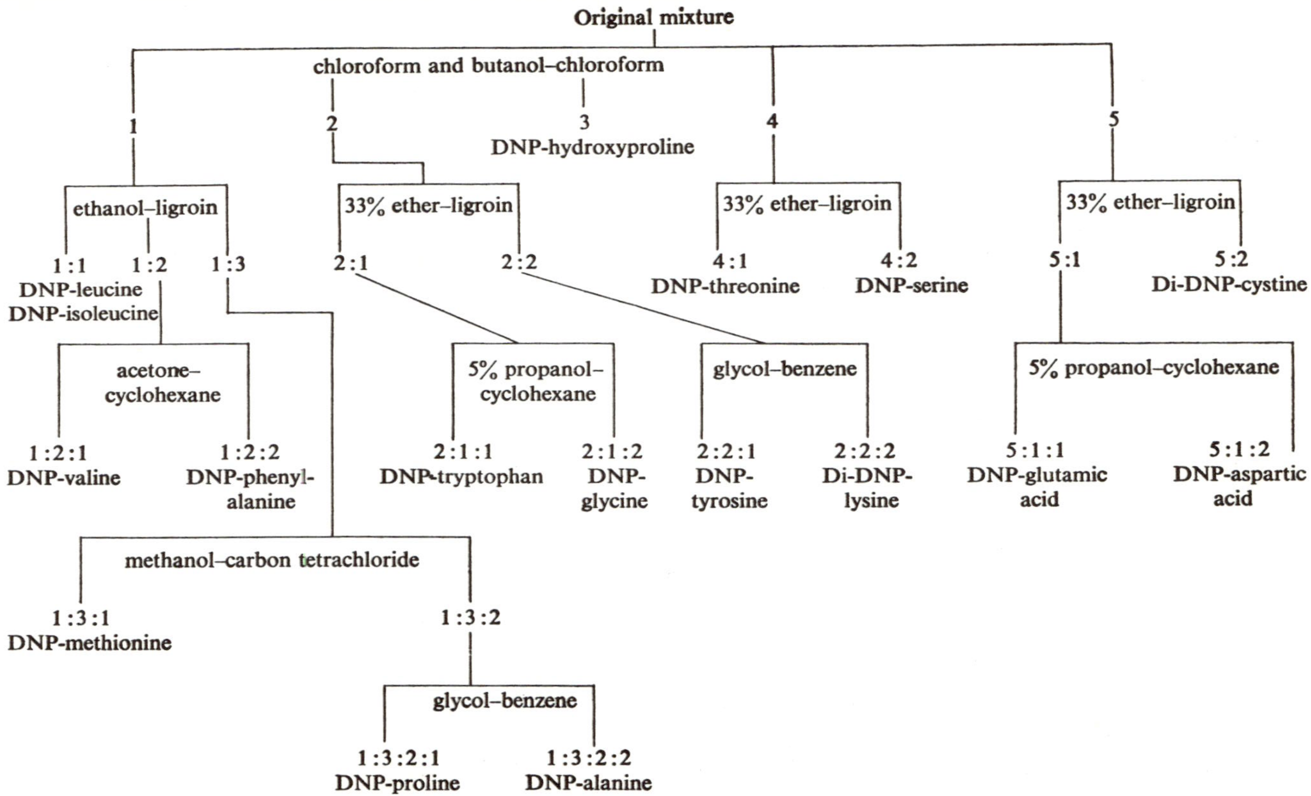

Fig. 64. A scheme of fractionation of ether-soluble DNP-amino acids (from Porter and Sanger[4]). (See text for composition of solvents.)

chloric acid being added if necessary to maintain a constant pH. The silica gel is filtered off, washed thoroughly with water, suspended in 0.5 *N* hydrochloric acid for 2 days, filtered and washed free from acid. It is finally dried in a current of warm air at 70°.

A glass tube 25 cm × 1 cm is suitable for most purposes and a perforated silver disc should be fitted on to a ledge at the lower end. A filter paper is layered on to the disc, the bottom of the column is immersed in solvent and a slurry of silica gel in solvent is then poured in. All air bubbles must be removed and the surface of the gel must be left perfectly level. A perforated stainless steel disc attached to a long rod is useful in the preparation of these columns.

Unbuffered columns are prepared from silica gel which has had half its weight of water ground in. When mixed organic solvents containing water miscible components are used about two thirds of the weight of the aqueous phase is ground in. High water content causes the bands to migrate too rapidly and might also cause cracking of the column.

Silica gels vary a great deal in their adsorption capacities. It might be expected that gels prepared from different sources would vary. However, matters are complicated further by the fact that different batches from the same source vary in behaviour even when prepared by a standardized procedure. A number of batches should be prepared and tried out.

An important factor in the separations is the load. Best results are obtained when the amounts of DNP-amino acid added are just enough to be seen (0.1–0.2 μmoles); amounts in excess of 2 μmoles will probably lead to 'tailing'. Although the principle governing the operation of the columns is mainly that of liquid–liquid partition it appears that adsorption plays a significant role.

A scheme for the fractionation of ether-soluble DNP-amino acids devised by Porter and Sanger[4] is shown in Fig. 64. The terminology indicates the percentage by volume of the first component in solution with the second component.

A. Ethanol–ligroin column. One vol of water, one vol of ethanol and 10 vol. ligroin (b.p. 80–100°) are mixed together. 1 ml of the lower aqueous phase is added to each 2 g dry silica gel, and the upper ethanol–ligroin phase is made the moving phase of the column.

B. Methanol–carbon tetrachloride column. Prepared similarly from 1 vol. water, 1 vol. methanol and 15 vol. carbon tetrachloride.

C. Acetone–cyclohexane column. Prepared from 1 vol. water, 1 vol. acetone and 10 vol. cyclohexane.

D. Glycol–benzene column. Glycol and benzene are shaken together, 1 ml of the glycol layer is added to each 1 g dry silica and the benzene layer is used as the moving phase.

In the operation of these columns the following artifacts should be recognized.

(1) A yellow band moving faster than any DNP-amino acid in chloroform.

(2) DNP-tyrosine and DNP-methionine on acid hydrolysis give rise to additional coloured bands stationary in chloroform.

(3) DNP-tryptophan after hydrolysis yields a brown compound which is not soluble in ether.

(4) Di-DNP-cystine yields a band which moves slowly in 3% butanol–chloroform.

(5) Certain stable DNP-peptides are easily mistaken for DNP-amino acids.

The acid soluble derivatives are eluted from butanol–chloroform and methyl-ethyl-ketone columns as indicated in Table 22.

TABLE 22

BAND RATES OF ACID-SOLUBLE DNP-AMINO-ACIDS

(Figures are values of *R*—Gordon *et al.*[22])

DNP-amino-acids	*Developing solvents*		
	Methyl ethyl ketone-ether	*Butanol-chloroform*	
	66%	*17%*	*30%*
Di-DNP-histidine	Fast	Fast	–
S-DNP-cysteine	Fast	0.08	Fast
α-DNP-arginine	0.35	–	0.2
ε-DNP-lysine	0.20	–	0.15
α-DNP-lysine	0.11	–	–

References p. 218

In the presence of formaldehyde the ionization of the α-amino group of ε-DNP-lysine is suppressed and a clearer separation from α-DNP-arginine is made possible. The column is prepared from 100 g of silica (9% moisture) mixed with 40 ml of 1:9 (by vol) formaldehyde–water which has been neutralized to pH 7 with 0.5 *M* disodium hydrogen phosphate[23].

Useful separations may be carried out on buffered columns as has been shown by Blackburn[24] who points out that the variations in behaviour of different batches of gel might well be attributed to variations in the pH of the aqueous phase in contact with the gel. In Blackburn's method the gel is treated for 3 h with excess hydrochloric acid after precipitation. The gel is then washed and allowed to stand in distilled water for 3 days. Table 23 gives the R_F's of DNP-amino acids run on buffered columns. Buffer A is a 2 *M* solution of sodium dihydrogen phosphate, buffer B is 0.5 *M* phosphate, pH 6.61 and buffer C is 0.25 *M* phosphate, pH 5.95. The acid soluble DNP-derivatives may be fractionated on a buffered column using buffer A and the solvent systems used by Sanger (see above).

TABLE 23

BAND RATES (*R* VALUES) OF DNP-AMINO ACIDS IN CHLOROFORM AND BUTANOL–CHLOROFORM[24]

DNP amino acid	*R values*			
	$CHCl_3$ buffer A	*$CHCl_3$ buffer B*	*3% BuOH $CHCl_3$, buffer C*	*17% BuOH $CHCl_3$, buffer B*
DNP-leucine	Fast	0.7	–	–
DNP-valine	Fast	0.5	–	–
DNP-phenylalanine	Fast	0.7	–	–
DNP-methionine	Fast	0.5	–	–
DNP-proline	Fast	0.3	–	–
DNP-alanine	Fast	0.2	–	0.5
DNP-tryptophan	0.6	0.4	Fast	–
DNP-glycine	0.9	0.03	0.5	0.2
Di-DNP-lysine	0.7	0.06	Fast	Fast
Di-DNP-tyrosine	Fast	0.07	0.9	Fast
DNP-threonine	0.3	Slow	0.18	0.15
DNP-serine	0.11	Slow	0.08	0.07
DNP-glutamic acid	0.17	Slow	0.2	Slow
DNP-aspartic acid	0.06	Slow	Slow	Slow

Chromatography on kieselguhr

In order to avoid the serial use of silica gel columns Mills[25] employed kieselguhr columns on which the DNP-derivatives of all the amino acids usually found in a protein hydrolysate could be resolved. 'Acid washed' kieselguhr (British Drug Houses, Ltd.) is activated before use by shaking in a large bottle with a lump of ammonium carbonate enclosed in a small cloth bag. The pH of an aqueous extract of this adsorbent should be between 6 and 7. The solvents are different mixtures of methyl ethyl ketone with chloroform. A column (1.5 × 12 cm) is prepared from 6.5 g of dry kieselguhr compressed in 45% methyl ethyl ketone-chloroform by means of a stainless steel ramrod. Amino acid N (2–3 μg) is suitable for each band and in general a column of the size described will carry 30–40 μg of protein N. Table 24 gives the order of elution of DNP-amino acids from kieselguhr.

Acid soluble derivatives are removed from contaminating salts after evaporation to dryness by extraction into a 10% solution of

TABLE 24

ORDER OF ELUTON OF DNP-AMINO ACIDS FROM KIESELGUHR[25]

DNP-amino acid	*Solvent*
Di-DNP-cysteine DNP-aspartic acid DNP-glutamic acid	Methyl ethyl ketone
DNP-serine Di-DNP-histidine	75% methyl ethyl ketone–chloroform
DNP-methionine sulphone DNP-threonine DNP-glycine DNP-alanine DNP-proline Di-DNP-tyrosine Di-DNP-lysine Di-DNP-cystine DNP-valine DNP-leucine DNP-isoleucine DNP-phenylalanine Dinitrophenol	30% methyl ethyl ketone–chloroform to resolve DNP-threonine and DNP-glycine followed by 45% methyl ethyl ketone–chloroform.

References p. 218

6 *N* hydrochloric acid in methyl ethyl ketone. The residue after evaporation of this extract is then taken up in pure methyl ethyl ketone saturated with water and transferred to a column prepared in 75% methyl ethyl ketone–chloroform. DNP-arginine travels quickly and DNP-cysteic acid is removed slowly using pure methyl ethyl ketone. Cyteine is best determined as DNP-cysteic acid in this method.

The effect of buffering kieselguhr columns is shown in Table 25. The buffers are 1 *M* sodium dihydrogen phosphate, pH 4.03, 0.25 *M* phosphate, pH 6.5, and 0.4 *M* phosphate, pH 7.0. Each buffer (66% by weight) is ground into the dry kieselguhr and a suspension in the chosen solvent is packed down to give a column 1 cm × 12 cm. A preparative column 3 × 20 cm will take 350 μg of DNP derivatives (Perrone[26]).

It is commonly assumed that only the non-ionized form of DNP amino acids is partitioned between the organic and aqueous phases, and pH affects the gross partition coefficient by changing the position of equilibrium in the aqueous layer. Matheson[29] found that

TABLE 25

BAND RATES FOR DNP-AMINO ACIDS ON BUFFERED 'CELITE 545' COLUMNS[26]

(Figures are values of *R*)

	Ether			*Chloroform*	
	pH 4	*pH 6.5*	*pH 7*	*pH 4*	*pH 7*
DNP-tryptophan	fast	fast	0.9	fast	–
DNP-isoleucine	fast	fast	0.8	fast	0.2
DNP-leucine	fast	fast	0.8	fast	0.2
DNP-phenylalanine	fast	fast	0.6	fast	0.15
Di-DNP-lysine	fast	fast	0.5	fast	–
DNP-valine	fast	0.7	0.3	fast	0.07
DNP-methionine	fast	0.3	0.2	fast	–
DNP-alanine	fast	0.1	0.07	fast	–
DNP-proline	fast	0.1	0.07	fast	–
Di-DNP-tyrosine	fast	fast	0.1	0.5	–
DNP-glycine	fast	0.05	0.02	0.5	–
DNP-threonine	1.0	slow	slow	0.07	–
DNP-glutamic acid	0.9	slow	slow	0.03	–
DNP-serine	0.6	slow	slow	slow	–
DNP-aspartic acid	0.5	slow	slow	slow	–

almost any one of the ether-soluble DNP amino acids can be separated from a mixture when the derivatives are partitioned as ions between aqueous buffers and ethyl acetate. The R_F values are then much less dependent on pH and the bands are unusually narrow. The procedure is as follows. Hyflo Super-Cel is washed with 3 *N* HCl until no more colour is extracted, washed with water, and powdered (minus 100 mesh). Ethyl acetate (200 ml) and the appropriate buffer (40 ml, containing 35 g/l NaCl) are shaken together at room temperature and the resulting phases separated. Hyflo Super-Cel (4 g) is slurried in 50 ml of top phase, and 2.5 ml of lower phase is added dropwise. Columns of 15–19 cm are then prepared taking the usual precautions to remove lumps and bubbles. The sample mixture is loaded on to a column equilibrated at low pH (for example, 0.1 *M* Tris-maleic acid-NaCl, pH 5.4) which resolves the slowest amino acids. The fast-moving compounds are eluted, concentrated, and transferred to a column of high pH. The process is repeated until the whole mixture is completely resolved. A similar procedure was used by Portugal *et al.*[30] for the separation and estimation of DNP-glutamic and DNP-aspartic acids.

Chromatography on silicic acid–celite

This rapid technique of Green and Kay[27] enables an analysis to be completed in under two hours. Reproducibility is extremely good as most samples of silicic acid available commercially have been found satisfactory without special pretreatment.

To allow a convenient flow rate the silicic acid is mixed intimately with half its weight of celite filter aid which has been passed through a 60-mesh sieve to remove large particles. Columns of 1 cm or 1.4 cm diameter are packed under gravity to a height of about 17 cm and then strong suction is applied to contract the adsorbent into position in about 30 seconds. The surface of the column is levelled carefully and a circle of fine-mesh wire screen placed on top will prevent subsequent disturbance during addition of solvents[31].

Prewashing with redistilled solvents has a profound effect on the performance of the columns[32]. This is illustrated very clearly in Fig. 65 which shows the course of development of a zone of 4-nitrodiphenylamine (0.5 mg) with a developer of constant type (1:1 benzene–ligroin) in a column 1.9 × 15 cm. '*V* ml' is defined as the

References p. 218

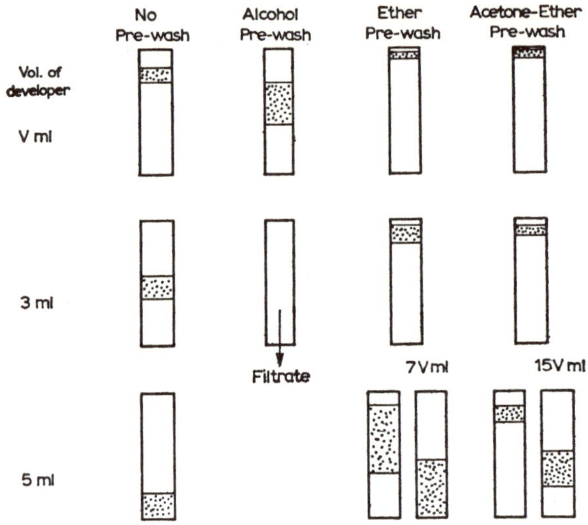

Fig. 65. The course of development of a zone of 4-nitrodiphenylamine with a developer of constant type from columns which have been given different types of pre-wash. Schroeder[32].

volume of solvent which is required to wet a column of adsorbent completely; the zones are idealized. It will be apparent that the adsorption increases in the following order: alcohol pre-washed adsorbent < unpre-washed adsorbent < ether pre-washed adsorbent < acetone-ether pre-washed adsorbent. It has been suggested that the effect of pre-washing is to modify the water content of the adsorbent.

In the procedure of Green and Kay the columns are pre-washed by successively passing through the column 0.2 V ml of ether, V ml of 1:1 acetone-ether, 0.8 V of ether, V ml of ligroin, and V ml of the developing solvent. During operation of the columns suction is maintained to give a partial vacuum of 70–78 mm. The sample is applied in acetic acid diluted with ligroin or acetone diluted with ligroin. The scheme of development is shown in Fig. 66.

The significance of the numbers is best explained by an example. 8AA-4A-L indicates 8 ml of acetic acid, 4 ml of acetone and enough ligroin to bring the total volume to 100 ml. The amount of development will depend on the mixture to be chromatographed and if there are members of Group III and Group IV only, the de-

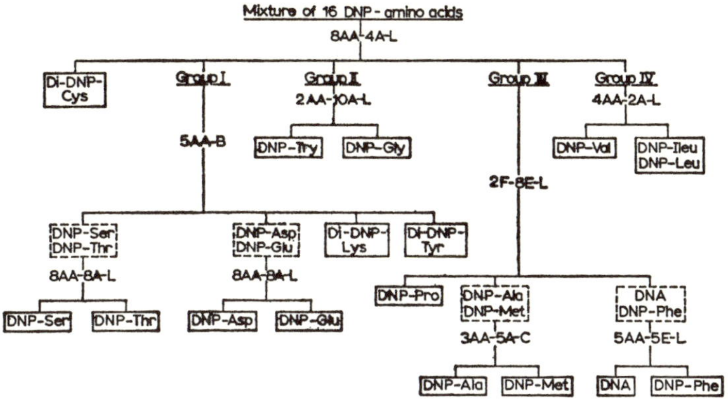

Fig. 66. Chromatographic separation of DNP-amino acids on silicic acid–celite columns. Development scheme of Green and Kay[27]. (AA, acetic acid; A, acetone; L, ligroin; F, formic acid; E, ethyl acetate; B, benzene; C, cyclohexane. DNA indicates 2,4-dinitroaniline.)

velopment can be stopped while the zones are still on the column. If members of Group I and Group II are present it will be necessary to elute the fast running zones off the column. After development is complete the column is extruded and cut into sections containing the slow-moving groups. The zones can then be eluted and prepared for rechromatography.

Some DNP-peptides and artifacts adsorb in the region of Group I. In Group III DNP-proline gives a zone considerably paler than

TABLE 26

EXTINCTION VALUES OF $20\mu M$-SOLUTIONS OF DNP-AMINO ACIDS AND PEPTIDES[2]

DNP-derivative	*Solvent*	*Extinction*	
		350 mμ	*390 mμ*
DNP-glycine	1% $NaHCO_3$	0.309	0.210
DNP-phenylalanine	1% $NaHCO_3$	0.313	0.214
DNP-glycylglycine	1% $NaHCO_3$	0.316	0.173
DNP-phenylalanylvaline	1% $NaHCO_3$	0.310	0.178
ε-DNP-lysine	N HCl	0.296	0.204
O-DNP-tyrosine	*N* HCl	0.058	0.0

most other DNP-amino acids and might be overlooked. The zones in Group IV are rather pale and no solvent has been found to resolve DNP-isoleucine from DNP-leucine. Dinitrophenol, which gives a nearly colourless zone is less strongly adsorbed than any of the DNP-amino acids.

The behaviour of a known mixture of DNP-amino acids in this scheme must be examined closely before any conclusions can be drawn regarding the identity of unknowns.

Table 26 shows the extinction values for 20 μM-solutions of DNP-amino acids and peptides. Agreement between readings taken at 350 mμ and 390 mμ is a useful criterion of purity. Similar values have been found at 340 mμ for solutions in glacial acetic acid. Beer's law is obeyed generally in solutions of concentrations below 50 μM.

Thin layer chromatography

Two-dimensional chromatography of DNP-amino acids on thin layers of Kieselgel G is a simple and speedy process which is used routinely in many laboratories. The method of Brenner *et al.*[33] requires the following solvent systems:

(1) Toluene–pyridine–ethylene monochlorhydrin–ammonia 0.8 *N* (100:30:60:60). This is similar to the toluene solvent described on p. 185; the upper phase is used for chromatography, the lower phase for treating the adsorbent.
(2) Chloroform–benzyl alcohol–acetic acid (70:30:3).
(3) Chloroform–*tert.*-amyl alcohol–acetic acid (70:30:3).
(4) Benzene–pyridine–acetic acid (80:20:2).
(5) Chloroform–methanol–acetic acid (95:5:1).

A two-dimensional chromatogram of a standard mixture of DNP amino acids (0.2 μg of each) using solvent systems 1 and 2 is shown in Fig. 67. Di-DNP-lysine and Di-DNP-tyrosine can be resolved in solvent systems 1 and 5; the leucines separate best in solvents 1 and 4 (Figs. 68 and 69).

In a recent application of the method dinitrophenylation of human urine gave rise to 45 ether-soluble DNP compounds, 24 of which were shown to be derived from authentic amino acids[34].

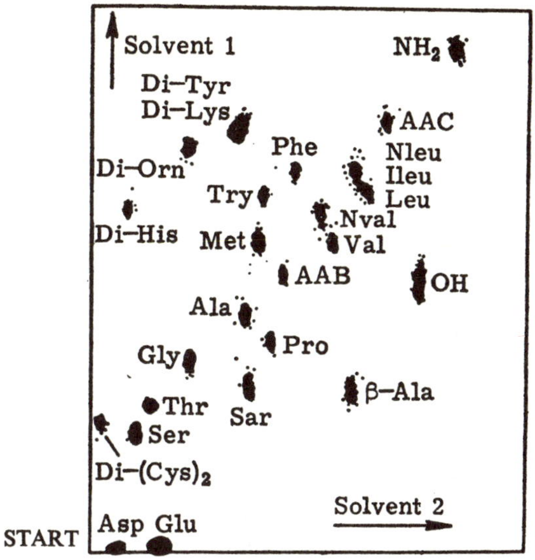

Fig. 67. Two-dimensional chromatogram of a reference mixture of 0.2 μg of each DNP amino acid in solvents 1 and 2; OH: 2,4-dinitrophenol; NH_2: 2,4-dinitroaniline; AAB: α-aminobutyric acid; AAC: α-aminocaprylic acid. Ultraviolet photocopy. Brenner *et al.*[33].

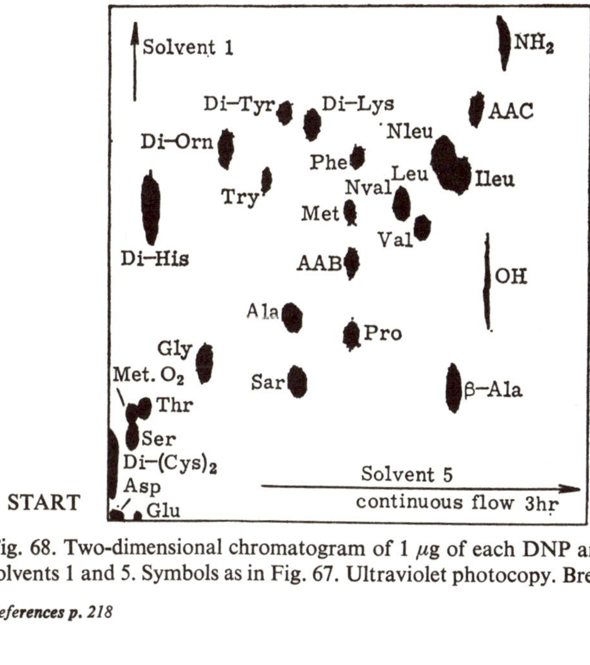

Fig. 68. Two-dimensional chromatogram of 1 μg of each DNP amino acid in solvents 1 and 5. Symbols as in Fig. 67. Ultraviolet photocopy. Brenner *et al.*[33].

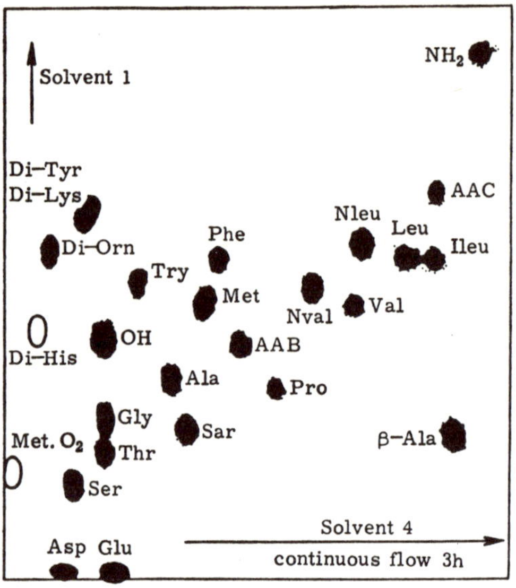

Fig. 69. Two-dimensional chromatogram of 1 μg of each DNP amino acid in solvents 1 and 4. Di-His and Met.O_2 are not present on the chromatogram illustrated. Symbols as in Fig. 67. Ultraviolet photocopy. Brenner *et al.*[33].

PAPER CHROMATOGRAPHY OF DNP-AMINO ACIDS

The particular circumstances of any given problem will largely determine the choice of the many solvent systems that have been proposed for the separation of DNP-amino acids on paper. The precautions to be taken in order to obtain good separations are much the same as have been discussed already in connection with chromatography of the amino acids themselves. An important additional precaution, however, is to run the chromatograms in the dark. Appropriate equilibration of the paper with the vapour phase of the system and maintenance of temperature control are important factors. Many grades of paper have been found equally satisfactory and detection of the resolved components is a simple matter. A useful property is the strong absorption of the spots in UV light.

Some of the drawbacks are frequent 'tailing' of the spots and the masking effect of a number of artifacts. Much attention has been given to the problem of eliminating dinitrophenol and a few of the methods will be discussed.

Removal of dinitrophenol

(1) A large quantity of dinitrophenol can be removed by vacuum sublimation. A cold finger is filled with crushed solid carbon dioxide and the flask is connected to a good high vacuum pump. Dinitrophenol will sublime on to the cold finger when the flask is warmed to about 100°[25].

(2) A short column of acid-washed alumina will retain DNP-amino acids in 2% aqueous acetic acid allowing the dinitrophenol to pass through. The former can then be eluted with 1% sodium bicarbonate solution[35].

(3) A short run on Whatman No. 3 paper by the ascending method in the solvent decalin–glacial acetic acid (1:1) will remove dinitrophenol to the solvent front. The spot is cut out and the slow-moving DNP-amino acids eluted from the paper strip with 1% sodium bicarbonate solution. The method will deal with up to 20 μg of dinitrophenol.

(4) The DNP-amino acids are applied in chloroform to a 1 × 30 cm column of silicic acid equilibrated with 0.067 *M* Na_2HPO_4 and chloroform; the artifacts are washed off with chloroform equilibrated with the same buffer; the DNP-amino acids are then eluted with chloroform containing 1% (v/v) glacial acetic acid[36].

Quantitative estimation of DNP-amino acids by method of Levy

A quantitative method based upon the formation of N-2,4-dinitrophenyl-derivatives of the amino acids was developed by Levy[37]. A protein hydrolysate equivalent to about 2–3 mg of protein is treated in 3 ml of water with a slight excess of FDNB for 80 min at pH 9.0 and 40°. The pH may be maintained constant by addition of 2 *N* sodium hydroxide from a micrometer burette or by means of an autotitrator. The slight excess of FDNB is extracted with ether and the solution acidified with 6 *N* HCl. The precipitated DNP-derivatives are extracted into 5 × 5-ml portions of ether and a 2-ml aliquot of the final solution is applied to a corner of a Whatman No. 1

References p. 218

paper ($18\frac{1}{4} \times 22\frac{1}{2}$ in.). The aqueous solution remaining after the ether extraction which contains DNP-arginine and α-DNP-histidine, is diluted to 10 ml and a 1-ml aliquot applied to an adjacent corner. As the area of the spots must be kept to a minimum it is usual to add the solutions by means of a fine capillary while maintaining a warm air stream underneath the horizontal sheet. The chromatogram is irrigated by the ascending method with toluene–chloroethanol–

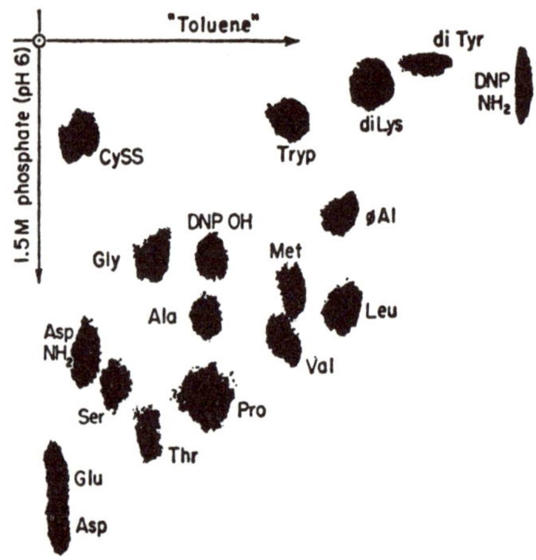

Fig. 70. Two-dimensional chromatogram of a synthetic mixture of dinitrophenyl amino acids (approx. 0.02 μM of each) according to the method of Levy[37].

pyridine–0.8 N ammonia (see p. 185), dried for 3–4 h at 40° and the spots due to DNP-arginine and α-DNP-histidine cut out. The paper is then run in the other dimension by the descending method using 1.5 M aqueous phosphate buffer (M NaH_2PO_4 + 0.5 M Na_2HPO_4). The pattern obtained is shown in Fig. 70. The spots are cut out and the colour eluted by warming each with 4 ml of water for 15 min at 55–60°. A few blanks cut from each sheet are treated similarly and the extinction values of all the solutions are determined at 360 mμ or 385 mμ in the case of DNP-proline. To determine the molar ratios of the DNP-amino acids the ratios of extinction values are multiplied by the following factors:

Asp,	0.99	Val,	0.99
Glu,	0.94	Met,	1.21
Cys,	0.56	Leu and Ileu,	1.10
Ser,	0.97	Phe,	1,03
Thr,	1.02	Tyr,	1.54
Gly,	1.03	Lys,	0.64
Ala,	1.09	His,	1.62
Pro,	0.93	Arg,	1.06

Dinitrophenol does not interfere with the chromatography provided the solutions are approximately 0.01 *M* with respect to amino acids.

Method of Blackburn and Lowther[38]

Strips of Whatman No. 4 paper 12 × 35 cm are dipped in phthalate buffer, pH 6.0. A mixture containing about 2–5 μg of each DNP-amino acid is applied to the paper and developed with one of the following solvents at room temperature.

(A) *Tert.* amyl alcohol.

(B) 30% Propanol–cyclohexane. Gives very good resolution of DNP-leucine, DNP-valine, DNP-alanine and DNP-glycine. Light petroleum (b.p. 100–120°) may be substituted for cyclohexane.

(C) 10% Ethanol–benzyl alcohol saturated with phthalate buffer.

Approximate R_F values of DNP-amino acids in these solvents are given in Table 27.

Method of Biserte and Osteux[39]

These authors stress the necessity for adequate temperature control (20° ± 1°) and the prior removal of dinitrophenol. The preparation of the toluene solvent calls for some care.

A. Toluene–pyridine–ethylene monochlorhydrin–ammonia 0.8 N (5:1:3:3).

The toluene, pyridine and ethylene monochlorhydrin are shaken together vigorously. The 0.8 *N* ammonia is added slowly *without shaking* and the two phases are allowed to equilibrate for 1 h. After running off the aqueous phase the solvent is filtered through a Whatman No. 1 paper to remove droplets of water. The aqueous phase

TABLE 27

R_F VALUES OF DNP-AMINO ACIDS[38,39]

DNP-Amino acids	*Phenol-isoamyl alcohol*	*Toluene solvent*	*Pyridine–isoamyl alcohol–NH_3*	*30% Propanol–cyclohexane*	*Tertiary amyl alcohol*	*10% Ethanol benzyl alcohol*
2,4-Dinitroaniline	0.95–1.0	0.95–1.0	0.95–1.0	—	—	—
2,4-Dinitrophenol	0.25	0.42–0.45	0.62	—	—	—
DNP-aspartic acid	0.05	0.08	0.05	0.02	slow	0.03
DNP-glutamic acid	0.05	0.09	0.06	0.05	0.04	0.07
DNP-asparagine	0.20	0.24	—	—	—	—
DNP-glutamine	0.23	0.27	—	—	—	—
DNP-hydroxyproline	0.23	0.28	0.40	—	—	—
DNP-serine	0.23	0.30	0.45	0.05	0.21	0.18
Di-DNP-cystine	0.20	0.32	0.41	—	—	—
DNP-glycine	0.33	0.35	0.45	0.10	0.23	0.26
DNP-threonine	0.31	0.36	0.53	0.07	0.36	0.26
DNP-proline	0.50	0.43	—	—	—	—
DNP-alanine	0.44	0.45	0.55–0.60	0.15	0.46	0.36
DNP-valine	0.60	0.56	0.70	0.23	0.79	0.59
DNP-methionine	0.57	0.59	0.71	—	—	—
DNP-methionine-sulphoxide	0.23	0.35	0.40	—	—	—
DNP-methionine-sulphone	0.36	0.48	0.66	—	—	—
DNP-tryptophan	0.63	0.62	0.75	—	—	—
DNP-leucine	0.68	0.64	0.80	0.28	0.88	0.71
DNP-phenylalanine	0.68	0.66	0.80	0.22	0.74	0.63
Di-DNP-lysine	0.75	0.72	0.78	—	—	—
Di-DNP-tyrosine	0.70	0.74	0.83	—	—	—
ε-DNP-lysine	0.65	0.40	0.40	—	—	—
α-DNP-arginine	0.74	0.48	0.57	—	—	—
Di-DNP-histidine	0.32–0.70	0.43–0.70	0.47–0.70	—	—	—
Ac-DNP-cysteic acid	—	0.02	—	—	—	—

and 300 ml of 0.8 *N* ammonia is placed at the bottom of the chromatography unit and the chromatogram paper allowed to equilibrate for 1 h. The DNP-amino acids separate into seven groups during the course of 6–12 h.

B. Phenol–isoamyl alcohol–water (1:1:1)

300 ml of 0.1% (v/v) ammonia is placed in the chromatography unit and the DNP-amino acid spots are first neutralized with ammonia vapour. Development takes 18–24 h.

C. Pyridine–isoamyl alcohol–ammonia 1.6 *N* (6:14:20)

200–300 ml of 1.6 *N* ammonia is placed in the chromatography unit. Development takes 12–14 h.

Table 27 gives the approximate R_F values of the DNP-amino acids in these solvents.

Other systems for DNP-amino acids

Rovery and Fabre[40] have described the behaviour of DNP-amino acids in the aqueous buffer 0.7 *M* citrate–hydrochloric acid, pH 6.2. The solvent is allowed to descend the paper a few centimetres below the origin. The DNP-amino acids are distributed in the unusual order: tyrosine, lysine, (tryptophan, cystine), glycine, phenylalanine, (alanine, methionine, serine), (valine, leucine), (glutamic acid, threonine, proline), aspartic acid.

The system *sec.*-butanol–methyl acetate–0.5 *N* hydrochloric acid–isopropyl chloride (50:30:16:4) gives a clear separation of the DNP-derivatives of serine, glycine, threonine, alanine, proline, valine and isoleucine in this order. It will also resolve the derivatives of arginine, lysine and histidine. All separations are carried out on paper buffered with phthalate, pH 6.

The acid-soluble DNP-amino acids chromatograph well in many solvents. The R_F values for these derivatives and those of leucine and valine in the system *n*-butanol–acetic acid–water (4:1:5) are given in Table 28.

Comment. Various methods for the fractionation of DNP-amino acids are described above; of these the two most generally followed are those of Green and Kay (p. 177) and Levy (p. 183).

TABLE 28

R_F VALUES OF ACID-SOLUBLE DNP-DERIVATIVES[41]

(Whatman No. 4 filter paper; butanol–acetic acid)

	R_F	*Colour of spot*	*Ninhydrin reaction*
O-DNP-Tyrosine	0.84	Colourless	Purple
α-N-DNP-Arginine	0.81	Yellow	No reaction
ε-N-DNP-Lysine	0.77	Yellow	Brown
Leucine	0.67	–	–
im-DNP-Histidine	0.57	Faint yellow	Brown
Valine	0.49	–	–
DNP-Cysteic acid	0.42	Yellow	–

CHROMATOGRAPHY OF DNP-PEPTIDES

DNP-peptides produced from partial hydrolysates of proteins have been fractionated on silica gel columns. As mentioned above it has been difficult to standardize on procedure owing to the variability in performance of the gels. It is good practice therefore to prepare a number of gels and calibrate them by means of synthetic DNP-amino acids. The factors governing the separation of amino acids on silica columns apply equally well to the fractionation of peptides. A preliminary extraction scheme[2] is as follows.

(1) An acidic solution of a partial hydrolysate of a DNP-protein is extracted three times with equal volumes of ethyl acetate.

(2) The bulked ethyl acetate extracts are extracted three times in countercurrent manner with 1% sodium bicarbonate solutions.

(3) The combined bicarbonate extracts are bulked, acidified with hydrochloric acid and extracted with ethyl acetate.

The final extract contains all acidic DNP-derivatives and a few non-acidic DNP-derivatives. The ethyl acetate extracts are evaporated and transferred to suitable columns. Useful solvents are chloroform *n*-butanol mixtures. Good results have been obtained with up to 30% (v/v) *n*-butanol. For rechromatography of fast-moving groups the concentration of *n*-butanol is lowered (*e.g.* to 15% or 5%). It is sometimes found that eluted materials become insoluble after being taken to dryness possibly due to salt formation. They may be brought into solution again by addition of a drop of concentrated hydrochloric acid followed by ether or ethyl acetate. The solution is then evaporated and quickly taken up in the new solvent for chromatography.

Early workers preferred water as the stationary phase and reported bad 'tailing' of bands on buffered columns. In a recent study of casein peptides Work and collaborators[42] successfully used columns buffered with phosphate (7.5 ml of 0.5 M phosphate buffer for 15 g silica). Desnuelle and Fabre[43] have worked out conditions for the separation of DNP-peptides containing aspartic acid residues. Basic DNP-peptides containing arginine have been resolved on buffered kieselguhr[44].

Basic peptides containing ε-DNP-lysine may be separated from colourless peptides by adsorption chromatography on talc. The DNP derivatives are adsorbed at the top of a short column of acid washed talc (5 g) and the contaminating materials washed through with N hydrochloric acid. A mixture of 4 parts ethanol and 1 part N hydrochloric acid will then bring off the DNP derivatives. A mixture of ε-DNP-lysine peptides may be fractionated on a column treated with N hydrochloric acid (0.5 ml per g silica gel) using the solvent 66% methyl ethyl ketone–ether or ethyl acetate.

DNP-peptides chromatograph well on the silicic acid–celite columns described on p.177. Samples are best applied to the column in a 5 volume per cent solution of glacial acetic acid in benzene. The peptides are first dissolved in the first solvent and the solution diluted with the latter solvent[45]. A preliminary development with 7 V ml of 8AA-4A-L (see p. 178) will remove most of the contaminating DNP-amino acids and weakly adsorbed DNP-peptides will begin to move down the column. Other solvents of the type used by Green and Kay should then be tried.

A procedure has been described by Keil[46] for the separation of DNP-derivatives of amino acids and peptides from free amino acids peptides using an anion exchanger.

'DNS' METHOD OF HARTLEY

'Dansyl' derivatives (DNS-X) prepared by coupling amino acids with 1-dimethylamino-naphthalene-5-sulphonyl chloride ('dansyl' reagent) are very resistant to acid hydrolysis and have an intense yellow fluorescence. They have been used by Gray and Hartley[47] in an end-group technique which is about 100 times more sensitive than the DNP method and therefore particularly suitable for very small amounts of material. Analysis of mixtures of DNS derivatives can be carried out by electrophoresis or by chromatography.

Electrophoretic method[47]

The peptide (10^{-3} μmole) dissolved in 15 μl of 0.1 M $NaHCO_3$ is coupled with DNS-Cl in acetone (1 mg/ml, 15 μl) at room temperature during 3 h. The sample is dried *in vacuo* and hydrolyzed with 6 N HCl (20 μl) at 105° for 6–12 h. A single electrophoresis in 0.8% acetic acid–0.4% pyridine, pH 4.40 (80 V/cm, 3 h) separates all the DNS amino acids obtained from performic acid-oxidized peptides, except the serine, proline and alanine derivatives. These three are resolved by electrophoresis in 0.1 *M* Na_3PO_4–0.1 *M* NaOH, pH 12.7 (30 V/cm, 2 h).

Chromatographic method[48]

Standards are prepared by mixing 1 ml of amino acid (6.5 μmoles in 0.1 *M* $NaHCO_3$) and 1 ml of DNS-Cl in acetone (6 mg/ml) and leaving overnight. $NaHCO_3$ is precipitated by the addition of acetone (8 ml) and the sample is then ready for chromatography. The following solvent systems are used.

(1) Light petroleum–acetic acid–water (100:90:10); separates Gly, Ala, Phe, Pro, Val, Leu and Ileu, in 12–14 h.

(2) Light petroleum–toluene–acetic acid–water (50:50:85:15); separates Gly, Ornithine, Try + Lys, in 15 h.

(3) Benzene–di-isopropyl ether–acetic acid–water (1:1:1:1); separates $AspNH_2$, $MetSO_2$, Ser, Asp, Glu, Tyr, $GluNH_2$ + citrulline, Thr + HyPro, in 14 h.

(4) Di-isopropyl ether–ethyl methyl ketone–acetic acid–water (8:2:5:5); separates Arg. His, taurine and cysteic acid in 14 h. Gray and Hartley[49] were able to carry out six consecutive steps in a sequence study utilizing the Edman degradation as a subtractive procedure, followed by investigation of the free amino groups of the residual peptide by the DNS method.

Heating the paper strips at 140° for 60 sec increases the fluorescent yield to about 350–400% as measured 30 sec after the heating. The fluorescence decays fairly rapidly so that it is about 225% at 10 min and 130% at 3 h[102].

PHENYLTHIOHYDANTOIN METHOD OF EDMAN

In 1927 Bergmann *et al.*[50] described the acid cleavage of phenylhydantoin from a phenylcarbamyl peptide. Abderhalden and Brockman[51] combined this method for the detection of N-terminal groups with one for splitting off a single residue at a time thus realizing a method for the stepwise degradation of a peptide. It was Edman[52,53], however, who extended the procedure further and evolved a simple technique which works smoothly for many types of peptide.

The first step is the formation of the phenylthiocarbamyl peptide (PTC-peptide) by reaction of phenylisothiocyanate with the peptide at pH 8–9 and at about 40°:

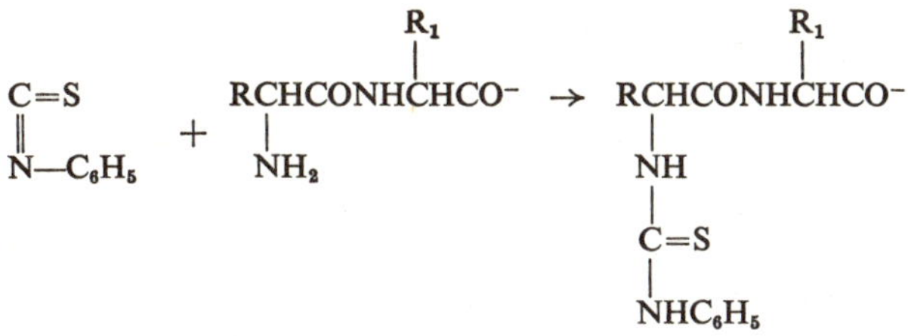

The second step is the cleavage of the N-terminal residue in the form of the phenylthiohydantoin. Edman[54] suggests that this cyclization reaction proceeds via a thiazolinone which may subsequently behave in one of two ways depending on the conditions as shown below.

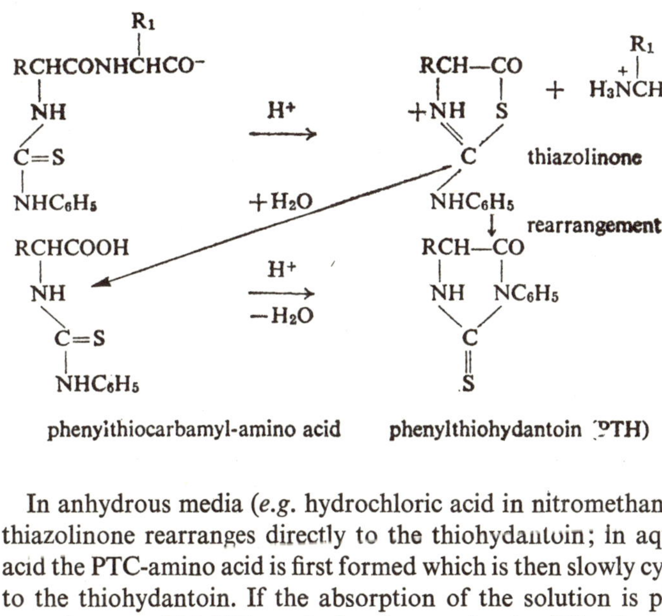

phenylthiocarbamyl-amino acid phenylthiohydantoin (PTH)

In anhydrous media (*e.g.* hydrochloric acid in nitromethane) the thiazolinone rearranges directly to the thiohydantoin; in aqueous acid the PTC-amino acid is first formed which is then slowly cyclized to the thiohydantoin. If the absorption of the solution is plotted over the range 235–275 mμ during the reaction it will be found that the absorption maximum shifts from a little below 240 mμ to about 265 mμ as the PTH is formed. For all practical purposes the method consists of the two steps (a) formation of PTC-peptide or protein and (b) splitting off the N-terminal amino acid as the thiohydantoin.

References p. 218

Preparation and properties of phenylthiohydantoins

In the original Edman procedure[52] 0.5–1 g of the amino acid is dissolved in 25 ml of aqueous pyridine (1:1) in a stoppered tube. The pH is adjusted to 9 with 1 *N* sodium hydroxide, the solution placed in a water bath at 40° and 1.2 ml of phenylisothiocyanate added with shaking. During a reaction time of about 30 min further alkali is added to maintain the pH at 9. The mixture is then extracted repeatedly with benzene to remove excess reagent and pyridine. When there is no further uptake of alkali a slight excess of 1 *N* hydrochloric acid is added to precipitate the PTC- amino acid. This is filtered off and warmed with 30 ml 1 *N* hydrochloric acid at 40°

TABLE 29

MELTING POINT DATA OF PHENYLTHIOHYDANTOINS OF AMINO ACIDS

Data taken from Edman[52,55]; those for serine and threonine from Ingram[56].

Amino acid	*M.p.*	*Solvent*
Alanine	185	acetic acid–water
Arginine	189	water
Asparagine	234	ethanol–water
Aspartic acid	229	ethanol–water
Cysteic acid	193–197 (dec.)	ethanol–water
Carbomethoxycysteine	164	ethanol–petroleum ether
Glutamine	201–211	acetic acid
Glutamic acid	166–167	ethanol–water
Glycine	245–248	acetic acid–water
Histidine	200–206	ethanol
Hydroxyproline	145, 171*	water
Isoleucine	173–175	ethanol
Leucine	176–179	ethanol
ε-Lysine	162–164	ethanol–water
Methionine	132	acetic acid–water
Phenylalanine	187	acetic acid–water
Proline	179	acetic acid–water
Serine	176–178	ethanol
Threonine	194	ethanol
Δ-Threonine	233	acetic acid–water
Tryptophan	177	acetic acid–water
Tyrosine	216	acetic acid–water
Valine	206–208	acetic acid–water

* Fraenkel–Conrat[57].

for 2 h. The PTH derivatives crystallize on cooling and further yields are obtained by concentration of the mother liquors. Most of the derivatives are recrystallized from aqueous acetic acid or ethanol; melting point data are given in Table 29.

The PTH derivatives of serine, threonine and cystine are extremely labile. Threonine loses a molecule of water during the reaction and the dehydro-derivative usually crystallizes out. Serine is similarly dehydrated and a very low yield of PTH is obtained. Cysteine and cystine lose H_2S during the procedure.

The derivatives of the labile amino acids are prepared under milder conditions[56]. The condensation with phenylisocyanate is done at room temperature and the cyclization allowed to proceed at pH 1 at room temperature for two days when the PTH derivatives crystallize out. Tryptophan PTH is decomposed in hydrochloric acid and the PTC-derivative must be cyclized in glacial acetic acid.

The PTH derivatives of all amino acids are readily extractable from aqueous solution by ethyl acetate with the exception of the derivatives of arginine and histidine. The latter may be obtained by evaporation of the solution and extracting with hot absolute ethanol. Arginine PTH cannot be extracted and tends to decompose in both acid and alkaline solutions.

A micro method for the preparation of PTH amino acids has been described by Sjöquist[58]. 250 μl of a solution containing 6 μl phenyl isothiocyanate per ml of acetone is added to 5 μmoles amino acids dissolved in 250 μl buffer, pH 10.1, in a stoppered quartz tube. The buffer prepared from 2.00 ml 2 N acetic acid, 1.20 ml triethylamine and distilled water to 25.0 ml has an apparent pH of 10.1 after dilution with an equal volume of acetone. The PTC-amino acids are formed at 25° during a reaction period of 2.5 h after which the tube is transferred to a desiccator (*high vacuum*) containing phosphorous pentoxide. The solution is evaporated carefully in a low vacuum and last traces of solvent are removed overnight in a vacuum of 0.001 mm Hg.

The dry residue is taken up in 100 μl of glass distilled water and 200 μl of acetic acid saturated with hydrogen chloride is added. Cyclization to the PTH amino acids is carried out at 25° for 6 h and the solvent is removed by evaporation in a desiccator over potassium hydroxide.

Only pure solvents should be used; acetone is refluxed with

References p. 218

potassium permanganate and stored over the same agent; triethylamine is refluxed for 3 h with phthalic anhydride (5 g/100 ml) and redistilled before use. The usual precautions to exclude moisture must be taken in the preparation of the acetic acid–hydrogen chloride reagent.

PTH-amino acids absorb in the region 245–275 mμ and the average molar extinction coefficient is 16,000. To determine the purity of a sample the extracted PTH-amino acid is taken to dryness and redissolved in ethanol. The ratio of the extinction at 245 mμ to that at 269 mμ should be close to 0.40 for a pure derivative. Serine, threonine, lysine and proline give values of 0.35. 0.46. 0.72 and 0.67 respectively[59] (see absorption curves in Fig. 71).

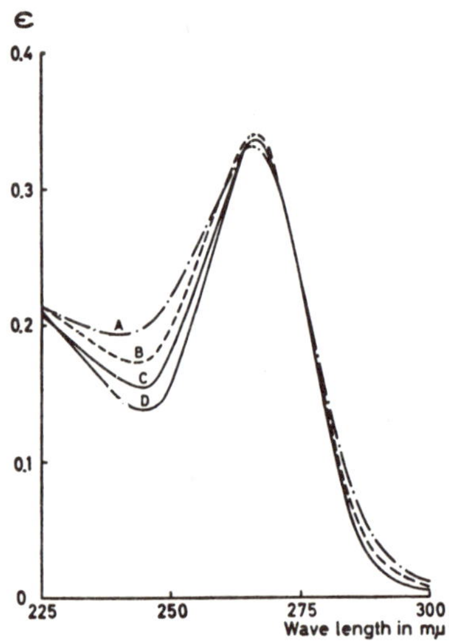

Fig. 71. U.V.-absorption during cyclisation of PTC-glycine after 1 h (A), 2 h (B), 3 h (C), and 5 h (D). Temperature 25°. Sjöquist[59].

Regeneration of the amino acid can be carried out by hydrolysis at a high temperature except in the case of the PTH-derivatives of serine, threonine and cystine. In one method[60] about 1 μM is heated with 0.1 ml redistilled hydrochloric acid (density 1.7) at 140° for 2 h in a sealed tube. The hydrolysate is then evaporated to dryness in a desiccator over potassium hydroxide. In the procedure of Fraenkel-Conrat *et al.*[57] hydrolysis is carried out in 1 ml of 5.7 *N* hydrochloric acid at 150° for 16–20 h. Yields of amino acids up to 70% are obtained if the tubes are evacuated (20–30 mm Hg) before hydrolysis. If tryptophan is suspected acid hydrolysis cannot be used; a useful recovery of this amino acid is obtained using 0.25 *N* barium hydroxide at 105° for 48 h.[52]

Quantitative paper chromatography of phenylthiohydantoins

The following procedure was worked out by Sjöquist[58].

Separations are carried out by the descending method in four one-dimensional chromatograms during a period of 2.5–3.5 h at 24°. The composition of each solvent is shown in Table 30. The upper phase of systems I, II and III are used; system IV is a single phase. Propionic acid is purified by boiling with chromic acid (5 g/150 ml) for 3 h. The liquid is poured off and distilled (boiling range 140–141°).

A correct equilibration procedure must be worked out for each individual chromatographic unit; the following details apply to glass tanks 20 × 30 cm and 60 cm high. Moving phase is poured into the jars to a depth of about 1 cm. Filter papers soaked in this solution are fixed to the short sides of the jar and wetted before each run for systems I, II and III. A dish containing stationary phase is placed at the bottom of each jar and a filter paper 15 cm wide allowed to hang into the solution. The stationary phase in jars I and II consists of 90% formic acid, in jar III of 75% formic acid and in jar IV of 25% formic acid. Before each run the hanging strips are sprayed with their stationary phase in jars I, II and III. Jar IV is allowed to equilibrate with moving and stationary phase for 2–3 days before chromatography and the solvents are replenished as they are used up. The solvents in jars I, II and III are changed each week.

TABLE 30

COMPOSITION BY VOLUME OF THE SOLVENT SYSTEMS

	I	*II*	*III*	*IV*
Heptane	60	60	30	40
Ethylene chloride	5	30	60	–
Propionic acid	5	–	–	–
n-Butanol	–	–	–	30
75% Formic acid	5	5	5	9

References p. 218

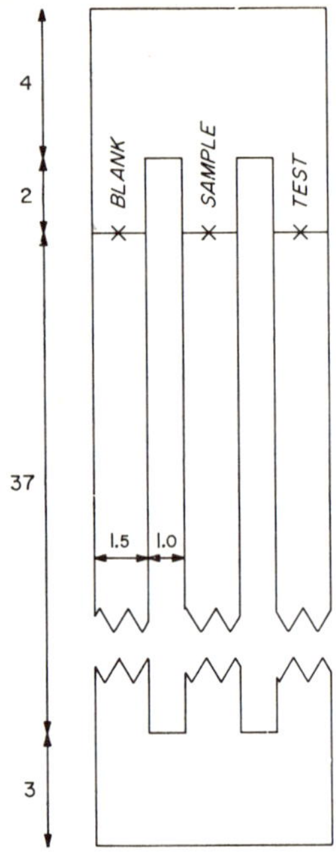

Fig. 72. Pattern for paper chromatogram with three strips. Dimensions in cm[58].

TABLE 31

COMPOSITION OF TEST SOLUTIONS

(The figures refer to the spots in Fig. 73)

A	*B*
1. PTH-leucine	2. PTH-isoleucine
3. PTH-valine	4. PTH-proline
5. PTH-phenylalanine	9. PTH-glycine
6. PTH-methionine	10. PTH-lysine
7. PTH-alanine	11. Phenylthiourea
8. PTH-tryptophan	12. PTH-tyrosine
9. PTH-glycine	14. PTH-glutamic acid
12. PTH-tyrosine	15. PTH-aspartic acid
13. PTH-threonine	17. PTH-arginine
16. PTH-serine	18. PTH-histidine
	19. PTH-cysteic acid

Whatman No. 1 papers are cut out according to the pattern in Fig. 72, dipped in a 0.1% aqueous solution of EDTA-disodium salt and dried at 100°. The paper may contain as many as seven or nine strips: each spot of test solution is flanked by a spot of the sample solution and a spot of the blank solution.

The test solutions A and B (see Table 31) contain 5 μmoles of each PTH amino acid dissolved in 2 ml of 90% acetic acid. They are stable for several months at 5°. Solution A (10 μl) is applied to chromatograms I and II and 10 μl of solution B to chromatograms III and IV.

The papers are allowed to equilibrate for 15 min in jars I, II and III, and for 30 min in jar IV. The solvent systems are freshly prepared before each run and the papers are irrigated to a depth of about 35 cm. Papers I, II and IV are dried at 100° for exactly 30 min and paper III for *exactly* 1.5 h.

Fig. 73 is a diagrammatic representation of UV photocopies of strips prepared by Sjöquist, showing the separation of the PTH derivatives of a synthetic amino acid mixture. To locate the spots use is made of their very strong UV absorption when viewed against a fluorescent screen (transparent perspex evenly sprayed with a suspension of fluorescent zinc silicate in chloroform). If the UV light is filtered to remove all visible light except red and red glasses are worn, the PTH amino acids appear as black spots on a red background.

The excised spots and 'blanks' are extracted in 2 ml of 95% ethanol for 1 h at 40° in glass stoppered tubes. The extinction values of the PTH amino acids are determined at 269 mμ. The PTH derivatives of serine and threonine undergo dehydration during the drying of the paper and the eluted decomposition products show an additional absorption peak at 320 mμ. PTH threonine is therefore estimated at 320 mμ but in the case of PTH serine the extinction at 269 mμ must be determined and substracted from the sum of PTH aspartic acid and PTH serine at 269 mμ. The ratio E^{269}/E^{320} is 0.43 and this constant is used for the conversion.

PTH leucine and PTH isoleucine overlap and they are determined together. PTH proline travels with PTH phenylalanine in system I and with PTH valine in system II. Since PTH valine and PTH phenylalanine are resolved completely in systems I and II respectively, a value for PTH proline can be calculated. PTH lysine is

References p. 218

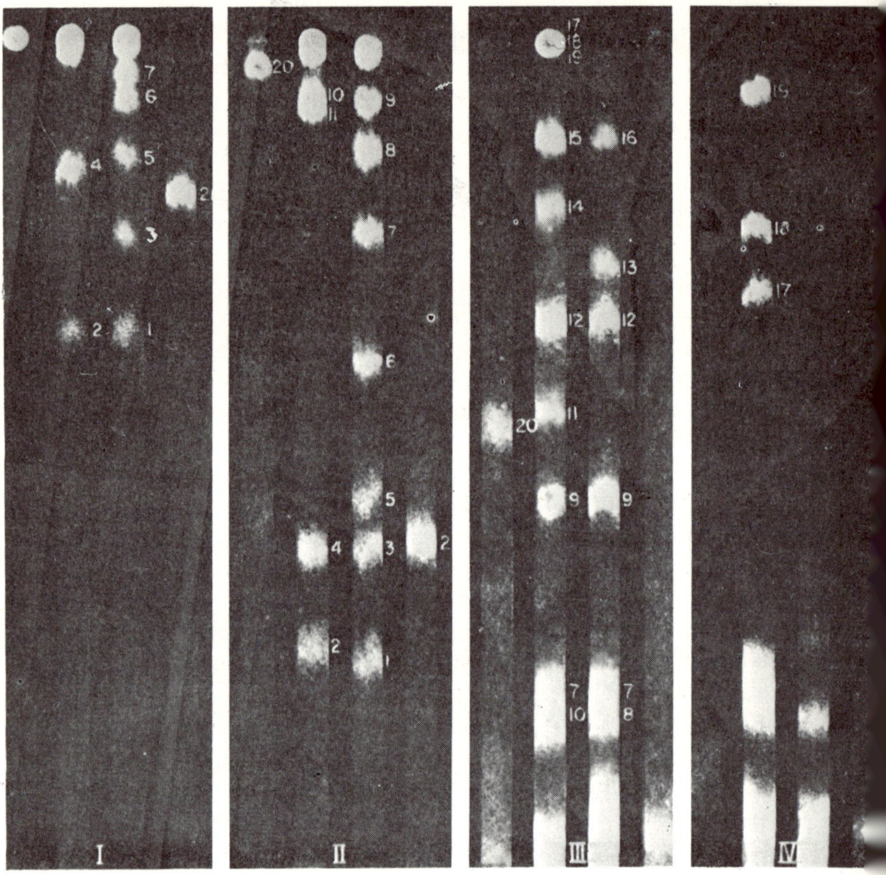

Fig. 73. U.V. photocopies (negatives) of strips run in systems I,II,III, and IV, showing the separation of the phenylthiohydantoins of a synthetic amino acid mixture. The figures to the right of the spots refer to the derivatives given in Table 31. In addition, numbers 20 and 21 represent PTH hydroxyproline and diphenylthiourea respectively. Sjöquist[58].

determined similarly from chromatogram II. The appearance of a phenylthiourea spot indicates ammonium ions present as contamination or released from amide groups. PTH hydroxyproline will interfere with ammonium determination. On blank strips an artifact sometimes appears between PTH phenylalanine and PTH valine in system I. It has the same R_F as diphenylthiourea and appears to be connected with the quality of the triethylamine used for the buffer.

The extinction values for 0.04 mole of each applied PTH amino acid at 269 mμ are shown in Table 32. Percentage yields are overall values; the factor C is the ratio of the amount of sample applied to the extinction. The amount A of amino acid in the sample in μmoles/mg is calculated from

$$A = \frac{C \times \varepsilon \times a}{b \times w}$$

where C is the coefficient in Table 32; ε, extinction value at 269 mμ or 320 mμ; a, μl 90% acetic acid in which the sample is dissolved before application to the chromatogram; b, μl of solution a applied to the chromatogram; and w, weight of sample in mg. A linear relationship exists between the amounts of amino acids applied to the procedure and the extinction values of the PTH derivatives. Lysine is the only exception and the extinction of PTH lysine should be compared with a standard curve. The reproducibility in each determination is $\pm$ 4%.

Partition column chromatography

Fairly quantitative recoveries of PTH amino acids from celite columns were obtained by Sjöquist[61]. Celite is purified by standing in concentrated hydrochloric acid for 3 days, washed with water to pH 4, then with 0.1% EDTA, water, and finally ethanol. The powder is dried at 100°.

A column 0.8 $\times$ 50 cm is prepared from 10 g celite mixed intimately with 6 ml of stationary phase and suspended in 100 ml of mobile phase. The sample is applied and washed in with the minimum volumes of solvent. Elution proceeds in two stages and the fractions are monitored at 269 mμ.

References p. 218

TABLE 32

EXTINCTION VALUES OF ELUTED PHENYLTHIOHYDANTOINS AT 269 mμ AND 320 mμ[58]

Amino acid	*ε* *Found*	*ε* *Calculated*	*Yield in %*	*C*
Alanine	0.334	0.320	104.2	0.120
(Ammonium chloride)*	0.121			0.330
Arginine	0.284	0.318	89.4	0.141
Aspartic acid	0.306	0.322	95.0	0.131
Cysteic acid	0.305	0.332	92.0	0.131
Glutamic acid	0.304	0.318	95.6	0.132
Glycine	0.290	0.298	97.4	0.138
Histidine	0.250	0.310	80.6	0.160
Hydroxyproline	0.254	0.272	93.4	0.157
Isoleucine, Leucine	0.308	0.337	91.5	0.130
Lysine**	0.455	0.580	78.5	0.088
Methionine	0.345	0.342	100.8	0.116
Phenylalanine	0.294	0.310	94.8	0.136
Proline	0.274	0.286	95.8	0.146
Serine***	0.221			0.181
Threonine***	0.300			0.133
Tryptophan	0.340	0.394	86.8	0.118
Tyrosine	0.297	0.312	95.2	0.135
Valine	0.300	0.330	90.9	0.133

* Determined as phenylthiourea.

** Dtermined from chromatogram II.

*** Measured at 320 mμ.

(1) A. Heptane, 250 ml; propionic acid, 100 ml; ethylene chloride 5 ml; water 30 ml.

B. Heptane, 85 ml; propionic acid, 5 ml; ethylene chloride 10 ml. Homogeneous system.

A is passed through the column at 15 ml per h. When PTH phenylalanine has travelled through the column elution is continued with solvent B. PTH amino acids migrate in the order, Leu, Val, Pro, Phe, Met, Ala, Try, Lys, OH-Pro. Fig. 74 illustrates results obtained after reaction of about 0.1 μM of each amino acid.

(2) After PTH phenylalanine has moved off the column a mixer is applied. The upper reservoir contains heptane, 58 ml; proprionic acid, 17 ml; ethylene chloride 25 ml; the lower chamber of the

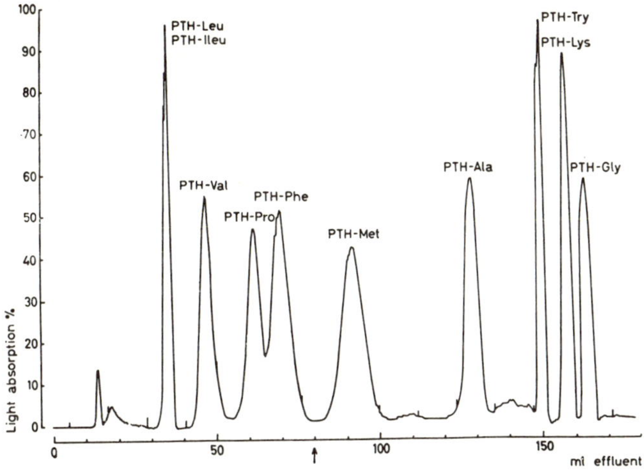

Fig. 74. Chromatographic separation of PTH-derivatives of amino acids on a Celite column (0.8 × 50 cm). The arrow indicates change of solvent (see text). Sjöquist[61].

mixer contains 150 ml of solvent A. Tyr, Glu, Thr, Asp, and Ser are then eluted in that order. PTH derivatives of Arg and His cannot be eluted from the column.

Losses of Ser and Thr are 60% and 30% respectively. A few artifacts appear between Ala and Lys.

PTC-proteins

Brilliant sequence studies have been made on peptides using the Edman method but until quite recently use of the method for quantitative end group analysis of high molecular weight proteins have been only partially successful. A very interesting adaptation of the Edman method by Eriksson and Sjöquist[62] has resulted in quantitative recoveries of the N-terminal amino acids of various serum proteins.

The protein (5–10 mg) is dissolved in 1 ml of saline and mixed with 2 ml of a solution of pyridine–triethylamine–phenyl isothiocyanate (100:3:1). The coupling is done at 40° during 1.5 h and the solution washed five times with 5-ml aliquots of benzene–ethylene chloride (3:1) previously saturated with 0.1 *N* sodium hydroxide.

Last traces of solvent are removed at 40° in a gentle stream of nitrogen. Cyclization is then brought about by treatment with 1 ml water and 2 ml hydrogen chloride-acetic acid at 40° for 2 h. The mixture is then freeze-dried to *near* dryness and the residue suspended in 2 ml water previously saturated with ethyl acetate–methyl ethyl ketone (2:1). After extraction with four 2-ml portions of the latter solvent the PTH amino acids are dissolved in 0.1 ml 90% acetic acid and chromatographed by the method described above. It is recommended that the cyclization, freeze-drying and extraction of the PTH derivatives should be carried out during the same day.

Step-wise degradation of peptides

During the early investigations difficulties were encountered with the Edman method owing to the insolubility of large peptides in anhydrous media and the generally poor yields of PTH derivatives obtainable. Two important findings led to rapid improvements in the procedure. In the first place it was discovered that aqueous media could be used for the degradation without appreciable hydrolytic splitting of the peptide bond[63]. Ottesen and Wollenberger[64] then reported a successful sequence determination on the hexapeptide released during the transformation of ovalbumin to plakalbumin. Second, it soon became clear that less damage to the labile PTH derivatives occurred in rather strong acid and the ideal procedure would be one using strong acid for the minimum time necessary to produce high yields of PTH derivatives. The elegant technique now described was developed by Fraenkel–Conrat and Harris.[65].

The peptide (0.2–0.3 mg) is dissolved in 4 ml of 50% aqueous dioxan, the solution brought to pH 8.7–9.0 with 0.01 *N* sodium hydroxide and the mixture stirred for 1.5 h at 40° with 0.1 ml phenyl isothiocyanate keeping the pH constant. The reaction mixture is then extracted seven times with benzene and the aqueous solution concentrated to dryness *in vacuo* over sodium hydroxide.

The sodium salt of the PTC-peptide is redissolved in water (2–10 ml) and aliquots corresponding to 0.2–1.0 μM are made 3 *N* with respect to hydrochloric acid and $0.2–1.0 \times 10^{-4}$ *M* with respect to peptide by addition of the correct amounts of water and 5.7 *N* hydrochloric acid. The rate of release of phenylthiohydantoin can be determined by following the change of the absorption maximum of the solution from 240 mμ or lower to 265–270 mμ during a period of about 2 h. If the transformation takes place too slowly for a given peptide the effect of increasing the temperature to 40–45° should be tried. Spectrophotometric observation has been found useful for the first two or three stages, but thereafter accumulated artifacts having high UV absorption might tend to interfere with the readings[66,67].

The PTH amino acids are extracted into ethyl acetate (with exception of PTH arginine and PTH histidine) and the residual peptide recovered by concentration of the aqueous solution as before. The residue is redissolved in 50% aqueous dioxan and submitted to the same cycle of operations.

Identification of the phenylthiohydantoin has usually been accomplished by the early procedures of Sjöquist. The solvent systems are the following: (a) Heptane–pyridine (70:30); (b) Upper phase of heptane–*n*-butanol–formic acid (40:20:40); (c) Heptane–*n*-butanol–formic acid (40:40:20).

It is essential to run a known mixture of amino acids simultaneously for comparison. At the end of a run the paper is dried for some hours at room temperature and dipped in a freshly prepared mixture of iodine and azide solutions (0.01 *M* iodine in 0.5 *M* potassium iodide and 0.5 *M* sodium azide, 1:1 v/v). The PTH derivatives appear as bleached areas on a brown background. If the paper is sprayed with 0.5% starch solution before chromatography the spots will appear on a purple background. Additional confirmation of identity is obtained by hydrolysis of the phenylthiohydantoins to the corresponding amino acids.

Subtractive method for peptides

Modern methods of amino acid analysis are far more reliable and accurate than any of the procedures for end-group determination. It is not surprising therefore that the subtractive technique has been favoured in recent sequence studies. The amino acid composition of the residual peptide is determined after each application of the Edman method, and identification of the PTH amino acids themselves is of secondary importance. The procedure of Konigsberg and Hill[68] illustrates this new approach.

References p. 218

Coupling with phenylisothiocyanate is carried out as follows. The peptide is dissolved in 2 ml of buffer containing N-ethylmorpholine (60 ml), acetic acid (1.5 ml), 95% ethanol (500 ml) and water (440 ml). Phenyl isothiocyanate (0.10 ml) is added and coupling allowed to proceed at 37° for at least 2½ h. The mixture is then concentrated to dryness and the residue extracted twice with benzene. Last traces of solvent are removed from the residue by exposure to a high vacuum for 5 min at 60°.

Cyclization is effected in anhydrous trifluoroacetic acid (2 ml) during 1 h at 25°, after which the reagent is removed by evaporation. The residue is dissolved in 0.2 *M* acetic acid (4 ml) and the solution extracted twice with an equal volume of benzene to remove the non-polar PTH derivatives and PTC amino acids. The aqueous phase is then transferred to a 0.3 × 6 cm column of Dowex 50-X2 (80 mesh) in the H^+ form. The resin is then washed successively with three 4-ml portions of 0.2 *M* acetic acid, three 4-ml portions of 0.17 *M* pyridine acetate buffer (pH 4.8), 4 ml of 1.07 *M* pyridine acetate, and finally with two 4-ml portions of 8.5 *M* pyridine acetate. Aliquots of the fractions are then spotted on paper and developed with ninhydrin. Neutral or acidic peptides usually appear in the second fraction of the 0.17 *M* buffer; basic peptides and those containing aromatic residues are eluted later. A portion of the peptides is analyzed, usually with automatic apparatus; the remainder is submitted to further degradation. The more polar phenylthiohydantoins and PTC amino acids are eluted in the initial fractions. Overall yields of released amino acids are 60–80%.

Degradation of proteins

A major difficulty arises when the Edman procedure is applied to proteins in that PTC-proteins and large PTC-peptides are generally insoluble in the anhydrous acid media used for their cyclization. To overcome this problem Fraenkel–Conrat[69] devised the ingenious technique of performing all operations on a thin layer of protein adsorbed on filter paper.

Small rectangular strips of Whatman No. 1 paper (1 × 5 cm) are marked at one end and suspended from this end by means of clips and a glass rod in a gentle draught. About 1–3 mg (0.05 ml) protein is deposited evenly on each of six or more strips to cover the bottom four-fifths. Several applications of more dilute solutions can be made with intermittent drying; proteins insoluble in water should be dissolved in salt solution or in weak acid or alkali. When the papers are dry 0.05 ml of 20% phenylisothiocyanate in peroxide-free dioxan is added and the strips allowed to stand for 2–3 h at 40° in an atmosphere saturated with pyridine, dioxan and water. They are then washed by shaking gently with benzene for 2–3 h using three changes of solvent. Finally the strips are washed repeatedly with ethanol–ether (1:1) until the washings show an extinction at 270 mμ of less than 0.025 (usually after 24 h).

Cyclization is brought about by exposing the strips to acid vapours in the following way. The dry strips are laid in small dishes on the bottom of a desiccator into which small beakers are placed containing 6 *N* hydrochloric acid and glacial acetic acid respectively. The desiccator is evacuated to about 100 mm Hg, closed, and kept at room temperature for 4–16 h depending on the particular protein. It is then necessary to remove the acid from the strips in an air stream before immersion in the alcohol–ether mixture to extract the phenylthiohydantoin. Extraction is usually complete after 4 h and the extinction of the combined extract and washings is determined over the range 260–275 mμ. An assessment of the purity of the released PTH amino acids is made as explained above and if all

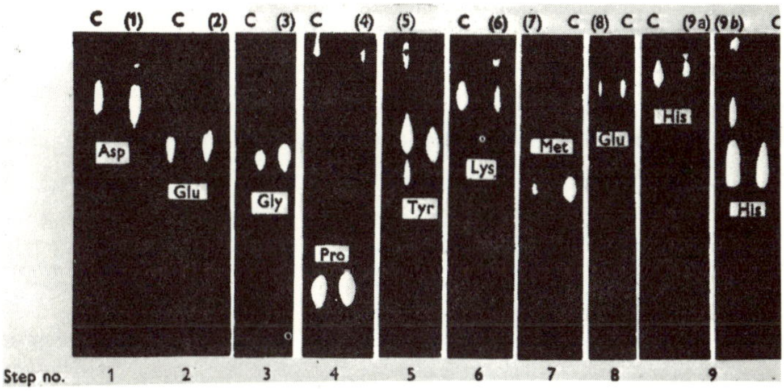

Fig. 75. Stepwise degradation of β-melanocyte-stimulating hormone by the phenyl isothiocyanate method. Amino acids identified as PTH derivatives by chromatography in heptane–pyridine (70:30 v/v) (1 to 9a) and in heptane–butanol–formic acid (40:20:40, by vol) (9b). C, control. Sequence established is Asp.Glu.Gly.Pro.Tyr.Lys.Met.Glu.His (from Harris and Roos[70]).

is well a further treatment with phenylisothiocyanate is carried out on the dried strips. The average yields from each successive step is 80% based on the maximum extinction.

Where the Edman method has been applied successfully 5–7 residues have been split off with little non-specific splitting. In several instances a peptide has been degraded with astonishing ease. Notable applications have been the determination of the decapeptide sequence Asp.Glu.Gly.Pro.Tyr.Lys.Met.Glu.His.Phe in β-melanocyte-stimulating hormone[70] (Fig. 75); heptapeptide sequence Ser.Tyr.Ser.Met.Glu.His.Phe in α-corticotropin[71]; a pentapeptide sequence Thr.Pro.Val.Thr.Pro in prolactin[72]; and sequences of a number of fragments derived from fibrinopeptides[73].

The efficiency of peptide mapping (see p. 55) might be improved by the application of the Edman method *in situ*[75]. N-terminal analyses of the PTC peptides would give a fair indication of the number of peptides in each spot.

Comments

The development of the pH-stat in the Carlsberg Laboratory[74] has made available a useful tool for the study of reactions involving peptides on a micro scale. The recent work of Hirs, Moore and Stein[67] illustrates the use of this refinement. 10–20 μM of peptide is mixed with 50 mg of sodium chloride in a water-jacketed test tube and suspended in 3–5 ml of either 30% aqueous dioxan or 50% aqueous pyridine. Water at 40° is circulated through the jacket and while vigorous magnetic stirring is maintained, N sodium hydroxide is added from a microburette until the pH is stable at pH 8.8. At this stage the solution is homogeneous and the microburette is replaced by the motor driven unit of the pH-stat which contains 0.20 N sodium hydroxide. The tube is flushed with CO_2-free nitrogen, the recorder of the pH-stat started and the instrument allowed to operate for 5–10 min. An uptake of alkali of less than 2 μl indicates an effective nitrogen barrier and 0.20 ml of redistilled phenylisothiocyanate is then added. Reaction is complete and uptake of alkali is usually constant after 2 h. An additional refinement is the use of cyclohexane in a small extractor for continuous removal of the PTC derivative.

As mentioned above the spectrophotometric examination of the reaction mixture may be complicated by UV absorbing contaminants. A number of workers have therefore used the Edman degradation in a subtractive manner; the reaction is then monitored by quantitative amino acid analysis.

Although much is known about the precedure there is considerable scope for further experimentation. High conversion to phenylthiohydantoins as indicated by spectrophotometry is not always followed by high recoveries from the reaction mixture. It is the opinion of a number of workers that the conditions of cyclization in acid are possibly still too vigorous and that it would appear desirable to extract the intermediate thiazolinone from its destructive environment as quickly as possible[76,77]. It is interesting to note that the modification of Eriksson and Sjöquist[62] employing hydrogen chloride in 50% acetic acid at 40° followed by freeze-drying offers a great improvement over earlier methods as judged by the recovery of N-terminal aspartic acid from human albumin. Anhydrous trifluoroacetic acid which is a good solvent for proteins, is a promising medium for cyclization. Even with this reagent the recoveries of phenylthiohydantoins are not as good as might be expected; perhaps a substitution reaction of the Friedel–Craft type is partly responsible.

OTHER CHEMICAL METHODS

Dihydroquinoxalines

The series of reactions leading to the formation of dihydroquinoxalines are the following:

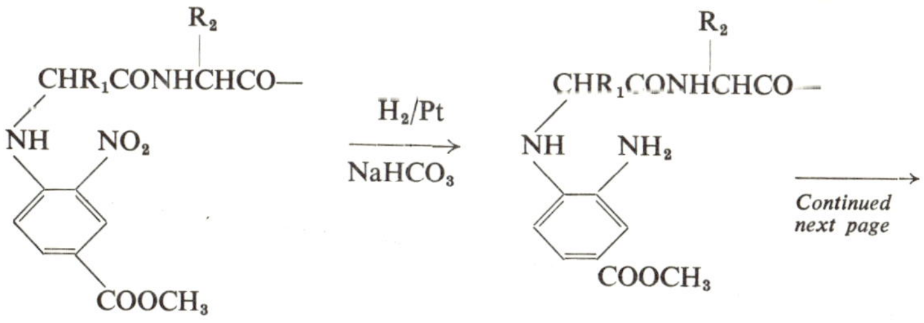

Continued next page

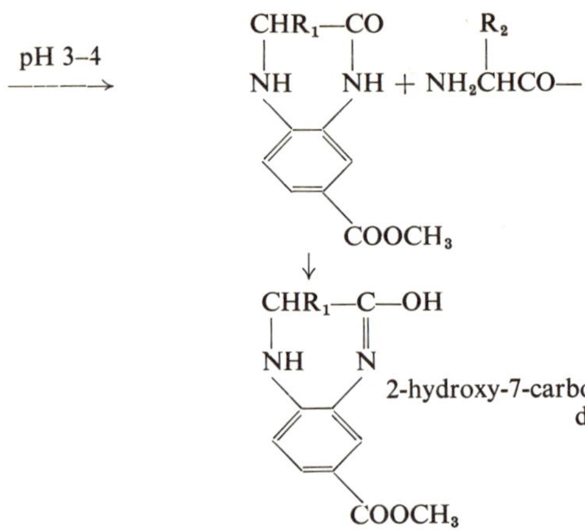

2-hydroxy-7-carbomethoxy-dihydroquinoxaline

As developed by Holley and Holley[78] the method employed 2-nitro-4-carbomethoxybenzene as the coupling agent. Later workers[79,80], have used DNP-peptides. In this modification the 2,4-diaminophenylpeptides produced by catalytic reduction must be protected from atmospheric oxidation and the lactam formation is carried out under hydrogen. Further treatment of the cyclic derivative with FDNB under nitrogen yields DNP-aminodihydroquinoxalines which are stable to air and can be extracted from the solution and identified by chromatography. DNP-amino acids and peptides undergo hydrogenolysis in methanol in the presence of a large excess of platinum catalyst. The free amino acids and peptides can be recovered in 90% yield (Fasold *et al.*[81]).

In another procedure[82] the formation of the sensitive 7-amino compounds is avoided by selective reduction of the 2,4-dinitrophenyl derivative with ammonium sulphide to give the 2-amino-4-nitrophenyl peptide. The 2,4-dinitrophenyl peptide is suspended in 95% ethanol (6 ml) and treated with a concentrated solution of ammonium sulphide (6 ml) at 45–50° during 30 min. The product is cyclized at pH 2–3 by heating at 70° for 15 min to give 7-nitro-3,4-dihydro-2-hydroxyquinoxaline.

Degradation using carbon disulphide

In 1950 Cook, Heilbron and Levy were studying the properties of 2-thio-thiazolidones with a view to their possible use in peptide synthesis. They were successful with glycine peptides which could be prepared from the thiazolidone and an amino acid ester in the presence of a tertiary base. The method did not work when 4-substituted thiazolidones corresponding to other amino acids were tried. The acidification step which led to the synthesis of glycyl peptides caused reversion to the original thiazolidone in the other cases. This aspect was turned to advantage by Levy[83] in a method for the removal of terminal groups using carbon disulphide.

Titration of peptides with alkali at constant pH in the presence of carbon disulphide yields the salts of substituted dithiocarbonic acids. On acidification in cold aqueous solution, these cyclize to 2-thio-5-thiazolidones with fission of the terminal peptide link.

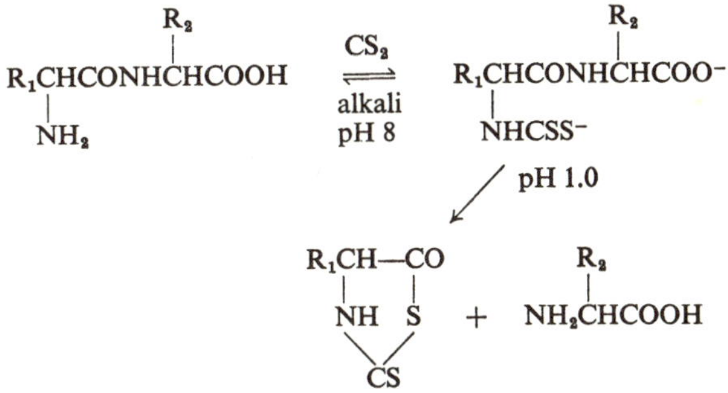

Details of the method are as follows. The peptide (0.2 mg per peptide bond for an end group; 1.0 mg per peptide bond for a sequence) is dissolved in 25% (v/v) dioxan and water (2–4 ml) and maintained at 40° in an atmosphere of purified nitrogen. The pH is adjusted to 8.0, carbon disulphide (0.25 ml) added to the stirred solution and the pH maintained constant by continuous small additions of 0.1 *N* sodium hydroxide. Control of pH is best done using an autotitrator and the flattening of the alkali uptake curve serves as an indication that the reaction is complete (2–3 h). The mixture is then continuously extracted with ether to remove dioxan and excess carbon disulphide. The pH is then adjusted to 1.0 and the mixture allowed to stand for 30 min at room temperature. The thiothiazolidone split off is then extracted with isopropyl ether (25 ml) and the aqueous layer returned for a second degradation cycle. The ether-extract is washed once with water (10 ml), dried over a little anhydrous magnesium sulphate, and the micromoles of thiothiazolidone present determined from the extinction at 280 mμ (ε = 13,500). The ether is then evaporated to dryness *in vacuo* and the residue refluxed gently with 0.25 ml dioxan + 0.25 ml 5 *N* HCl for 1.5 h. The solution, now containing the amino acid(s) is evaporated to dryness and the residue taken up in a small quantity of ethanol–0.1 *N* HCl (1:3 v/v) for identification by paper chromatography.

In their interesting account of the method, Leonis and Levy[84] were not too optimistic regarding application to proteins although considerable success was achieved with peptides. The presence of basic residues adjacent to the terminal peptide bond greatly diminishes the yield of fission as does elongation of the chain to long polypeptides. It has been suggested that in the one case the nitrogen atoms of basic residues compete with the terminal peptide bond for chelation with the dithiocarbamic acid, and in the other case external hydrogen bonding of the dithiocarbamate group to the helical backbone of the polypeptide chain inhibits fission.

A degradation of N-thioncarbalkoxy-peptides[85] is similar in many respects to the above procedure of Levy. N-thioncarbalkoxy derivatives of peptides are prepared by reaction with methyl ethyl xanthate in the presence of alkali, and cleavage of the cyclic intermediates is carried out in anhydrous nitromethane saturated with dry hydrogen chloride.

Condensation of peptides with N-acyldithiocarbamates provides another route to the labilization of the N-terminal residue[86]. This is illustrated in the following scheme.

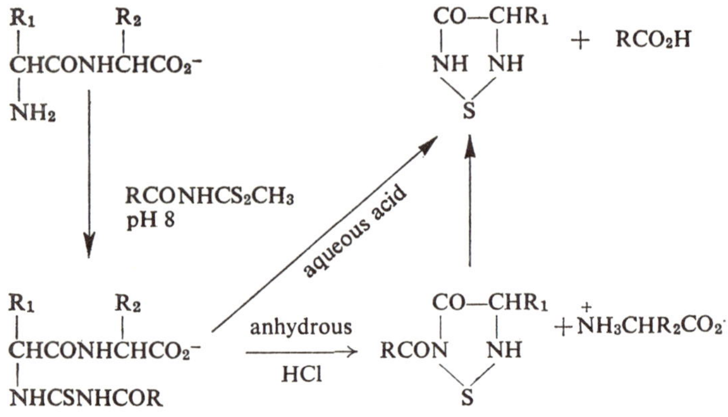

Methoxycarbonyl derivatives

These derivatives are formed according to the equation

$$NH_2CHRCO\text{—} + ClCOOCH_3 \rightarrow \underset{\displaystyle COOCH_3}{\underset{|}{N}}HCHRCO\text{—} + HCl$$

Examination of the properties of the methoxycarbonyl derivatives of amino acids and the conditions for their release intact from methoxycarbonyl-proteins was undertaken by Chibnall and Spahr[87]. They found that the amino acid derivatives are readily decomposed in hot strong hydrochloric acid but on hydrolysis with 8.6 *N* hydrochloric acid at 37° the loss of the methoxycarbonyl group is less than 14% in 10 days and 20% in 21 days. The methoxycarbonyl group has a definite labilizing effect on an adjacent peptide bond. The dipeptide leucylglycine is a particularly stable one, but the corresponding methoxycarbonyl derivative is completely split in 10 days under the above conditions.

The procedure for proteins may be illustrated by the reaction with insulin. The insulin (0.96 g) is dissolved in a mixture of aqueous 60% (v/v) ethanol (60 ml) and *N* sodium hydroxide (5.2 ml). Sodium bicarbonate (0.4 g) is then added followed by 0.4 ml of methoxycarbonyl chloride (methyl chloroformate) in 0.05-ml portions over a period of 3 h at room temperature. During the reaction the pH is maintained at 9 with *N* sodium hydroxide. The mixture is dialyzed against distilled water to remove salts and then against *N* hydrochloric acid which causes the methoxycarbonyl-insulin to precipitate. The product is centrifuged and washed with alcohol and ether.

The method chosen for the separation of the methoxycarbonyl derivatives was chromatography on a column of celite buffered at pH 4.5 and gradient elution with ethanol–chloroform. The zones were located by titrating each fraction with 0.005 *M* potassium hydroxide using bromothymol blue as indicator. After hydrolysis of the methoxycarbonyl-insulin with 8.6 *N* hydrochloric acid for 21 days at 37° and subsequent chromatography of the concentrated hydrolysate, Chibnall and Spahr were able to account for 0.98 group/mole of N-terminal phenylalanine and 0.77 group/mole of N-terminal glycine.

The procedure was applied to β-lactoglobulin, rabbit tropo-

References p. 218

myosin, and ovalbumin, and the results confirmed those obtained by the dinitrophenyl method of Sanger.

Carboxymethyl derivatives

These derivatives may be prepared by reaction of free amino groups with bromoacetate, iodoacetate or chloroacetate; bromoacetate is much more reactive than the others and is the reagent of choice. Korman and Clarke[88] used magnesium bromoacetate in an aqueous suspension of magnesium hydroxide and found that the majority of those amino acids studied formed the dicarboxymethyl derivative.

$$H_2NCHRCO\text{—} + (BrCH_2COO)_2Mg \rightarrow (HOOC.CH_2)_2NCHRCO\text{—}$$

Proline gave the mono-derivative, glutamic acid and phenylalanine gave some and aspartic acid a trace of mono-derivative. The guanidino group did not react and the reaction with serine, threonine, cystine and tryptophan is not fully understood. Lysine gave the tetra-derivative; the ε-di-carboxymethyl derivative could be isolated from a peptide such as alanyllysylalanine. Tyrosine gave two derivatives in both of which the phenolic hydroxyl group had also been carboxymethylated. Cysteine and histidine likewise react in their specific functional groups.

With the exception of the derivatives of cysteine, cystine, methionine, histidine, tryptophan and aspartic acid the amino acids give products which are precipitated by mercuric nitrate. The exceptions are precipitated by mercuric acetate in the presence of 0.1 *M* sodium chloride which distinguishes them from the α-carboxymethyl amino acids. The mercuric derivatives of the latter melt with decomposition at definite temperatures.

The derivatives of a few simple peptides were shown to yield insoluble mercury salts and after acid hydrolysis gave the carboxymethyl derivative of the N-terminal amino acid in yields, in most cases, of over 90%.

In general the amino acid derivatives may be prepared by gentle agitation with magnesium bromoacetate (8 moles) and excess magnesium oxide at 35° for 3 days. When the reaction is complete the solution is filtered, acidified to Congo red and treated with saturated mercuric nitrate solution. The mercuric derivatives are purified by solution in 0.05 *M* sodium chloride and reprecipitation with mercuric nitrate. The precipitate is washed free from nitrate, decomposed with H_2S and the resulting solution concentrated under reduced pressure to about 1 molar. The mercury-free carboxymethyl derivatives may be extracted with ethyl acetate after adjusting the pH of the concentrated solution to 1 to 1.5 and crystallized from the same solvent, ethano.l or water. Korman and Clarke isolated the carboxymethyl peptides as the mercuric derivatives and after decomposition with H_2S converted the products to the calcium salts. These were hydrolyzed with 6 *N* hydrochloric acid under reflux for periods of 18–40 h when high yields of the N-terminal residues were obtained. Carboxymethyl amino acids have been separated by paper chromatography in the solvent system *n*-butanol–ethanol–ammonia (3 *N*) (2:1:2) and the R_F values at 29–31° for Whatman No. 1 paper are shown in Table 33. The reduced

TABLE 33

R_F VALUES (± 0.03)) OF CARBOXYMETHYLAMINO ACIDS[88]

Di-Cm-glycine	0.37	Di-Cm-serine	0.33
Di-Cm-alanine	0.48	O-mono-Cm-tyrosine	0.44
Di-Cm-valine	0.48	Tri-Cm-tyrosine	0.36
Di-Cm-leucine	0.73	Di-Cm-lysine	0.32
Di-Cm-isoleucine	0.54	Tetra-Cm-lysine	0.27
Mono-Cm-phenylalanine	0.62	Di-Cm-arginine	0.28
Di-Cm-phenylalanine	0.62	Di-Cm-histidine	0.27
Cm-proline	0.45	Tri-Cm-histidine	0.32
Mono-Cm-aspartic acid	0.35	Di-Cm-methionine	0.37
Mono-Cm-glutamic acid	0.38		

ninhydrin reagent of Long *et al.*[89] is used for detection purposes. After spraying, the paper is heated at 110–115° for 1–3 min and allowed to stand at room temperature overnight. Spots of the dicarboxymethyl amino acids appear purple, those of the monoderivatives pink, and those of the free amino acids dark bluish brown.

An example of the carboxymethylation of proteins may be illustrated with reference to bovine serum albumin[90]. A solution containing 500 mg of crystalline protein, 6 mmoles of bromoacetate and excess of magnesium oxide is gently agitated at 35°. The course

References p. 218

of the reaction is followed by removing aliquots of the mixture for analysis by various methods. Loss of amino nitrogen and failure to give diazo coupling of imidazole and phenolic residues are useful criteria; in the case of bovine serum albumin the disulphide linkages become more accessible to reduction as the carboxymethylation proceeds and this can be assessed by treatment with sulphite followed by Folin phosphotungstic acid solution[91]. After 27–72 h the mixture is centrifuged and the clear supernatant acidified to pH 2 with hydrochloric acid. A practically quantitative yield of the carboxymethyl-protein is obtained after the solution has been kept for several hours at 0°.

Release of the N-terminal residue is brought, about by hydrolysis with 6 *N* hydrochloric acid under reflux for 20 h. An aliquot of the hydrolysate containing 15.0 mequivs. of N is evaporated under reduced pressure to remove the hydrochloric acid and the residue taken up in 10 ml of water and treated with 20 ml of 1 *M* mercuric acetate solution. The precipitate is filtered and washed and the supernatant and washings treated with saturated mercuric nitrate. In this way Korman and Clark were able to isolate a mercuric derivative from bovine albumin which behaved in many ways like mercuric carboxymethyl aspartate. Assuming the product was homogeneous the recovery amounted to 0.87 group/mole based on the total N of the isolated salt.

The carboxymethylation procedure offers a promising alternative to the dinitrophenyl method and a full assessment of its value in end-group determination must await further development.

Carbamyl derivatives

Hydantoins are formed from cyanate and amino acids according to the following scheme:

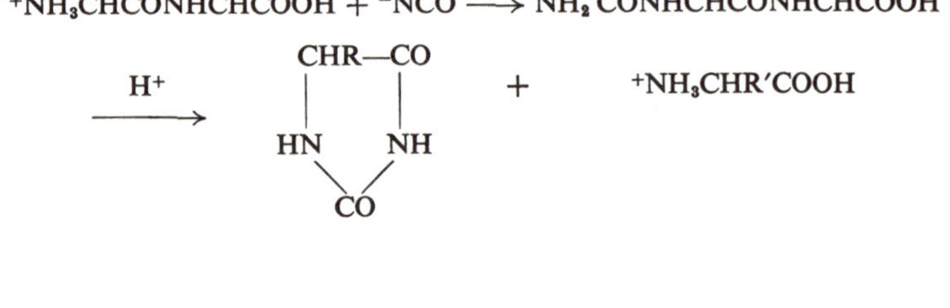

Carbamylations of proteins or large peptides is conveniently carried out by the method of Stark and Smyth[100]. N-Ethylmorpholine acetate buffer–8 *M* urea is prepared from the base (2 ml), water (2 ml), acetic acid and urea; 2.5 ml of this solution is added to 50 mg of protein or peptide dissolved in 2.5 ml of 8 *M* urea. KNCO (250 mg) is then added and the mixture allowed to stand overnight at 50°. Excess cyanate is decomposed with acetic acid (5 ml). Cyclization to hydantoins is performed on 0.5 μmole of carbamyl protein contained in 1–3 ml of 50% acetic acid. An equal volume of concentrated HCl is added and the mixture heated in an evacuated sealed tube at 100° for 1 h.

Neutral and acidic hydantoins are readily eluted with water from a 0.9 × 10 cm-column of Dowex 50-X2; the derivatives of histidine and arginine, together with amino acids and peptides formed during the cyclization procedure, are removed with 1 *M* NH_4OH and fractionated further on Dowex 50-X2 using 0.8 *M* HCl as eluent. After isolation most of the hydantoins can be hydrolyzed to amino acids in 6 *N* HCl at 100°. The derivatives of serine and threonine are, however, markedly unstable to acid hydrolysis.

Reductive methylation

N-Dimethyl amino acids are formed by the action of formaldehyde on the free amino acids followed by catalytic reduction with hydrogen. Originally studied by Bowman and Stroud[92] the principle was later developed by Ingram[93] as an end-group method. The catalyst is prepared from 1 g of palladous chloride in 1 ml 10 *N* hydrochloric acid diluted to 100 ml with water. Following the addition of 5 g of hydrated sodium acetate and 5 g of activated charcoal, the suspension is stirred in an atmosphere of hydrogen at room temperature and pressure until the absorption of H_2 has finished. The catalyst is ready after washing with water and drying in the air.

The formaldehyde treatment is a simple matter of adding one drop of 40% aqueous formaldehyde to about 0.5 ml of a 0.01 *M* solution of a peptide. The air in the container is then displaced by nitrogen, a little palladized charcoal added and hydrogen passed through the suspension for 4–12 h at 40°. The catalyst is removed by centrifugation, the formaldehyde evaporated and the residue hydrolyzed with hydrochloric acid. The dimethylamino acids are

References p. 218

fairly stable to hydrolysis, but the percentage loss for any given derivative must be ascertained for the particular conditions chosen.

Proline forms a mono-methyl derivative, cystine a tetramethyl derivative, and histidine the α-dimethyl derivative. Some R_F values for the dimethylamino acids in the systems *n*-butanol–acetic acid–water (25:6:25), *n*-butanol–pyridine–water (5:2:3) and phenol–water, are given in Table 34. The zones may be detected with the

TABLE 34

R_F VALUES FOR SOME DIMETHYLAMINO ACIDS[93]

Dimethyl derivative	*n-Butanol–acetic acid–water*	*n-Butanol–pyridine–water*	*Phenol–water*
Glycine	0.27	0.15	0.86
L-Alanine	0.35	0.20	0.88
L-Leucine	0.71	0.54	0.91
DL-Isoleucine	0.68	0.51	0.90
L-Tyrosine	0.53	0.48	0.88
L-Phenylalanine	0.69	0.57	0.88
L-Glutamic-acid	0.33	0.10	0.85
L-Ornithine	0.16	0.09	0.86
L-Lysine	0.14	0.07	0.79
L-Histidine	0.14	0.17	0.70
L-Cystine	0.40	0.38	0.94
L-Proline	0.38	0.21	0.93

indicator thymol blue (0.04%) in a mixture of *n*-butanol and ethanol (1:1) which is made 0.01 *N* with respect to H_2SO_4. Dimethylamino acids appear as yellow spots on a red background while the paper is damp. Alternatively the paper can be sprayed with methyl iodide which converts the dimethylamino derivatives to betaines. The excess methyl iodide is removed by evaporation and the paper held in an atmosphere of gaseous chlorine. The liberated iodine is then detected with starch solution[94].

LEUCINE AMINOPEPTIDASE

Leucine aminopeptidase was originally prepared from swine intestinal mucosa but a more convenient preparation from swine kidney has been described by Smith *et al.*[95,96]. For an excellent review of the properties of the enzyme see Smith and Hill[97].

TABLE 35

HYDROLYSIS OF AMINO ACID AMIDES BY LEUCINE AMINOPEPTIDASE[98]

(The value for L-leucinamide is given as 100, and for the other amides as the relative rate.)

Substrate	*Mn^{2+} activation*		*Mg^{2+} activation*	
	C_0	*Relative rate*	*C_0*	*Relative rate*
L-Leucinamide	14,000	100	6,600	100
DL-Norleucinamide	14,200	101	7,200	105
DL-Norvalinamide	11,800	84	7,200	109
DL-α-Amino-*n*-buryramide	5,100	36	5,100	77
L-Alaninamide	470	3.4	325	4.9
Glycinamide	18	0.13	7	0.10
L-Isoleucinamide	2,800	20	1,120	17
L-Alloisoleucinamide	1,000	7.1		
L-Valinamide	2,400	17		
L-Tryptophanamide	3,400	24		
L-Phenylalaninamide	3,600	26	1,140	17
L-Tyrosinamide	2,200	16		
L-Histidinamide	2,700	19.1	680	10
L-Lysinamide	1,000	7.1	800	12
L-Argininamide	1,000	7.1		
L-Isoglutamine	310	2.2		
L-Aspartic acid diamide	410	2.9	250	38
L-Serinamide	106	0.76		
L-Prolinamide	100	0.71		
Hydroxy-L-prolinamide	80	0.57		
DL-α-Aminocaprylamide	10,500	75		
DL-Tertiary leucinamide	16	0.11		
α-Aminoisobutyramide	0	0		
D-Leucinamide	0	0	0	0
β-Alaninamide	0	0	0	0

The enzyme is labile at pH values above 9 or below 7 and is optimally active in the pH region 7.8 to 9.3. The activity is stabilized by Mg^{2+} and Mn^{2+} and chelating agents which remove these ions may cause irreversible inactivation. Before use it is necessary to incubate the enzyme for a short period at 40° with either 0.002 M $MnCl_2$ or 0.04 *M* $MgCl_2$; the crude enzyme may require 3 h, purer samples 30 min to 1 h. Substances that contain large aliphatic groups act as inhibitors. The effectiveness of the straight chain aliphatic alcohols up to hexanol is in the order of increasing chain length;

References p. 218

power to inhibit decreases with the degree of chain branching. Diisopropylphosphofluoridate is not an inhibitor for the enzyme.

The relative rates of hydrolysis of amino acid amides by leucine aminopeptidase are listed in Table 35. Some compounds are hydrolyzed according to first-order kinetics, others to zero-order, and the less favourable ones according to mixed first-order and zero-order kinetics. For those substrates which do not follow zero-order kinetics a zero-order proteolytic coefficient, C°, is calculated from the initial rate of hydrolysis. For first-order reactions $C_1 = k_1/E$ where E is the enzyme concentration in mg protein per ml or mg protein N per ml, and $k_1t = \log(100/100 - x)$ where x is the per cent of substrate hydrolyzed at time t (Ref. [99]). For zero-order reaction, $C_1 = k_1/E$ and $k_1t = x$. The proteolytic coefficient is usually calculated for an initial substrate concentration of 0.05 M.

From Table 35 it is apparent that the action of the enzyme is greatest with those substances containing the largest side chains. Aromatic residues are less sensitive and residues bearing polar groups are hydrolyzed relatively slowly. The rates of hydrolysis of peptides are also governed by the nature of the amino acid adjacent to the N-terminal position. In several instances partial elucidation of the N-terminal sequence has been possible. Both the oxidized A and B chains of insulin are readily hydrolyzed to free amino acids. From kinetic studies it has been possible to deduce the sequence of the first 6 N-terminal residues of the B chain. Useful information has been obtained about the N-terminal sequence of a number of natural peptides and of peptide fragments derived from proteins by the action of leucine aminopeptidase. Certain proteins, *e.g.* ribonuclease, lysozyme, Zn-insulin and serum albumin, are resistant to attack but are readily hydrolyzed after oxidation with performic acid or after removal of the zinc in the case of insulin. Complete digestion of a peptide or protein would indicate the L-configuration for all the constituent amino acids. For a sensitive precise assay using a pH-stat see Bryce and Rabin[103].

REFERENCES

1 F. Sanger, *Biochem. J.*, 39 (1945) 507.
2 F. Sanger, *Biochem. J.*, 45 (1949) 563.
3 H. G. Cook and B. C. Saunders, *Biochem. J.*, 41 (1947) 558.

4 R. R. Porter and F. Sanger, *Biochem. J.*, 42 (1948) 287.
5 A. L. Levy and D. Chung, *J. Am. Chem. Soc.*, 77 (1955) 2899.
6 E. Margoliash, *Nature*, 175 (1955) 293.
7 F. Sanger, *Biochem. J.*, 40 (1946) 261.
8 K. R. Rao and H. A. Sober, *J. Am. Chem. Soc.*, 76 (1954) 1328.
9 R. L. M. Synge, and M. A. Youngson, *Biochem. J.*, 78 (1961) 708.
10 H. Davoll, R. A. Turner, J. G. Pierce and V. Du Vigneaud, *J. Biol. Chem.*, 193 (1951) 363.
11 R. R. Porter, *Methods Med. Res.*, 3 (1950) 256.
12 F. Sanger and E. O. P. Thompson, *Biochem. J.*, 53 (1953) 353.
13 B. J. Campbell, R. J. Schlüter, G. F. Weber and W. F. White, *Biochim. Biophys. Acta*, 46 (1961) 279.
14 F. Šorm, J. Korbl and L. Matousek, *Collection Czech. Chem. Commun.*, 15 (1950) 295.
15 B. Pollara and R. W. Von Korff, *Biochim. Biophys. Acta*, 39 (1960) 364.
16 D. W. Russell, *Biochem. J.*, 87 (1963) 1.
17 S. Akabori, S. Sakakibara and K. Sakakibara, *Bull. Chem. Soc. Japan*, 32 (1959) 312.
18 F. S. Scanes and B. T. Tozer, *Biochem. J.*, 63 (1956) 282.
19 D. M. P. Phillips, *Biochem. J.*, 68 (1958) 35.
20 R. R. Redfield and C. B. Anfinsen *J. Biol. Chem.*, 221 (1956) 387.
21 J. F. Roland and A. M. Gross, *Anal. Chem.*, 26 (1954) 502.
22 A. H. Gordon, A. J. P. Martin and R. L. M. Synge, *Biochem. J.*, 37 (1943) 79.
23 K. Bailey, *Biochem. J.*, 49 (1951) 23.
24 S. Blackburn, *Biochem. J.*, 45 (1949) 579.
25 G. L. Mills, *Biochem. J.*, 50 (1952) 707.
26 J. C. Perrone, *Nature*, 167 (1951) 513.
27 F. C. Green and L. M. Kay, *Anal. Chem.*, 24 (1952) 726.
28 F. S. Steven, *Anal. Biochem.*, 4 (1962) 316.
29 N. A. Matheson, *Biochem. J.*, 88 (1963) 146.
30 A. V. Portugal, R. Green and T. M. Sutherland, *J. Chromatog.*, 12 (1963) 183.
31 K. H. Trueblood and E. W. Malmberg, *J. Am. Chem. Soc.*, 72 (1950) 4112.
32 W. A. Schroeder, *Ann. N.Y. Acad. Sci.*, 49 (1948) 204.
33 M. Brenner, A. Niederwieser and G. Pataki, *Experientia*, 17 (1961) 145.
34 G. Pataki and M. Keller, *Helv. Chim. Acta*, 47 (1964) 787.
35 F. Turba and G. Gundlach, *Biochem. Z.*, 326 (1955) 322.
36 F. S. Steven, *J. Chromatog.*, 8 (1962) 417.
37 A. L. Levy, *Nature*, 174 (1954) 126.
38 S. Blackburn and A. G. Lowther, *Biochem. J.*, 48 (1951) 126.
39 G. Biserte and R. Osteux, *Bull. Soc. Chim. Biol.*, 33 (1951) 50.
40 M. Rovery and C. Fabre, *Bull. Soc. Chim. Biol.*, 35 (1953) 541.
41 F. Sanger and H. Tuppy, *Biochem. J.*, 49 (1951) 463.
42 B. A. Askonas, P. N. Campbell, G. Godin and T. S. Work, *Biochem. J.*, 61 (1955) 105.
43 P. Desnuelle and C. Fabre, *Biochim. Biophys. Acta*, 18 (1955) 49.

[44] R. Monier and M. Jutisz, *Biochim. Biophys. Acta*, 15 (1954) 62.
[45] W. A. Schroeder, *J. Am. Chem. Soc.*, 74 (1952) 5118
[46] B. Keil, *Collection Czech Chem. Commun.*, 23 (1958) 740.
[47] W. R. Gray and B. S. Hartley, *Biochem. J.*, 89 (1963) 59P.
[48] A. A. Boulton and I. E. Bush, *Biochem. J.*, 92 (1964) 11P.
[49] W. R. Gray and B. S. Hartley, *Biochem. J.*, 89 (1963) 379.
[50] M. Bergmann, E. Kann and A. Mieckely, *Ann.*, 458 (1927) 56.
[51] E. Abderhalden and H. Brockman, *Biochem. Z.*, 225 (1930) 386.
[52] P. Edman, *Acta Chem. Scand.*, 4 (1950) 277, 283.
[53] P. Edman, *Acta Chem. Scand.*, 7 (1953) 700.
[54] P. Edman, *Acta Chem. Scand.*, 10 (1956) 761.
[55] P. Edman and K. Lauber, *Acta Chem. Scand.*, 10 (1956) 466.
[56] V. M. Ingram, *J. Chem. Soc.*, (1953) 3717.
[57] H. Fraenkel-Conrat, J. I. Harris and A. L. Levy, *Methods Biochem. Anal.*, 2 (1955) 383.
[58] J. Sjöquist, *Biochim. Biophys. Acta*, 41 (1960) 20.
[59] J. Sjöquist, *Arkiv Kemi*, 11 (1957) 129.
[60] D. F. Elliott and W. S. Peart, *Biochem. J.*, 65 (1957) 246.
[61] J. Sjöquist, *Arkiv Kemi*, 11 (1957) 151.
[62] S. Eriksson and J. Sjöquist, *Biochim. Biophys. Acta*, 45 (1960) 290.
[63] H. Fraenkel-Conrat and J. Fraenkel-Conrat, *Acta Chem. Scand.*, 5 (1951) 1409.
[64] M. Ottesen and A. Wollenberger, *Compt. Rend. Trav. Lab. Carlsberg, Ser. Chim.*, 28 (1953) 463.
[65] H. Fraenkel-Conrat and J. I. Harris, *J. Am. Chem. Soc.*, 76 (1954) 6056.
[66] D. F. Elliott and W. S. Peart, *Biochem. J.*, 65 (1957) 246.
[67] C. H. W. Hirs, S. Moore and W. H. Stein, *J. Biol. Chem.*, 235 (1960) 633.
[68] W. Konigsberg and R. J. Hill, *J. Biol. Chem.*, 237 (1962) 2547.
[69] H. Fraenkel-Conrat, *J. Am. Chem. Soc.*, 76 (1954) 3606.
[70] J. I. Harris and P. Roos, *Biochem. J.*, 71 (1959) 434.
[71] J. I. Harris and C. H. Li, *J. Am. Chem. Soc.*, 76 (1954) 3607.
[72] R. D. Cole, I. I. Geschwind and C. H. Li, *J. Biol. Chem.*, 224 (1957) 399.
[73] J. Sjöquist, B. Blombäck and P. Wallen, *Arkiv Kemi*, 16 (1960) 425.
[74] C. F. Jacobsen, J. Leonis, K. Linderstrøm-Lang and M. Ottesen, *Methods Biochem. Anal.*, 4 (1957) 171.
[75] I. Sjöquist and J. Sjöquist, *Biochim. Biophys. Acta*, 71 (1963) 210.
[76] E. O. P. Thompson, *Advan. Org. Chem.*, 1 (1960) 149.
[77] F. Sanger, *J. Polymer Sci.*, 49 (1961) 3.
[78] R. W. Holley and A. D. Holley, *J. Am. Chem. Soc.*, 74 (1952) 5445.
[79] V. M. Ingram, *Biochim. Biophys. Acta*, 20 (1956) 577.
[80] M. Jutisz and W. Ritschard, *Biochim. Biophys. Acta*, 17 (1955) 548.
[81] H. Fasold, G. Steinkopff and F. Turba, *Z. Physiol. Chem.*, 335 (1961) 1.
[82] E. Scoffone, E. Vianello and A. Lorenzini, *Gazz. Chim. Ital.*, 87 (1957) 354.
[83] A. L. Levy, *J. Chem. Soc.*, (1950) 404.
[84] J. Leonis and A. L. Levy, *Compt. Rend. Trav. Lab. Carlsberg, Ser. Chim.*, 29 (1954) 57, 87.

85 G. W. Kenner and H. G. Khorana, *J. Chem. Soc.*, (1952) 2076.
86 D. T. Elmore and P. A. Toseland, *J. Chem. Soc.*, (1954) 4533.
87 A. C. Chibnall and P. F. Spahr, *Biochem. J.*, 68 (1958) 135.
88 S. Korman and H. T. Clarke, *J. Biol. Chem.*, 221 (1956) 113.
89 A. G. Long, J. R. Quayle and R. J. Stedman, *J. Chem. Soc.*, (1951) 2197.
90 S. Korman and H. T. Clarke, *J. Biol. Chem.*, 221 (1956) 133.
91 O. Folin, *J. Biol. Chem.*, 106 (1934) 311.
92 R. E. Bowman and H. H. Stroud, *J. Chem. Soc.*, (1950) 1342, 1349.
93 V. M. Ingram, *J. Biol. Chem.*, 202 (1953) 193.
94 H. Kiessling and J. Porath, *Acta Chem. Scand.*, 8 (1954) 859.
95 R. L. Hill and E. L. Smith, *J. Biol. Chem.*, 231 (1958) 117.
96 R. L. Hill, D. H. Spackman, D. M. Brown and E. L. Smith, *Biochem. Prep.*, 6 (1958) 35.
97 E. L. Smith and R. L. Hill, *Enzymes*, 4 (1960) 37.
98 E. L. Smith and R. L. Hill, In: *The Enzymes*, Vol. 4, P. D. Boyer, H. Lardy and K. Myrbäck (*Eds.*) Academic Press, 1960, p. 37.
99 G. W. Irving, Jr., J. S. Fruton and M. Bergmann, *J. Biol. Chem.*, 138 (1941) 231.
100 G. R. Stark and D. G. Smyth, *J. Biol. Chem.*, 238 (1963) 214.
101 R. J. Pollitt, *Chem. Comm.*, (1965) 262.
102 A. A. Boulton, *Federation of European Biochemical Societies Meeting, Abstracts*, Vienna, 1965, A 265.
103 G. F. Bryce anf B. R. Rabin, *Biochem. J.*, 90 (1964) 509.

CHAPTER 7

C-TERMINAL SEQUENCE DETERMINATION

CLEAVAGE OF C-TERMINAL GROUPS WITH CARBOXYPEPTIDASE

Carboxypeptidase has been prepared from beef pancreatic glands by a somewhat lengthy process based on the original procedure of Anson[1]. Neurath *et al.*[2] have recently described an improved method for the preparation of carboxypeptidase A from a stable acetone powder of pancreas glands. This enzyme is considerably more soluble in 1 *M* NaCl at neutral pH than is the enzyme isolated by the Anson[1] procedure. The chemical relationships among the various forms of bovine pancreatic carboxypeptidase A have been discussed by Sampath Kumar *et al.*[3]. Commercial samples usually require treatment with diisopropylphosphofluoridate (DFP) to inhibit traces of chymotrypsin, trypsin or subtilisin. Porcine pancreatic carboxypeptidase A exists in three electrophoretically distinct active forms essentially identical in enzymic properties[4,5].

Carboxypeptidase is practically insoluble in water and in dilute salt solutions. Before use it is usually washed 3 or 4 times with ice-cold water. A convenient way of preparing solutions for end group analysis[6] is to mix about 0.25 ml of the aqueous suspension of the enzyme with 6 ml of 2 *M* sodium chloride and gradually raise the pH of the solution to 10 by addition of 0.01 *N* NaOH from a microburette. As soon as the enzyme dissolves, 0.6 ml of *M* sodium phosphate buffer at pH 8.0 is added bringing the pH to 8.3. Any inactive protein that separates on standing the solution at 4° for 2 h is centrifuged off in the cold. The enzyme concentration may be estimated spectrophotometrically from the data

$$E^{280}_{1\,\mathrm{cm}}/0.0123 = \text{concn. of carboxypeptidase in } \mu\text{g protein N/ml}$$

$$E^{280}/E^{260} = 1.86$$

The *p*H optimum of carboxypeptidase appears to vary considerably with the nature of the C-terminal residue exposed to its action and values from pH 5.5 to 9.2 have been reported. It has even been found possible to utilize the differential action of the enzyme at different pH values to prepare modified proteins. Thus Davie *et al.*[7] were able to isolate in crystalline form β-lactoglobulin derivatives from which the two terminal iso-leucines had been removed by the enzyme at pH 7.6. Further action of carboxypeptidase at pH 9.2 led to another crystalline derivative from which the two terminal histidines had been removed.

An absolute requirement for attack by the enzyme is a free α-carboxyl group; a chain terminating in an amide group will not be susceptible. A bond involving an aromatic residue such as phenylalanine is readily split but C-terminal arginine, lysine or proline blocks the action of the enzyme. Leucine and valine are split less readily than phenylalanine but more rapidly than residues with smaller or more polar side chains. Terminal residues carrying a free —SH group or an —SO_3H group do not appear susceptible but terminal cystine may be liberated.

It is sometimes possible to obtain further information about the C-terminus by plotting the rates of release of the different amino acids released by the action of the enzyme. Fig. 76 shows the data

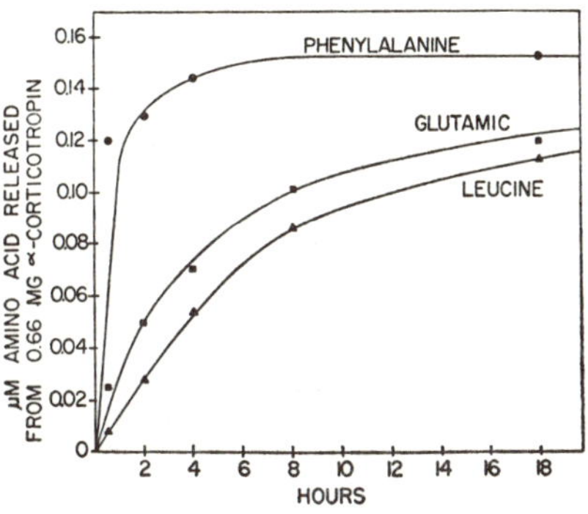

Fig. 76. Rate of release of amino acids from α-corticotropin by the action of carboxypeptidase, estimated as their DNP derivatives. Harris and Li[8].

for α-corticotropin obtained by Harris and Li[8], at an enzyme:substrate ratio of 1.25 and at pH 8.0–8.5 and 37°. Aliquots of the reaction mixture were removed at intervals, the enzyme destroyed with sufficient hydrochloric acid to bring the pH to 3.0–3.5, and the amino acids isolated as their DNP-derivatives at pH 9.0. The C-terminal sequence –Leu.Glu.Phe was established, the enzyme action having ceased at a prolyl residue.

Another carboxypeptidase of pancreatic origin has been describes by Folk and Gladner[9,10] and designated carboxypeptidase B. It acts specifically on peptides containing arginine, lysine or ornithine as the C-terminal group. Table 36 shows the relative rates of release

TABLE 36

AMINO ACIDS RELEASED FROM THE α-CHAIN IN μMOLES/μMOLE α-CHAIN[11]

(CBP-B added initially; CBP-A added after 65 min.)

Time (min)	*CBP-B*		*CBP-B+CBP-A*	
	35	65	105	255
Arginine	0.91	0.92	0.97	0.96
Tyrosine	0.86	0.86	0.85	0.89
Lysine	0.47	0.77	0.90	1.02
Serine	0.04	0.10	0.85	1.42
Threonine	—	0.03	0.35	1.40
Leucine	—	0.02	0.41	1.20
Valine	—	—	0.29	0.93
Alanine	—	—	0.15	0.28
Phenylalanine	—	—	0.12	0.14
Histidine	—	—	0.11	0.10

of amino acids from normal adult human haemoglobin α-chain in the presence of carboxypeptidase B alone, and combined with carboxypeptidase A[11]. The conditions were 1.4 mM substrate, 0.1 mM carboxypeptidase-A, 130 units/ml carboxypeptidase-B (assayed against hippuryl-lysine), 0.2 M phosphate buffer, pH 7.65, 0.056 M sodium lauryl sulphate (to solubilize the chain), 25°. The sequence –Ser.Lys.Tyr.Arg. was established for the α-chain and under similar conditions the sequence for the β-chain was found to be –(His.Ala) Lys.Tyr.His. Carboxypeptidase B from porcine pancreas has also been described[12,13]; its specificity differs from that recently reported for the analogous bovine enzyme which was found active towards basic and aromatic compounds[14].

A C-terminal sequence for carboxypeptidase B was determined by Folk *et al.*[15]. The enzyme was first denatured with 0.2% sodium dodecylsulphate, treated with the inhibitor *o*-phenanthroline (0.005 *M*) and dialyzed exhaustively against 0.02 *M* veronal buffer, pH 7.5, containing 0.1 *M* sodium chloride. The denatured enzyme was incubated at 25° with carboxypeptidase A (0.5% substrate; 80:1 enzyme substrate ratio) in the presence of 0.001 *M* diisopropylphosphofluoridate. The sequence –(Val,Ser).Asp(NH_2).Thr. was established. The same material treated with hydrazine followed by dinitrophenylation (see p. 226) afforded 0.9 mole DNP-threonine and 0.10–0.15 mole of DNP-serine per mole.

Difficulties in interpretation arise when it is not known for certain that a protein is a single peptide chain. Also there is always the possibility that the C-terminal sequence consists of two similar residues (*e.g.* –Phe.Phe in bovine growth hormone)[16]. A number of proteins exist which are not attacked by the enzyme and the presence of a C-terminal residue has then been established using another reagent. Yeast triosephosphate dehydrogenase is stable to the enzyme but is digested in strong urea solution[17].

Unambiguous identification of the C-terminal residue should be possible with the new method of Kowalsky and Boyer[18] which is based on carboxypeptidase digestion in $H_2{}^{18}O$. It will be observed from the scheme below that amino acids released from the C-terminus do not contain excess ^{18}O, whereas those from interior positions contain excess ^{18}O derived from the $H_2{}^{18}O$.

Hydrolysis of a peptide chain in $H_2{}^{18}O$

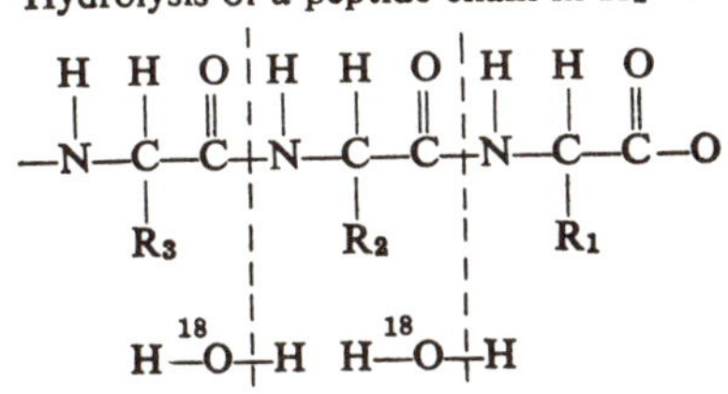

The technique has the advantage that quantitative recovery of the released amino acid is not necessary.

HYDRAZINOLYSIS METHOD OF AKABORI

The method is based on the discovery of Akabori *et al.*[19] that the C-terminal amino acids of a protein are liberated by treatment with anhydrous hydrazine. All other amino acids in peptide linkage are converted to the amino acid hydrazides which may be condensed with benzaldehyde to form water-insoluble dibenzal derivatives. The C-terminal amino acids remaining in the aqueous solution can then be identified by any suitable method.

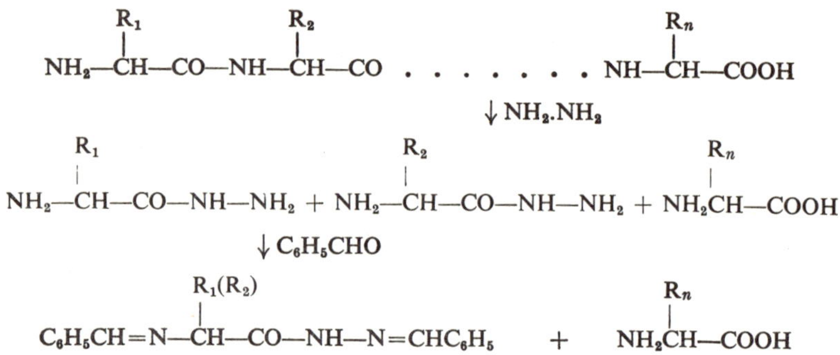

Anhydrous hydrazine may be prepared by distilling 80% hydrazine hydrate with calcium oxide[20]. Kusama[21] has described the following preparation: 50 g of 80% hydrazine hydrate, 500 g toluene and 500 g calcium oxide are shaken together and allowed to stand overnight. The mixture is then refluxed for 3 h with protection from atmospheric moisture, and distilled as the azeotropic mixture boiling at 93–94°. The lower layer of the distillate is used as anhydrous hydrazine without further purification. The reagent can be checked and amino acid hydrazides estimated by titration with bromate[22] according to the equation

$$\mathrm{RCH(NH_2)CONHNH_2} + \tfrac{2}{3}\,\mathrm{KBrO_3} \rightarrow \mathrm{RCH(NH_2)COOH} + \mathrm{N_2} + \mathrm{H_2O} + \tfrac{2}{3}\,\mathrm{KBr}$$

The hydrazides give a red-purple colour with ninhydrin but can be distinguished from amino acids on a paper chromatogram by the dark brown or black spots produced after spraying with *n*-butanol saturated with ammoniacal silver nitrate solution. The solvent used by Akabori for the separation of mixtures of amino acids and their hydrazides was pyridine bases (b.p. 130–150°)–aniline–water (9:1:4 by vol.).

In an improved procedure[23] dinitrophenylation converts the amino acids to the corresponding DNP-derivatives and the hydrazides to the corresponding di-DNP-derivatives. Thus 1 μM of protein is hydrolyzed with about 1 g of anhydrous hydrazine (99–100%) at 100° for 5–10 h. The mixture is evaporated carefully *in vacuo* over concentrated sulphuric acid, the residue dissolved in 15 ml of 2% sodium bicarbonate solution or 0.2 *M* phosphate buffer, pH 8, and a solution of 0.3 ml FDNB in 30 ml of ethanol added. The reaction mixture is shaken for 2 h in the dark at room temperature, 45 ml of water are added and the solution acidified with dilute hydrochloric acid. The DNP-derivatives are extracted into ethyl acetate and the acidic DNP-derivatives extracted in turn from the ethyl acetate with 2% bicarbonate solution. C-terminal derivatives and those of glutamic acid and aspartic acid are precipitated from the aqueous solution by dilute acid, extracted again into ethyl acetate and chromatographed on silica gel. The DNP-derivatives of the acidic amino acid mono-hydrazides can be characterized as the DNP-amino acids after hydrolysis with 6 *N* HCl for 5 h at 100°.

An alternative method after the removal of the excess hydrazine is to dissolve the residue equivalent to 0.5 μM protein in 3 ml 0.2 *M* phosphate buffer, pH 7.3, and treat with 0.2 ml isovaleraldehyde for 30 min at 0° with shaking. The precipitate is filtered off and washed well. The filtrate and washings, which are now acid, are retreated with the aldehyde at pH 7–7.5, and the final filtrate is extracted with ether and treated with sodium bicarbonate and FDNB. Kusama has recommended the use of enanthaldehyde which does not usually form crystalline hydrazides and permits cleaner separation of the upper layer from the lower aqueous layer.

The recovery of C-terminal residues as the free amino acids or as the DNP-derivatives varies considerably with the conditions of hydrazinolysis. In a survey of the end-groups of samples of serum albumin obtained from different species of animal the recovery of DNP-alanine (horse, bovine, goat and sheep) was 40–60% and that of DNP-leucine (human, dog, rabbit) was 45–65%[21].

In a further extension of the method Ohno[24] and Niu and Fraenkel-Conrat[25,26] introduced the principle of partial hydrazinolysis which provides additional information about the C-terminal sequence. The procedure leads to a mixture of di-DNP-peptide hydrazides and DNP-peptides; acidic DNP-derivatives are formed from both the C-terminal peptide groups and from non-C-terminal

References p. 248

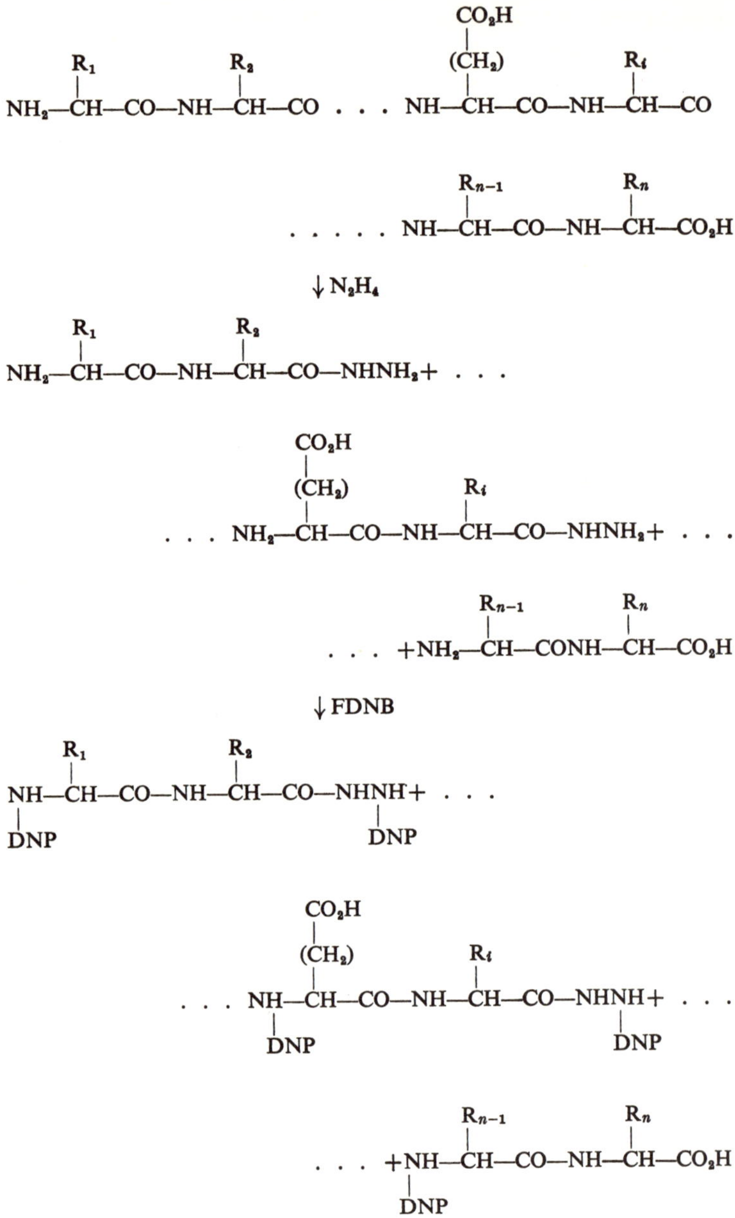

peptide fragments containing the acidic amino acids. This is illustrated in the scheme on p. 228.

Haruna and Akabori[27] isolated 35 acidic DNP-peptide derivatives from a partial hydrazinolysis of ovalbumin (20 g) conducted in hydrazine (100 g) at 100° for 3 h. Under these conditions only 0.3 mole of C-terminal amino acid appeared in the free state. It was found convenient to use the counter-current distribution method for the bulk separation of the large amounts of DNP-peptide mixtures and then to proceed to column chromatography. Pretreatment of the mixture was avoided to preserve the acidic peptide hydrazides. An interesting finding is the apparent lability of the peptide bonds linked with the acidic amino acids and giving rise to a high proportion of peptides containing N-terminal aspartic or glutamic acids.

Hydrazinolysis as an end-group method suffers from the serious disadvantages of low recoveries and the complete loss, if C-terminal, of the residues cystine, cysteine, asparagine and glutamine[28]. Methionine is converted to methionine sulphoxide, arginine to ornithine in moderate yield, and aspartic acid and glutamic acid are partially destroyed. Table 37 gives the recovery of amino acids as DNP-derivatives after a 10 hour hydrazinolysis as determined by Niu and Fraenkel-Conrat[25] who also report that addition of water up to 10% to the hydrazine failed to produce spurious end groups from a number of proteins.

The severity of the treatment with hydrazine can be overcome to a certain extent by the use of an acid catalyst. In 1 *M* hydrazine sulphate at 60° the reaction is about five times as fast as in anhydrous hydrazine and less destruction of some amino acids occurs. Bradbury[29] has compared the relative rates of acid-catalyzed hydrazinolysis and hydrolysis reactions for various peptides, and it is clear that the wide range of reactivity of different peptide bonds towards hydrazine might lead to a simpler mixture of peptide fragments than would be obtained after partial hydrolysis.

REDUCTION OF C-TERMINAL GROUPS TO AMINO ALCOHOLS

Karrer *et al.*[30], in a series of papers, first demonstrated the elegance of using lithium aluminium hydride in the preparation of amino alcohols from amino acid esters. The ability of this powerful re-

TABLE 37

RECOVERY OF AMINO ACIDS AS DNP-DERIVATIVES AFTER HYDRAZINOLYSIS (10 h PERIOD)[25]

	Average	*Range*
Alanine	0.45	0.38–0.53
Arginine!	0.25	0.18–0.32
Aspartic acid	0.19	0.13–0.25
Glutamic acid	0.16	0.11–0.23
Glycine	0.37	0.25–0.47
Histidine	0.43	0.31–0.47
Hydroxyproline	0.50	0.43–0.51
Isoleucine	0.54	0.49–0.60
Leucine	0.54	0.46–0.58
Lysine	0.47	0.35–0.56
Methionine[b]	0.56	0.50–0.61
Phenylalanine	0.59	0.54–0.64
Proline	0.52	0.43–0.61
Serine	0.39	0.34–0.51
Threonine	0.52	0.45–0.59
Tryptophan	0.39	0.30–0.50
Tyrosine	0.44	0.35–0.49
Valine	0.76	0.72–0.82

[a] Converted to and estimated as di-DNP-ornithine.

[b] Recovered as DNP-methionine sulphoxide when benzaldehyde treatment was involved.

ducing agent to convert free carboxyl groups to primary alcohols was made the basis of an attempt by Fromageot *et al.*[31,32] to determine the C-terminal end groups of insulin. A suspension of the protein in N-ethylmorpholine containing a cationic detergent was treated with an 8-fold excess of lithium aluminium hydride at 55° for a period of eight hours. Extraction of the neutralized acid hydrolysate with ether afforded ethanolamine and propanolamine, corresponding to glycine and alanine terminal residues; β-amino-γ-hydroxybutyric acid corresponding to an asparagine residue was later identified[33] and the ethanolamine was considered an artifact.

In an independent investigation Chibnall and Rees[34] studied the reduction of insulin methyl ester with lithium borohydride, a much milder reducing agent, and a true appraisal of the method was possible following the publication by the Cambridge school of precise analytical data drawn from a study extending over many years. It is obvious now that the determination of the end groups of proteins of molecular weight above 10,000 will be complicated by the appearance of non-terminal amino alcohols arising from reductive cleavage of peptide bonds. This will be seen from the following scheme:

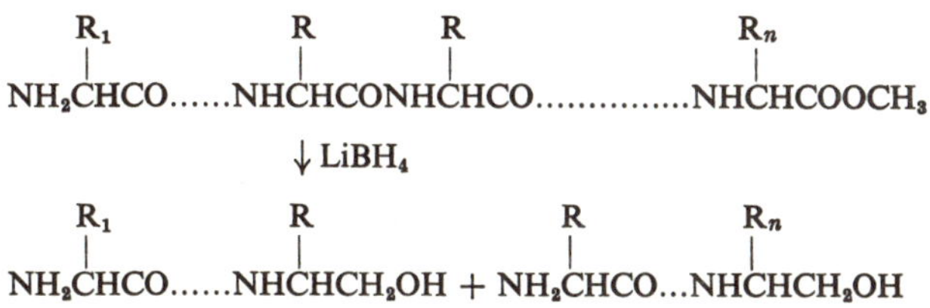

Reductive fission by lithium borohydride is selective. In the insulin molecule one glycine residue is particularly susceptible. Peptide links at the glutamine and asparagine residues appear to be labile in a number of proteins. Table 38 gives a list of the various dicarboxylic acid, dicarboxylic acid amide, and C-terminal residues present in an intact protein together with a list of the corresponding products present in the hydrolysate of the reduced protein ester. Table 39 shows an analysis of the products formed from insulin and from insulin esterified with methanolic 0.1 *N* HCl for 24 h. Table 39 does not include possible highly basic products that might arise from the complete reduction of a peptide bond.

When allowances are made for the small quantities of amino alcohols produced by reductive cleavage of intact insulin the true end groups are clearly recognizable as one residue of alanine and one residue of asparagine. Although the side reactions occur only to the extent of 1–2% of the total peptide bonds the interference is serious in the case of high molecular weight proteins. Nevertheless the procedure using lithium borohydride may well prove of value in protein chemistry and a few practical details of the technique are included in this section.

It will be observed from Table 38 that C-terminal asparagine and glutamine will undergo reduction at the esterified α-carboxyl group to give after hydrolysis, γ-β-HAB and δ-γ-HAV respectively. On the other hand the same residues occurring in the peptide chain will remain intact during the reduction and will appear as aspartic and

TABLE 38

COMPARISON BETWEEN THE VARIOUS DICARBOXYLIC ACID, DICARBOXYLIC ACID AMIDE, AND C-TERMINAL RESIDUES PRESENT IN THE INTACT PROTEIN, WITH THE CORRESPONDING PRODUCTS PRESENT IN THE HYDROLYSATE OF THE REDUCED PROTEIN ESTER[35]

Residue in protein		*Product expected in hydrolysate*
Glutaminyl,	N-terminal or in chain	Glutamic acid
Isoglutaminyl,	N-terminal or in chain	Glutamic acid
Glutamine,	C-terminal	δ-hydroxy-γ-aminovaleric acid (δ-γ-HAV)
Isoglutamine,	C-terminal	δ-hydroxy-α-aminovaleric acid (δ-α-HAV)
α-Glutamyl,	N-terminal or in chain	δ-hydroxy-α-aminovaleric acid (δ-α-HAV)
γ-Glutamyl,	N-terminal or in chain	δ-γ-HAV
Glutamic acid,	C-terminal	Glutamidiol
Asparaginyl,	N-terminal or in chain	Aspartic acid
Isoasparaginyl,	N-terminal or in chain	Aspartic acid
Asparagine,	C-terminal	γ-hydroxy-β-aminobutyric acid (γ-β-HAB) or lactone
Isoasparagine,	C-terminal	γ-hydroxy-α-aminobutyric acid (γ-α-HAB) or lactone
α-Aspartyl,	N-terminal or in chain	γ-hydroxy-α-aminobutyric acid (γ-α-HAB) or lactone
β-Aspartyl	N-terminal or in chain	γ-hydroxy-β-aminobutyric acid (γ-β-HAB) or lactone
Aspartic acid	C-terminal	Aspartidiol
Monoamino or basic amino acid (terminal)		β-amino alcohol
Monoamino or basic amino acid amide		Monoamino or basic amino acid

TABLE 39

C-TERMINAL RESIDUES IN INSULIN (MOLES/MOLE PROTEIN)[35]

Material	*Leucinol*	*Tyrosinol*	*Alaninol*	*Asparti-diol*	*Ethanol-amine*	*γ-β-HAB*	*δ-γ-HAV*
Insulin methyl ester	0.1	0.027	1.110	0.08	0.19	0.89	0.18
Insulin	0.048	0.023	0.156	—	0.161	0.08	0.16
Difference	0.052	0.004	0.955	0.08	0.029	0.81	0.02

glutamic acids after hydrolysis. Aspartyl and glutamyl residues in peptide chains however will undergo reduction at esterified ω-carboxyl groups to give γ-α-HAB and δ-α-HAV respectively. Finally, C-terminal aspartic and glutamic acids will give diols on reduction. Chibnall *et al.*[36] have devised a procedure for the determination of amide distribution in proteins based on the use of lithium borohydride and utilizing the periodate oxidation procedure for 1,2-hydroxyamines and diols. The four products aspartic and glutamic acids, δ-γ-HAV and γ-β-HAB can be separated on a column of Dowex-50 by elution with N-hydrochloric acid. δ-α-HAV and γ-α-HAB are computed by difference. Table 40 gives an analysis of the asparagine and glutamine residues in different proteins expressed as residues/mole of protein, allowance having been made for amide loss during esterification.

TABLE 40

NUMBER OF ASPARAGINE AND GLUTAMINE RESIDUES IN DIFFERENT PROTEINS[36]

Residue	*β-Lactoglobulin (mol.wt.37,000)*	*β_1-Lactoglobulin (mol.wt. 37,000)*	*Lysozyme (mol. wt. 14,700)*	*Ox insulin (mol. wt. 5,732)*
Asparaginyl	10	10	13	3
Glutaminyl	18	18	5	3
Aspartyl	21	21	7	-
Glutamyl	32	32	—	4

A step-wise degradation of peptides from the carboxyl end based on N,O migration of acyl groups in conjunction with a reduction procedure was envisaged by Bailey[37]. In this scheme the selective removal of C-terminal amino alcohols from peptides requires two steps.

References p. 248

N-Peptidylamino alcohols formed from the corresponding peptide esters by reduction with a metal hydride are first treated with an acid or acid chloride to bring about N,O acyl migration. The resulting β-amino esters are then treated with a metal hydride to liberate the C-terminal amino alcohols.

$R_1CONHCHR_2CONHCHR_3COOCH_3$	Peptide ester
$\downarrow$ Reduction	
$R_1CONHCHR_2CONHCHR_3CH_2OH$	N-peptidylamino alcohol (β-hydroxyamide)
$\downarrow$ Acid chloride	
$R_1CONHCHR_2COOCH_2CHR_3NH_3^+$	O-peptidylamino alcohol (β-amino ester)
$\downarrow$ Reduction	
$R_1CONHCHR_2CH_2OH + HOCH_2CHR_3NH_2$	

N,O-acyl migration in β-hydroxyamides derived from N-toluene-*p*-sulphonylpeptide esters has been achieved in 85–90% yield using phosphorus oxychloride at room temperature. Reductive cleavage of the β-amino esters with lithium borohydride has afforded termi-

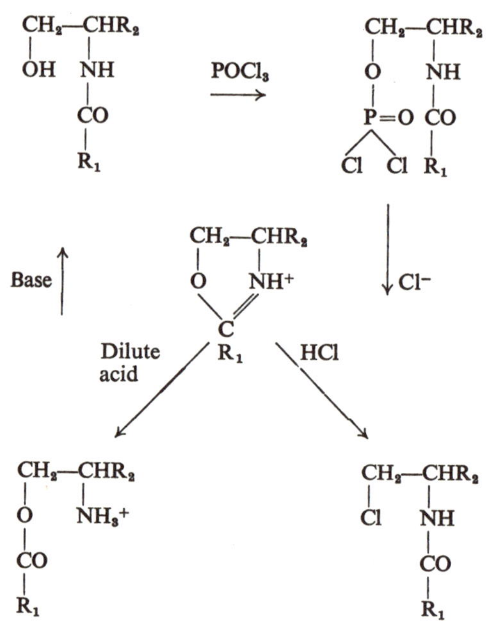

nal amino alcohols in 85–90% yield. The residual peptide at this stage is simultaneously converted into a new β-hydroxyamide capable of rearrangement with phosphorus oxychloride.

The mechanism of the cyclization of hydroxyamides is discussed in another connexion on page 151; the scheme shown on page 234 illustrates the relationship between the different compounds and also the origin of the small amounts of β-chloroethylamides that might contaminate the hydroxyamides after each cycle.

Procedure for the reduction of esters of proteins and peptides

Tetrahydrofuran

The solvent is refluxed over lithium aluminium hydride, distilled and stored in a dark bottle. It rapidly peroxidizes and samples that have been kept longer than two weeks should be redistilled over lithium aluminium hydride before use. It has a boiling point of 65°.

Lithium aluminium hydride

Great care must be taken with this reagent as it is spontaneously inflammable in moist air. Transfer of the reagent should be carried out in a dry-air cabinet and large lumps are best left alone or broken up very gently with the aid of a steel needle.

Lithium borohydride

Solutions between 1 and 2.5 *M* are prepared by refluxing the reagent (rough weight judged by bulk) with 30–40 ml of anhydrous tetrahydrofuran for 2 h. With the aid of a filter-stick the cooled solution is sucked into a bottle from which the air has been displaced by dry nitrogen. The apparatus and containers must be protected from atmospheric moisture. If the sample of lithium borohydride is grossly impure it is best to extract with ether in a centrifuge bottle and centrifuge off the insoluble material. On removal of most of the solvent the reagent crystallizes out and will then dissolve in tetrahydrofuran to give a clear solution. The reducing power of the solutions may be assessed by titration of the lithium with standard acid, or more accurately by the manometric method of Krynitsky *et al.*[38] depending on the volume of hydrogen liberated on decomposition with acid.

Esterification of proteins and peptides[39]

The protein (250 mg) is placed in a 50-ml flask and dried overnight *in vacuo* at room temperature over phosphorous pentoxide and sodium hydroxide. About 20 ml of anhydrous methanol is added followed by methanolic hydrogen chloride dropwise until the protein dissolves. Sufficient excess of both methanolic hydrogen chloride and methanol is then added to bring the final volume to 25 ml and the acid concentration to 0.1 *N*. The protein will sometimes precipitate at this stage but this is not disadvantageous. The flask is immersed in a water bath at 25° and shaken slowly during 24 h. About 60 ml of ether is then added and after standing for a short time the precipitated ester hydrochloride is removed by centrifuging, washed twice with ether (50 ml), dried *in vacuo* over P_2O_5 and NaOH and then equilibrated in the air at room temperature.

During the incubation period of 24 h at 25° some alcoholysis of the amide groups occurs; with insulin this was found to be about 6.6% of the amide nitrogen. Also there is a tendency during the esterification for the serine and threonine residues to participate in N,O acyl migration. The latter phenomenon does not appear to take place when the protein ester is made with diazomethane. Peptides of high molecular weight are best treated as for proteins. The methyl esters of simple peptides are conveniently prepared by dissolution of the free peptides in saturated methanolic hydrogen chloride and concentration of the solution *in vacuo* at room temperature[40].

Reduction of protein esters

The following is the procedure worked out by Chibnall and Rees[35]. The protein ester (100 mg) is placed in an open reaction vessel (10 × 2 cm) and dried *in vacuo* for 2 h at 50°. A three-necked multiple adapter is fitted to the reaction vessel, a mercury-sealed stirrer and a condenser with a soda-lime trap are placed in position and a standard solution of lithium borohydride introduced through the remaining free socket. An eight-fold molar excess of reagent is required for full reduction of protein esters that are insoluble in tetrahydrofuran. Allowance must be made also for consumption of reagent by bound HCl and by disulphide linkages present. The suspension of protein in about 2 ml of tetrahydrofuran (containing

10–20 mg reagent) is refluxed gently on an oil bath for 6 h, cooled, and treated with a slight excess of methanolic hydrogen chloride with stirring for 1 h. Acetone–ether (1:1, v/v; 8 ml) is then added and the reduced protein removed by centrifuging.

The β-amino alcohols present in an acid hydrolysate of the reduced protein are separated by passage through a column (1 $\times$ 10 cm) of Dowex 2-X10 which has been regenerated with carbonate-free *N* NaOH and then thoroughly washed with CO_2-free water. A sample containing about 30 mg of protein hydrolysate in a volume of 1 ml is transferred to the column and the amino alcohols eluted with 25 ml CO_2-free water. δ-γ-HAV and γ-β-HAB arising from C-terminal glutamine or asparagine are retained on the column.

Reduction of peptide esters

Peptide esters generally remain in solution provided the lithium borohydride solution is not less than 1 molar. The procedure is best illustrated by an example[41].

L-Leucylglycylglycine ethyl ester hydrochloride (160 mg) is dried over P_2O_5 and NaOH and then *in vacuo* at 70° for 3 h. Dry nitrogen is admitted into the flask followed by anhydrous tetrahydrofuran (0.4 ml) and 2.3 *M* lithium borohydride solution (0.47 ml; 2 mole proportions). The clear solution is left for 17 h at room temperature. The mixture is then cooled to $-10°$, decomposed with *N* HCl (1.1 ml), left for a few hours and neutralized to bromothymol blue with *N* lithium hydroxide. The solution is concentrated to dryness to expel all gaseous products, taken up in 1 ml of water, and applied to a short column (1.5 $\times$ 6.5 cm) of Dowex 50-X8 (200 mesh). On elution with 0.1 *N* ammonium hydroxide the β-hydroxyamide formed by the reduction of the peptide ester travels down the column as a visible band adjacent to the neutralization front. Ammonia is removed from the solution by concentration *in vacuo* and the free hydroxyamide converted to the hydrochloride by addition of the calculated amount of ethanolic hydrogen chloride.

The procedure for the same peptide protected by the toluene-*p*-sulphonyl group is as follows: 1.75 *M* lithium borohydride solution (0.34 ml; 2.25 molar proportions) is added to the derivative (100 mg) in anhydrous tetrahydrofuran (0.7 ml) and the clear solution left at room temperature overnight. The mixture is cooled, diluted with 50% ethanol (4 ml) and decomposed with dry Dowex-50

References p. 248

(200 mg) to give a solution of about pH 8. After allowing to stand at this pH for a few hours more resin is added to lower the pH to about 5 and the mixture kept at room temperature for a further 24 h. The resin is filtered off, the filtrate concentrated to dryness *in vacuo* and the boric acid removed as the volatile methyl ester by repeated evaporation with anhydrous methanol–benzene (2:1). A 94% yield of the crystalline β-hydroxyamide is obtained.

Rearrangement to an O-peptidylamino alcohol

A suspension of the above N-peptidyl derivative (114 mg) in acetonitrile (0.2 ml) gives a clear solution on addition of phosphorus oxychloride (0.2 ml). After 3 h at room temperature the solvent is removed *in vacuo*, the intermediate oxazoline decomposed with water, and the solution neutralized with solid barium carbonate. The filtered solution contains the hydrochloride of the N-toluene-*p*-sulphonyl derivative of O-leucylglycylaminoethanol. It crystallizes from ethanol–ethylacetate in 87% yield. On neutralization with alkali a quantitative conversion to the original hydroxyamide occurs.

Reductive cleavage of O-peptidylamino alcohols

Reduction of the O-peptidyl compounds proceeds smoothly under the same conditions as for the peptide esters described above. On addition of water to the reaction mixture the toluene-sulphonyl derivative of the new hydroxyamide separates out and the aqueous solution contains the liberated amino alcohol.

Partition chromatography of amino alcohols (Rees[42])

Useful separations and excellent recoveries of a number of β-amino alcohols have been obtained from a buffered column of silica gel by gradient elution with ethanol-chloroform. The buffer solution, pH 4.85, is made from 184.5 g $Na_2HPO_4.12H_2O$ and 51 g citric acid monohydrate dissolved in water and the volume made to one litre. Solvent A is the lower phase of the mixture, buffer solution–ethanol–chloroform (1:1:1); solvent B is the upper phase of the mixture, buffer solution–ethanol–chloroform (1:2:1). Equilibration of the solvent and operation of the column is carried out at $27.5 \pm 1°$. A column 9.5 cm high is prepared from silica gel (5 g) ground with 2.5 ml of the upper phase from solvent A in the form of a slurry made with 40 ml of the lower phase. Before use the column is washed with 8-hydroxyquinoline (25 mg in 1 ml of solvent A).

The sample containing the β-amino alcohol hydrochloride is dissolved in 1.5 ml of solvent A and applied to the column. Gradient elution is established by feeding solvent B (containing 15 ml ethan nol/100 ml solvent) into a 430-ml mixing flask containing solvent A. The separations obtained by Rees are shown in Fig. 77 and the method could no doubt be developed to separate many more β-amino alcohols.

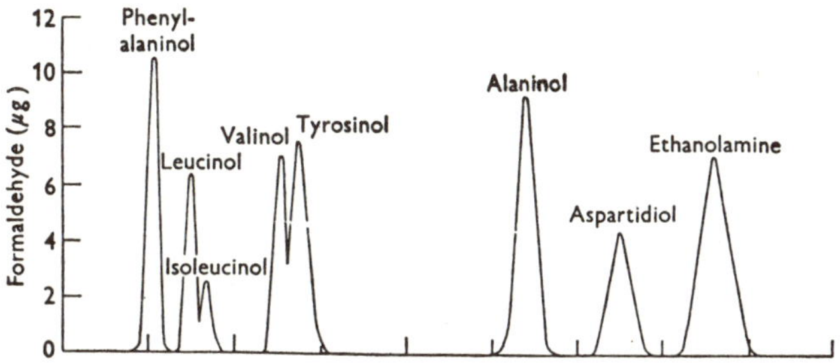

Fig. 77. Chromatographic separation of β-amino alcohols from a synthetic mixture on a column of buffered silica gel (1.3 × 9.5 cm), gradient elution with ethanol–chloroform being used. (From Rees[42].)

Estimation of β-amino alcohols

The method of Rees[42] determines the β-amino alcohol as formaldehyde formed after oxidation with periodate according to the equation

$$R.CHNH_2CH_2OH + HIO_4 \rightarrow NH_3 + HCHO + RCHO + HIO_3$$

A standard formaldehyde curve is first obtained with a series of solutions of serine of concentration 20–100 μg/ml. The following reagents are required.

(1) Periodate solution: 1.07 g sodium periodate is dissolved in 44.7 ml water and 5.3 ml of 2 *N* NaOH added.

(2) Chromotropic acid: 0.9 g of Judex special reagent and 0.5 ml of saturated sodium metabisulphite are adjusted to 25 ml and stored in a dark bottle.

The standard serine solution (0.2 ml) is mixed with 0.1 ml saturated K_2HPO_4 solution and 0.7 ml of water, and then oxidized

TABLE 41

COMPARATIVE R_F VALUES FOR AMINO ALCOHOLS AND AMINO ACIDS IN VARIOUS SOLVENTS[43–45]

	Solvent		
	1	*2*	*3*
Glycine	0.07	0.06	0.33
Ethanolamine	0.18	0.15	0.74
Alanine	0.13	0.12	0.56
Alaninol	0.25	0.23	0.83
Serine	0.06	0.06	0.32
Serinol	0.16	0.15	0.69
Threonine	0.10	0.09	0.48
Threoninol	0.17	0.31	—
Valine	0.30	0.24	0.77
Valinol	0.40	0.49	1
Leucine	0.46	0.44	0.81
Leucinol	0.53	0.63	1
Isoleucine	0.44	0.39	0.82
Isoleucinol	0.50	0.63	1
Proline	0.15	0.14	0.88
Prolinol	0.28	0.26	1

with one drop (approx. 0.02 ml) of periodate solution during 4 min. One drop of saturated metabisulphite solution is added to destroy the excess periodate followed by 0.1 ml of the chromotropic acid solution. After cooling in ice 4 ml of sulphuric acid (98–100%, N-free) is added slowly from a burette, the tubes shaken and the colour developed by heating in a boiling water bath for 10 min. The tubes are cooled again, diluted with 5 ml of water and the extinction values measured at 570 mμ.

The effluent fractions from the silica column are analyzed in the same way after rendering acid with one drop of 4% H_3PO_4 in ethanol and evaporation almost to dryness in a water bath.

TABLE 41 (continued)

	Solvent		
	1	*2*	*3*
Phenylalanine	0.39	0.44	0.82
Phenylalaninol	0.54	0.70	1
Tyrosine	0.21	0.24	0.59
Tyrosinol	0.40	0.54	0.81
Aspartic acid	0.05	0.01	0.17
α-Amino-γ-hydroxy-butyric acid*	0.10	—	0.57
β-Amino-γ-hydroxy-butyric acid*	0.10	—	0.57
Aspartidiol	0.19	0,23	0.82
Glutamic acid	0.09	0.01	0.23
α-Amino-δ-hydroxy-valeric acid*	0.14	—	0.65
γ-Amino-δ-hydroxy-valeric acid*	0.15	—	0.67
Glutamidiol	0.21	0.23	0.85
Lysine	0.02	0.01	0.53
Lysinol	0.08	0.07	0.78
Histidine	0.03	0.07	0.64
Histidinol	0.08	0.12	—
Arginine**			
Argininol**			

* Can be separated well on Whatman No. 4 paper (Rees).
** Arginine 0.12, Arginol 0.35 in pyridine–collidine–water (60:20:20).
Table 41 is reproduced, with permission, from *The Chemical Structure of Proteins, edited by* G. E. W. Wolstenholme and M. P. Cameron (Ciba Foundation Symposium, J. & A. Churchill Ltd., London, 1953), p. 89.

Paper chromatography of β-amino alcohols

When β-amino alcohols from a hydrolysate are to be identified by paper chromatography the removal of the relatively large amount of salts presents a problem. A very neat solution to this was devised by Rees. In phenol (70 g)–water (30 ml) to which has been added 5% (v/v) ammonium hydroxide(s.g. 0.880), the amino alcohols travel with the solvent front leaving a visible salt spot some distance behind. A 6 cm ascending run in phenol is sufficient and the paper can then be dried at 45° and the section containing the salt cut off. The paper sheet is then ready for chromatography in the usual way.

Useful solvents for the separation of amino alcohols are:

(1) *n*-Butanol–acetic acid–water (77:6:17)

(2) *n*-Butanol saturated with water +0.1% ammonia

(3) Phenol saturated with water +0.1% ammonia

(4) Propanol–water (80:20, v/v)

(5) Pyridine–isoamyl alcohol–water (35:35:27); for isoleucinol, leucinol and phenylalaninol.

Comparative R_F values for β-amino alcohols and amino acids in the first three of these solvents using Whatman No. 1 paper at 20–22° are listed in Table 41. In general the β-amino alcohols run in front of the corresponding amino acids and the ninhydrin colour yield is often only 10–20% of that given by the amino acids. Many of the β-amino alcohols can be oxidized back to the amino acids in fair yield by heating with a suspension of silver oxide for a number of hours at 100°. Silver ions can be removed as silver chloride and the amino acids formed can be adsorbed on a short column of Dowex 2 in the free base form. Elution with *N* HCl will provide a solution of the amino acids for subsequent identification[42].

Column chromatography of DNP-amino alcohols

Ether is the only solvent that can be used with any success for extraction of β-amino alcohols from aqueous mixtures containing amino acids and salts and even in this solvent ethanolamine and aminodiols are practically insoluble. A number of investigators therefore looked into the possibility of their isolation as the 2,4-dinitrophenyl derivatives. Sanger's reagent, 1-fluoro-2,4-dinitrobenzene may be used for this purpose but products other than the straightforward N-substituted derivatives may arise, and a welcome contribution to this problem was the novel use by Crawhall and Elliott[46] of 2,4-dinitrobenzenesulphonate (DNBS) in the presence of sodium carbonate or borax. A hydrolysate of a reduced protein (100 mg) is brought to pH 9.5–10.0 and treated with DNBS (1 g)

TABLE 42

COMPOSITION OF SOLVENT SYSTEMS (PARTS BY VOLUME) FOR ELUTION OF DNP-AMINO ALCOHOLS[46]

	Solvent system No.								
	1	*2*	*3*	*4*	*5*	*6*	*7*	*8*	*9*
Ethyleneglycol	2	2	Excess	2	2	2	—	—	—
Benzene	20	20	Excess	—	20	—	—	20	20
Cyclohexane	—	—	—	20	20	—	—	—	—
Acetone	2	1	—	1	—	—	—	1	1
Chloroform	—	—	—	—	—	12	Excess	—	—
Carbon tetrachloride	—	—	—	—	—	12	—	—	—
Glycerol	—	—	—	—	—	—	Excess	—	—
Trimethylene glycol	—	—	—	—	—	—	—	2	—
Tetramethylene glycol	—	—	—	—	—	—	—	—	2

in the presence of borax (2.5 g) in 5 ml of water at 95° for 20 h in a sealed tube. At the end of the reaction the DNP-amino alcohols are extracted with ether.

A number of good methods for the column chromatography of DNP-amino alcohols are available[47–49]; the following is that of Crawhall and Elliott[46] using Hyflo Super-Cel as a support and the solvent systems described in Table 42.

The Hyflo Super-Cel (Johns-Manville, Co., London) is made into a thin cream with 6 *N* HCl and the mixture allowed to stand overnight. It is then filtered, treated twice more in the same way, washed free from chloride and dried at 100°. Before use the solid is ignited at 600° for 8 h. A column (1.2 × 35 cm) is poured from a slurry of Hyflo Super-Cel (6 g) in the stationary phase (3 g) of the solvent and packed down with a steel plunger. The columns are allowed to

References p. 248

become just dry before application of the sample (1 ml) and are operated under slight positive pressure (3–5 cm Hg). If a fraction is to be transferred to another column for further resolution it is

TABLE 43

R_F VALUES OF DNP-AMINO ALCOHOLS IN VARIOUS SOLVENT SYSTEMS[46]

N-DNP-derivative	*Solvent system No.*								
	1	*2*	*3*	*4*	*5*	*6**	*7*	*8*	*9*
Serinol	0.24	0.11	0.02	—	—	—	0.25	0.15	—
Aspartidiol	—	0.14	—	—	—	—	0.58	0.18	0.14
Glutamidiol	0.20	0.13	—	—	—	—	0.64	0.14	0.11
Threoninol	0.56	0.23	—	—	—	—	0.72	0.25	—
Tyrosinol	0.70	0.25	—	—	—	—	0.58	—	—
Lysinol (mono-)	—	—	0.21	—	—	—	—	—	—
Ethanolamine	—	0.88	0.50	—	—	—	—	—	—
Alaninol	—	—	0.70	—	—	0.37	—	—	—
Ammonia	—	—	0.70	—	—	0.27	—	—	—
Lysinol (di-)	—	—	0.90	—	—	—	—	—	—
Prolinol	>1.20	>1.20	1.04	0.10	0.35	—	—	—	—
Valinol	—	—	1.05	0.33	0.55	—	—	—	—
Leucinol	—	—	1.20	0.60	0.78	—	—	—	—
Isoleucinol	—	—	—	0.60	0.80	—	—	—	—
Phenylalaninol	—	—	1.10	0.23	0.55	0.87	—	—	—
O-Methyltyrosinol	—	—	—	0.20	0.52	1.00	—	—	—

* Separates DNP-alaninol from dinitroaniline.

necessary to remove the glycol first by heating at 50°/0.01 mm Hg for a few minutes. The eluted fractions are evaporated on the water pump at 30–40°, and the residue dissolved in chloroform for determination of the extinction values at 350 mμ. A standard curve is constructed for ethanolamine and alaninol; the other DNP derivatives are estimated by reference to the alaninol curve, due allowance neing made for molecular weight. Table 43 lists the R_F values of DNP-amino alcohols in various solvent systems using as stationary phases ethylene glycol, glycerol, trimethylene glycol or tetramethylene glycol.

OTHER CHEMICAL METHODS

Isothiocyanate degradation

Under somewhat vigorous conditions a peptide in which the free amino group must be protected is converted to an acylthiohydantoin by the simultaneous action of acetic anhydride and ammonium thiocyanate. The thiohydantoin corresponding to the C-terminal residue is released by mild treatment with alkali.

This reaction has been studied off and on for many years since

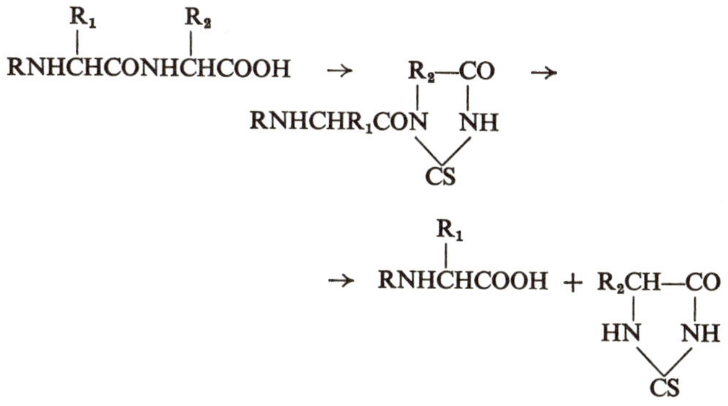

its introduction by Schlack and Kumpf[50] in 1926. Waley and Watson[51] introduced modern chromatographic methods but with insulin as a model concluded that some interaction between the liberated thiohydantoin and the protein prevented the extraction of the derivative from neutral or acid solution. From studies on wool keratin and insulin, Tibbs[52] also concluded that further modification of the procedure was necessary.

High yields of the intermediate acylthiohydantoins under mild conditions using the reagent diphenyl phosphoroisothiocyanatidate were obtained by Kenner *et al.*[53]. This reagent is made simply from diphenyl phosphorochloridate[54] and potassium isothiocyanate in acetonitrile at room temperature and distils at 105°/0.1 mm. The degradation of peptides is carried out in a sealed tube at room temperature during a period of two days according to the scheme on page 246:

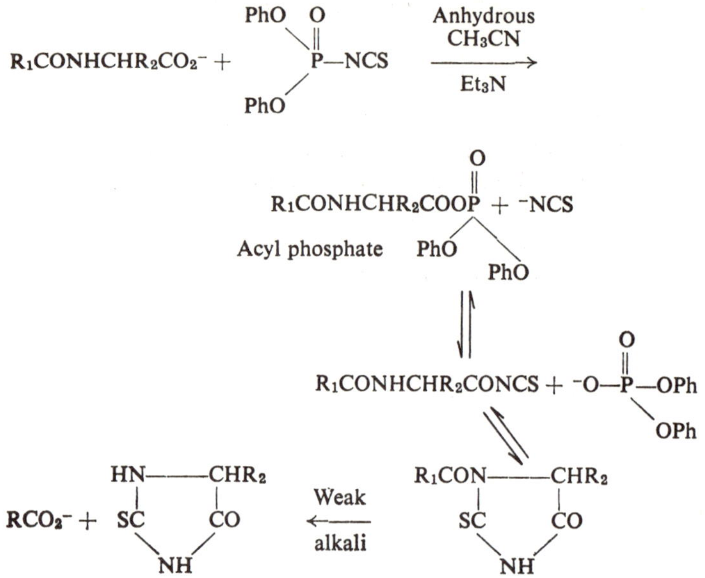

Paper chromatography of the 2-thiohydantoins has been described by Edward and Nielsen[55] who recommend the solvents butanol saturated with water or petroleum ether (b.r. 100–200°)–propionic acid (1:1). Dautrevaux and Biserte[56] employed isoamyl alcohol saturated with water (20–25 cm run at 20°) in one dimension and benzene–acetic acid–water (1:1:1, 30 cm run) in the second dimension. The zones are identified by spraying with Folin and Denis reagent[57]. On exposure of the dried paper to ammonia fumes the thiohydantoins of tyrosine and tryptophan give green spots, those from other amino acids give blue spots. The isothiocyanate degradation has been carried out on protein adsorbed on filter-paper[56] (0.75 mg protein per cm band) in a manner similar to the technique employed by Fraenkel-Conrat for the Edman degradation (see p. 204). In this way the C-terminal alanine was confirmed for horse serum albumin. Low recoveries of the 2-thiohydantoins due to side reactions are very often the rule and perhaps the procedure is best used as a subtractive method. Fox *et al.*[58] have demonstrated that the amino acids of lysozyme other than the C-terminal residue, are not destroyed by treatment with the Schlack and Kumpf reagents; after acid hydrolysis an accurate amino acid analysis revealed the loss of only one leucine residue per mole of 14,700.

Other methods

A novel approach to the selective removal of C-terminal residues was the carbodi-imide method of Khorana[59]. Under very mild conditions carbodiimides (RN:C:NR) react with N-acylpeptides to form acylureas which undergo an interesting degradation in the presence of alkali.

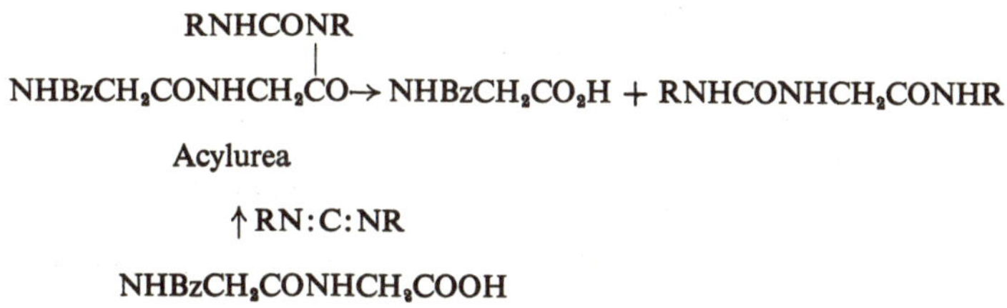

The C-terminal residue may be released from the cleavage product by hydrolysis with acid or alkali. A serious disadvantage is the contamination of the degraded peptide with the starting materials regenerated from the acylureas by simple hydrolysis.

Anodic oxidation of a protected peptide has been described by Boissonnas[60]. The derivative (N-carbobenzoxy- or DNP-) is dissolved in a little anhydrous methanol, cooled in ice and submitted to the oxidation for a period of 1 to 3 h using a current of 50–150 mA at 200–600 V. The product may be hydrolyzed completely with 20% hydrochloric acid in a sealed tube at 100° or submitted to a mild hydrolysis to preserve the remaining peptide.

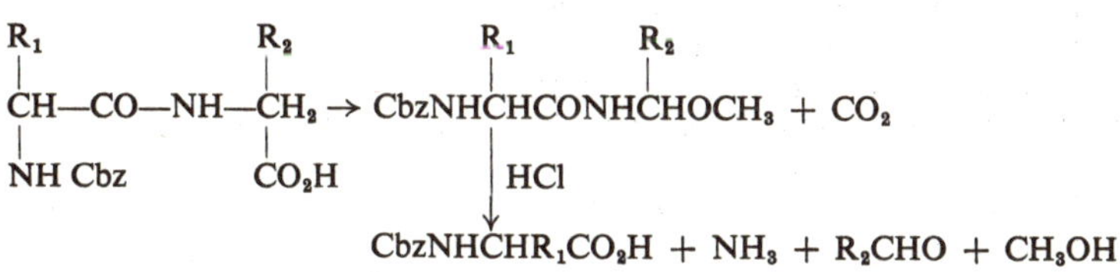

References p. 248

REFERENCES

[1] M. L. Anson, *J. Gen. Physiol.*, 20 (1937) 663, 777.
[2] B. J. Allan, P. J. Keller and H. Neurath, *Biochemistry*, 3 (1964) 40.
[3] K. S. V. Sampath Kumar, K. A. Walsh, J. Bargetzi and H. Neurath, *Biochemistry*, 2 (1963) 1475.
[4] J. E. Folk and E. W. Schirmer, *J. Biol. Chem.*, 238 (1963) 3884.
[5] J. E. Folk, *J. Biol. Chem.*, 238 (1963) 3895.
[6] M. A. Mitz, quoted in C. H. W. Hirs, S. Moore and W. H. Stein, *J. Biol. Chem.*, 235 (1960) 633.
[7] E. W. Davie, C. R. Newman and P. E. Wilcox, *J. Biol. Chem.*, 234 (1959) 2635.
[8] J. I. Harris and C. H. Li, *J. Biol. Chem.*, 213 (1955) 499.
[9] J. E. Folk, *J. Am. Chem. Soc.*, 78 (1956) 3541.
[10] J. E. Folk and J. A. Gladner, *J. Biol. Chem.*, 231 (1958) 379, 393.
[11] G. Guidotti, *Biochim. Biophys. Acta*, 42 (1960) 177.
[12] J. E. Folk, K. A. Piez, W. R. Carroll and J. A. Gladner, *J. Biol. Chem.*, 235 (1960) 2272.
[13] E. C. Wolff, E. W. Schirmer and J. E. Folk, *J. Biol. Chem.*, 237 (1962) 3094, 3100, 3105.
[14] E. Wintersberger, D. J. Cox and H. Neurath, *Biochemistry*, 1 (1962) 1069, 1078.
[15] J. E. Folk, R. C. Braunberg and J. A. Gladner, *Biochem. Biophys. Acta*, 47 (1961) 595.
[16] C. H. Li, *Advan. Protein Chem.*, 12 (1957) 269.
[17] Y. D. Hasley and H. Neurath, *J. Biol. Chem.*, 217 (1955) 247.
[18] A. Kowalsky and P. D. Boyer, *J. Biol. Chem.*, 235 (1960) 604.
[19] S. Akabori, K. Ohno and K. Narita, *Bull. Chem. Soc. Japan*, 25 (1952) 214.
[20] A. Stähler, *Ber.*, 42 (1909) 3018.
[21] K. Kusama, *J. Biochem. (Tokyo)*, 44 (1957) 375.
[22] I. M. Kolthoff, *J. Am. Chem. Soc.*, 46 (1924) 2009.
[23] S. Akabori, K. Ohno, T. Ikenata, A. Nagata and I. Haruna, *Proc. Japan Acad.*, 29 (1953) 561.
[24] K. Ohno, *J. Biochem. (Tokyo)*, 42 (1955) 615.
[25] C. I. Niu and H. Fraenkel–Conrat, *J. Am. Chem. Soc.*, 77 (1955) 5882.
[26] C. I. Niu and H. Fraenkel–Conrat, *Biochim. Biophys. Acta*, 16 (1955) 597.
[27] I. Haruna and S. Akabori, *J. Biochem. (Tokyo)*, 47 (1960) 513.
[28] H. Locker, *Biochim. Biophys. Acta*, 14 (1954) 533.
[29] J. H. Bradbury, *Biochem. J.*, 68 (1958) 475, 482.
[30] P. Karrer *et al.*, *Helv. Chim. Acta*, 31 (1948) 1617, 2088; *Helv. Chim. Acta*, 32 (1949) 1034, 1156, 1934.
[31] C. Fromageot, M. Jutisz, D. Meyer and L. Penasse, *Compt. Rend.*, 230 (1950) 1905.
[32] C. Fromageot, M. Jutisz, D. Meyer and L. Penasse, *Biochim. Biophys. Acta*, 6 (1950) 283.
[33] P. Jolles and C. Fromageot, *Biochim. Biophys. Acta*, 9 (1952) 416.

[34] A. C. Chibnall, and M. W. Rees, In: G. E. W. Wolstenholme and M. P. Cameron (*Eds.*), *The Chemical Structure of Proteins*, Churchill, London, 1953, p. 70.
[35] A. C. Chibnall and M. W. Rees, *Biochem. J.*, 68 (1958) 105.
[36] A. C. Chibnall, C. Haselbach, J. L. Mangan and M. W. Rees, *Biochem. J.*, 68 (1958) 122.
[37] J. L. Bailey, *Biochem. J.*, 60 (1955) 173.
[38] J. A. Krynitsky, J. E. Johnson and H. W. Carhart, *Anal. Chem.*, 20 (1948) 311.
[39] A. C. Chibnall, J. L. Mangan and M. W. Rees, *Biochem. J.*, 68 (1958) 114.
[40] E. J. Wilson and E. Pacsu, *J. Org. Chem.*, 7 (1942) 126.
[41] J. L. Bailey, *Biochem. J.*, 60 (1955) 170.
[42] M. W. Rees, *Biochem. J.*, 68 (1958) 118.
[43] P. Jolles and C. Fromegeot, *Bull. Soc. Chim. France*, (5), 18 (1951) 862.
[44] P. Jolles and C. Fromageot, *Biochim. Biophys. Acta*, 9 (1952) 287.
[45] A. Rhein and M. Jutisz, *Biochim. Biophys. Acta*, 9 (1952) 645.
[46] J. C. Crawhall and D. F. Elliott, *Biochem. J.*, 61 (1955) 264.
[47] M. Jutisz, M. Privat de Garilhe, M. Suquet and C. Fromageot, *Bull. Soc. Chim. Biol.*, 36 (1954) 117.
[48] W. Grassman, H. Hörmann and H. Endres, *Z. Physiol. Chem.*, 296 (1954) 208.
[49] J. M. Luck and A. Wilcox, *J. Biol. Chem.*, 205 (1953) 859.
[50] P. Schlack and W. Kumpf, *Z. Physiol. Chem.*, 154 (1926) 125.
[51] S. G. Waley and J. Watson, *J. Chem. Soc.*, (1951) 2394.
[52] J. Tibbs, *Nature*, 168 (1951) 910.
[53] G. W. Kenner, H. G. Khorana and R. J. Stedman, *J. Chem. Soc.*, (1953) 673.
[54] P. Brigl and H. Muller, *Ber.*, 72 (1939) 2121.
[55] J. T. Edward and S. Nielsen, *Chem. Ind.* (*London*), (1953) 197.
[56] M. Dautrevaux and G. Biserte, *Compt. Rend.*, 240 (1955) 1153.
[57] O. Folin and W. Denis, *J. Biol. Chem.*, 14 (1913) 95.
[58] S. W. Fox, T. L. Hurst, J. F. Griffith and O. Underwood, *J. Biol. Chem.*, 77 (1955) 3118.
[59] H. G. Khorana, *J. Chem. Soc.*, (1952) 2081.
[60] R. Boissonnas, *Helv. Chim. Acta*, 35 (1952) 2226.

CHAPTER 8

DIALYSIS AND GEL FILTRATION

In the present phase of biochemical endeavour when a great number of workers are preoccupied with the task of fractionating complex mixtures, it is curious that the molecular-sieve properties of membranes have been so little exploited. The physicochemical principles governing the rates of migration of solutes through semi-permeable membranes, whether by free diffusion or under the influence of an electric field, have not been completely formulated but interesting information of great practical value has already emerged from recent advances in the subject. Some of these practical aids will be considered as well as the new developments in gel filtration.

DIALYSIS

The technique is used mainly today for the removal of small molecules from aqueous solutions of proteins and other high molecular weight solutes and for the concentration of protein solutions. A common practice is to have both ends of a dialysis sac tied during the dialysis of concentrated solutions but as will be seen below, the high osmotic pressures developed may be disadvantageous. Also the testing of dialysis tubing for pin-hole leaks under pressure is not recommended.

The concentration of aqueous solutions of proteins may be effected in three ways. One method is to suspend the tube in a cold air stream; depending on the diameter of the tubing and with the upper end open to the air, water will be lost at 4° at a minimum rate of about 2–3 ml/h. In another method the dialysis sac is filled with a water soluble high polymer *e.g.* Carbowax polyethylene glycol 20 MM or Ficoll, and placed in the solution to be concentrated. Water then passes through into the sac by a process of forced osmosis. A third technique is the very old one of ultrafiltration whereby water is forced through the pores of the membrane under high pressure or by suction. The membrane must be suitably supported and clogging of the pores tends to slow up the process of filtration.

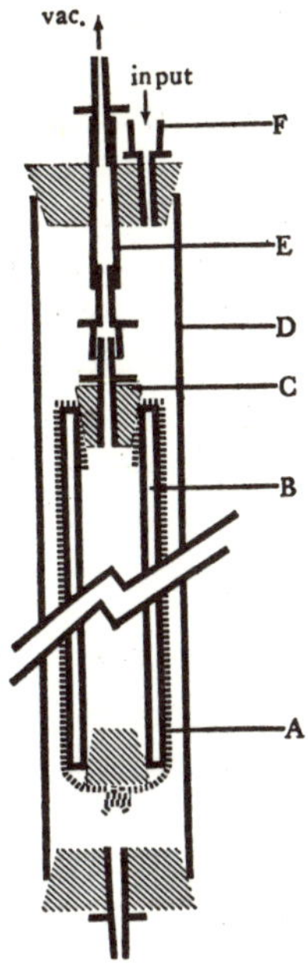

Fig. 78. Ultrafilter for concentrating protein solutions. Siegelman and Firer[1]. See text for explanation.

A convenient ultrafilter[1] for the concentration of large volumes of protein solutions is illustrated in Fig. 78. An essential part of the apparatus is a 3-ft. length of linear porous polyethylene tubing *B* (external diam. 1 in., internal diam. 0.5 in.; Porous Plastics Ltd., Dagenham, Essex). closed at one end with a rubber stopper. Dialysis tubing *A* (36/100 Visking) tied at one end and about 8 in. longer than *B* is soaked in water, tested for leaks, and stretched over *B*. The free end is tucked into the tube to be closed with a silicone stopper *C*. The large glass tube *D* is filled with the protein solution to be concentrated before introducing the ultrafiltration assembly. A vacuum of 7–10 cm Hg is applied giving a concentration rate of about 1 ml/min. Several complete units can be connected in series.

There are a number of studies in the literature on the correlation of molecular size with the rates of diffusion of solutes through semipermeable membranes. Signer *et al.*[2] used rather complicated apparatus in their work on low molecular weight solutes; Craig *et al.*[3–5] have accumulated useful information with simple apparatus; Synge[6] has studied the electrical migration of solutes inside porous membranes. The preparation of collodion membranes has been described by Elford[7,8], Mould and Synge[9], Synge[6], Adair[10] and Carr *et al.*[11]. A survey of early work on the properties of membranes has been made by Stauffer.[12] Ackers and Steere[13] have devised a technique for measuring the pore size of agar gels in thin membranes.

Very simple apparatus is required for the determination of diffusion rates of solutes through membranes as is shown in Fig. 79.

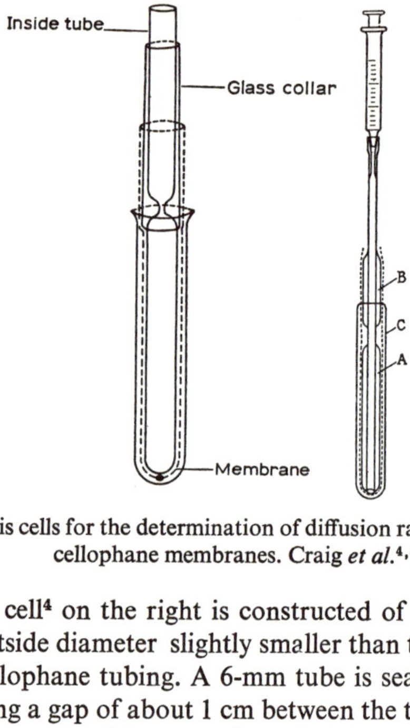

Fig. 79. Dialysis cells for the determination of diffusion rates of solutes through cellophane membranes. Craig *et al.*[4,5].

The dialysis cell[4] on the right is constructed of two glass tubes A and B of outside diameter slightly smaller than the diameter of the extended cellophane tubing. A 6-mm tube is sealed inside the two tubes allowing a gap of about 1 cm between the two sections A and B. The upper section B is approximately 0.5 mm larger in diameter than the lower section and is about 4 cm in length. The 6 mm tube is open at the lower end and fitted with an adapter to take a 2-ml syringe.

The dialysis membrane is soaked in water and one end is tied off with a single knot, care being taken not to stretch that part of the membrane which will be used for the actual dialysis. Sufficient membrane to cover A and part of B is pulled gently over the glass; that over A is kept immersed in water, the upper part over B is allowed to dry and contract on to the glass. The external vessel C is of a size such that volume of solvent 10–20 times that initially used inside the sac will provide a level as high as the upper part of A. The collapsed rim of C provides a snug fit between C, and A and B and when the apparatus is clamped just below the ground joint the arrangement facilitates the centering of A in C. Before the solution to be dialyzed is inserted a small hole is punched with a needle through the membrane just below the bottom of B. Depression of the plunger of the syringe forces liquid out around the annular space inside the membrane. Plastic cannular tubing is useful for placing and withdrawing solutions at the lower end of the inner tube. The solution to be dialyzed must be of sufficient volume to be in contact with the membrane without allowing any liquid to remain

TABLE 44

COMPARATIVE ESCAPE RATES OF A SERIES OF POLYPEPTIDES AND PROTEINS[4]

Solute	*Mol. wt.*	*50% escape time, min*		
		0.1 N HAc 20/32	*0.01 N HAc 20/32*	*0.01 N HAc 18/32*
Tryptophan	204	......	4	6
Bacitracin	1,422	......	15	21
Salmine	2000–8000	......	40	150
Subtilin	3,300	......	54	138
Insulin	5,733	1.5 h		Will not pass
Cytochrome *c*	12,000		60	
Ribonuclease	13,600	3.5 h	2 h	
Lysozyme	14,000	3.5 h	2.3 h	
Trypsin	20,000	......	4 h	
Trypsinogen	20,000	......	6.5 h	
Chymotrypsin	24,500	......	5 h	
Chymotrypsinogen	25,000	12 h	9 h	
Pit. lact. hormone	26,000	......	13 h	
Gliadin	27,000	......	29 h	
Ovomucoid	28,000	......	35 h	
Pepsin	35,500	......	80 h	
Ovalbumin	45,000	......	More slowly than pepsin	

References p. 275

in the glass tube in A. This volume is 0.6 ml for 20/32 and 18/32 Visking cellophane membranes when the diameter of A is 14.8 mm. The unit is hung in the solvent in C to a height so that the levels of liquid inside the membrane and the outside are equal. After arbitrary time intervals the outside solution is changed and at each change the solution inside the membrane is redistributed over the membrane by manipulation of the plunger. The rate of escape of the solute is calculated on the basis of total recovery of the solute from the external and internal solutions.

Table 44 lists the comparative escape rates of a series of polypeptides and proteins; about half of these solutes deviate from straight line behaviour—an indication of impurity, association or obscure membrane effect. Craig and collaborators conclude that molecular size is the main factor determining the escape rate and that charge plays a less significant role. The effect of various concentrations of salt on the escape rate of a protein is illustrated in Fig. 80. In general dialysis is more rapid at higher temperatures

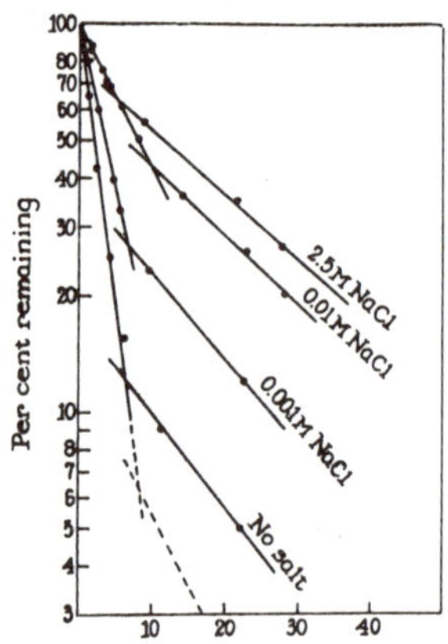

Fig. 80. Escape curves of ribonuclease through 20/32 cellophane in 0.01 *N* acetic acid at various salt concentrations (——) and in distilled water (---). Craig *et al.*[4].

provided the molecule has a stable configuration. Increased rates might not be manifest, however, with proteins that tend to denature at these temperatures. The addition of solvents or a denaturing agent such as urea invariably decreases the rates of diffusion.

Craig and Konigsberg[5] have modified the porosity of cellophane by various methods. A list of the porosities of several sizes of Visking cellophane tubing is given in Table 45.

TABLE 45

POROSITIES OF VISKING CELLOPHANE TUBING[4]

Size	*Thickness (mm)*	*Result with insulin*
8/32	0.064	Passes slowly
18/32	0.0175	Does not pass
20/32	0.020	Passes rapidly
23/32	0.0275	Does not pass
27/32	0.025	Slower rate than with 20/32

Cellophane tubing when full of water will stretch under its own weight and increase its porosity. Application of air pressure to the water will increase the porosity further. Acetylation with pyridine and acetic anhydride reduces the porosity of cellophane reproducibly, and differences in effective molecular diameters of the order

TABLE 46

RELATIVE DIALYSIS RATES OF VARIOUS COMPOUNDS[15]

	Molecular weight	*Half-time escape rates (min)*	
		copolyester membrane	*cuprophane membrane*[b]
Urea	60.8	58	68
Creatinine	113.1	117	123
Uric acid	168.1	217	320
Ascorbic acid	176.1	135	178
Glucose	180	268	223
Thiamine chloride	337.3	150	160
Sucrose	342	397	270
Raffinose	504.5	930	450
Bacitracin	1411	568	720
Insulin[a]	6000	26% in 30 h	None
Polysarcosine	6700	36% in 30 h	15% in 30 h

[a] In this experiment the insulin was probably present as the dimer (mol. wt. 12,000).

[b] Cuprammonium cellulose film.

References p. 275

of 2% or less can be detected by dialysis through thin films of cellophane thus modified[16]. A low porosity cellophane membrane was used by Craig and collaborators[14] to study the conformations in aqueous solutions of a number of small peptides of biological interest. The demands of medical research for membranes of critical performance have encouraged a search for new synthetic membranes[15]. The relative dialysis rates of various compounds through a low porosity film of a copolymer based on polyoxyethyleneglycol and polyethylene terephthalate[15] are shown in Table 46.

ELECTRODIALYSIS

This term described the removal of small molecules from larger molecules in aqueous solution under the influence of an applied electric field. The apparatus used consists of a glass or plastic cell which is divided into three compartments by means of suitable membranes[17]. The centre compartment contains the sample solution, the two outer compartments the electrode solutions. With this type of apparatus a protein solution may be desalted, metal ions removed from enzyme solutions in activation studies, or contaminating ions removed from isotopically labelled materials. Drawbacks to the method are occasional precipitation of protein at the membranes, and adsorption effects (see Haselbach and Piquet[18] and Synge[6]).

A cell containing a fourth compartment has been used by Synge[19], and Sanger and Tuppy[20] for the group separation of amino acids, amino acid derivatives and small peptides according to charge. The four compartment ionophoresis cell designed by Synge[19] is illustrated in Fig. 81. In operation the cathode compartment A was filled with 1% (w/v) aqueous NH_3; the solution to be analyzed was adjusted to pH 6 and placed in B; the third compartment was filled with 1% (w/v) aqueous acetic acid, and the final anode compartment was filled with 0.1 *N* H_2SO_4. A potential difference of 200 V was applied and during the experiment the pH of the specimen compartment was maintained at 6. When a mixture of alanine, ornithine and aspartic acid was fractionated the specimen compartment was found to be free of ornithine and aspartic acid which had migrated into the cathode and anode compartments respectively.

In the operation of the ionophoresis cells the arrangement of the

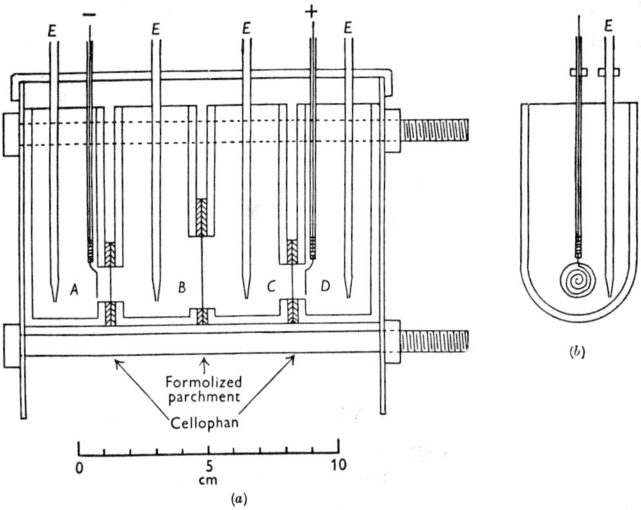

Fig. 81. Diaphragm cell assembled with four compartments: (a) sectional view from side. (b) end elevation of a compartment. A, cathode compartment; B, specimen compartment; C, acetic acid compartment; D, anode compartment; E, inlets for stirring gas. Synge[19].

membranes is very important. In Fig. 81, for example, the negatively charged cellophane membranes induce an endosmotic flow towards the cathode whilst the formolized parchment carrying a positive charge induces an endosmotic flow in the reverse direction. This ensures a minimum back diffusion into compartment B of the ions accumulated in compartments A and C.

A positive membrane is made by passing finely woven cloth through a warm 10% solution of gelatin, and after allowing the material to dry at room temperature for 24 h, by immersion in a 4% formaldehyde solution for several hours. The membrane is finally washed thoroughly to remove all soluble materials.

As already mentioned denaturation is the chief drawback to electrodialysis as a means of desalting proteins, and this arises partly from poor dissipation of coulombic heat and partly from ion depletion at the membranes (membrane polarization). Low D.C. voltages containing a minimum A.C. component are therefore recommended. Greatly improved electrodialysis rates[21] are attained when the conventional membranes are replaced by ion exchange membranes.

References p. 275

Suitable 'permselective' membranes available commercially (Ionics Inc., Mass., U.S.A.) are Nepton-AR-111A (anionic) and Nepton-CR-61 (cationic). Their use in apparatus designed for the preparation of isoionic protein is described by Katz and Ellinger[22].

GEL FILTRATION

An early observation of the molecular sieve effect in gels was made by Synge and Tiselius[23] who fractionated hydrolysis products of amylose in agar-agar jelly. Starch grains were used by Lathe and Ruthven[24] in a detailed study of the molecular-sieve mechanism. It was found that proteins do not penetrate ordinary hydrated starch but good resolution of substances in the molecular weight range 100–1000 is possible. Starch which has been swollen to about three times the original size by warming in water is permeable to larger molecules such as insulin, myoglobin and haemoglobin. Amino acids move through a starch column roughly in the order to be expected for a weak cation exchanger, but the ion-exchange effect can be counteracted by using eluents of appropriate ionic strength. Partridge[25] has drawn attention to the molecular-sieve properties of ion-exchange resins themselves and the increased selectivity obtainable when this additional mechanism operates.

Fractionation of biological materials has been accomplished on

TABLE 47

SEPHADEX TYPES AND SPECIFICATIONS (Courtesy of Pharmacia, Uppsala)

Type	*Approximate exclusion limit (molecular weight)*	*Water regain (g water/g dry gel)*	*Bed volume (ml)/g*	*Particle size (μ)*
G-10	up to 700	1.0 ± 0.1	2–3	40–120
G-15	up to 1500	1.5 ± 0.1	2.5–3.5	40–120
G-25	5000	2.5 ± 0.2	5	
fine				20–80
coarse				100–300
G-50	10,000	5.0 ± 0.3	10	
fine				20–80
coarse				100–300
G-75	50,000	7.5 ± 0.5	12–15	40–120
G-100	100,000	10.0 ± 1.0	15–20	40–120
G-150	150,000	15 ± 1.5	20-30	40-120
G-200	200,000	20.0 ± 2.0	30–40	40–120

Sephadex, granulated agar[26], polyacrylamide[27], cross-linked hyaluronic acid[28], polyvinyl ethyl carbitol and polyvinyl pyrrolidone[29]. Procedures for gel filtration in Sephadex will be discussed first.

Cross-linked dextrans (*Sephadex*)

Porath and Flodin[30] introduced particulate dextran gels for desalting proteins in 1959, and as a result of collaboration between the school at Uppsala and the Swedish organization AB Pharmacia a valuable range of cross-linked dextran gels are now available under the trade name *Sephadex*. Some 200 papers have appeared describing various techniques and applications, and useful surveys of the new developments are to be found in review articles by Porath and Flodin[31,32].

Dextran is a heterogeneous glucan of high molecular weight produced by many strains of *Leuconostoc*. Most of the glucosidic linkages are α-D-1→6 but a few are of the 1→3 type. A stable three dimensional polymer can be formed from dextran by reaction of the glucose hydroxyls with the cross-linking agent epichlorhydrin or with bifunctional epoxides. Two alcohol molecules are bridged in the following manner with little change in the total hydroxyl content.

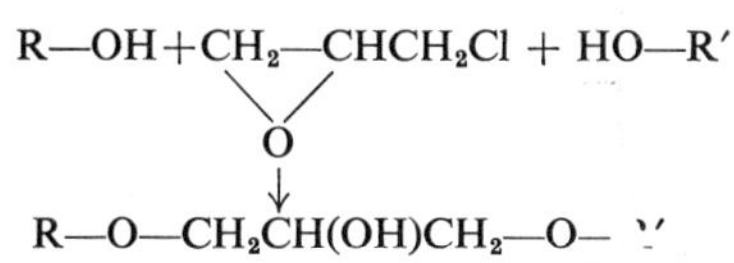

A few negatively charged sites exist in the products amounting to 20–30 μequiv./g of dry gel[33]. Sephadex gels are stable to acetic acid and 0.02 *N* HCl. The types and specifications of Sephadex are listed in Table 47. All types are available in bead form and the makers' estimates for the approximate exclusion limits are based on data obtained with carbohydrate macromolecules and not with proteins.

Before use the powders are suspended in dilute buffer solution and the finer particles removed by decantation. Columns are packed in much the same way as described for cellulose columns (p. 333) and values of 10:1 and 20:1 are preferred height-to-diameter ratios. After use the columns may be washed by causing an upward flow of water from the bottom through the column. The columns exhibit a high capacity and the greatest resolution is obtained with particle

References p. 275

sizes in the range 250–400 mesh. Resistance to flow in columns packed with the fine grades of Sephadex sets a limit to the length of column that may be profitably operated. The gravity flow-rate of a 2 × 70 cm column of Sephadex G-50 (100–250 mesh) is 40 ml/h. Properties of Sephadex in the bead form are discussed later.

Some remarkable separations of proteins, peptides and amino acids were achieved by Porath[34] utilizing dextran gels of different degrees of cross-linking. The elution curves for an artificial mixture of proteins and amino acids separated on a gel of low cross-linking are shown in Fig. 82. R_F values of various substances on a highly cross-linked gel are listed in Table 48, from which it will be seen

TABLE 48

R_F VALUES OF VARIOUS SUBSTANCES ON A GEL OF A HIGH DEGREE OF CROSS LINKING[34]

Substance	*Buffer system*			
	1 M HAc	*0.3 M Py+0.1 M HAc*	*1 M Py+0.05 M HAc*	*1 M Py*
β-MSH	0.81	0.80	—	—
Viscotoxin	0.94	0.96	0.70	0.15
Vasopressin	0.56	0.55	0.51	0.30
Polymyxin	0.86	—	0.65	0.15
Bacitracin A	—	0.63	0.70	0.75
Oxidized glutathione	0.61	0.66	—	0.99
Ferriporphyrin-*c* peptide	0.53	0.70	0.80	0.95
Phthaloyl-γ-amino butyryl-histidine	0.50	—	0.57	—
Histidine	0.53	0.50	0.50	0.49
Isoleucine	0.52	0.51	0.50	0.50
Glycine	0.51	—	0.50	0.48
Arginine	0.52	—	0.41	0.30
Glutamic acid	0.51	—	0.48	0.60
Phenylalanine	0.45	0.45	0.47	—
Tyrosine	0.40	0.39	0.39	0.39
Tryptophan	—	0.27	0.27	0.27
DNP-glycine	0.26	—	0.34	0.77
DNP-isoleucine	0.28	—	0.36	0.77
DNP-tryptophan	0.09	—	0.21	0.75
Na-phosphate	—	0.45	—	—
Na-borate	—	0.40	—	—

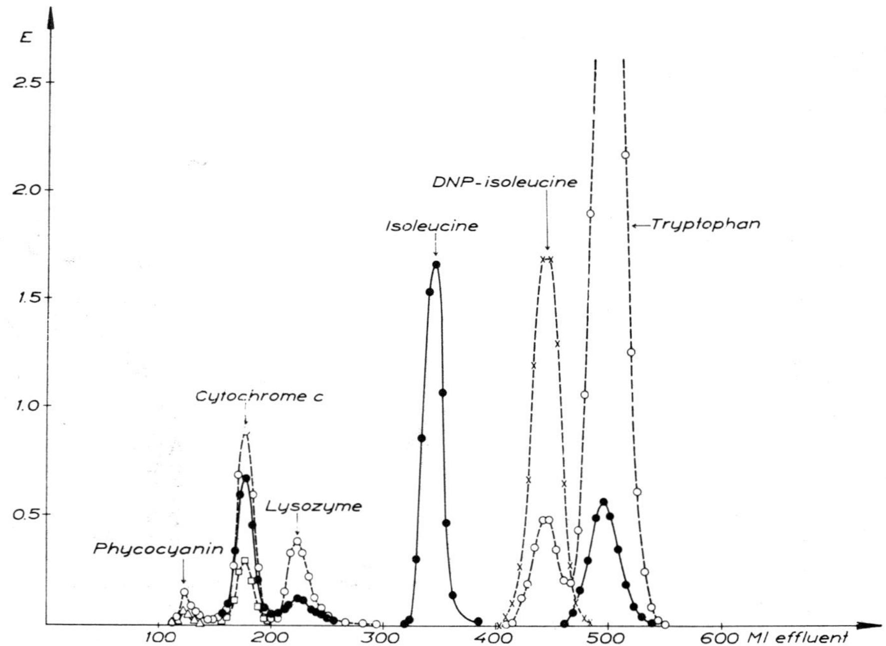

Fig. 82. Filtration diagram of an artificial mixture of selected substances. Filtration was carried out on a gel of low cross-linking in phosphate buffer, pH 6.8, ionic strength 0.05 (△——△) extinction at 615 mμ; (○——○) extinction at 280 mμ; (□——□) extinction at 550 mμ; (×——×) extinction at 360 mμ; (●——●) ninhydrin colour (from Porath[34]).

that basic peptides and amino acids move slowly through the gels in 1 *M* pyridine and faster in 1 *M* acetic acid.

An extensive study of the chromatographic behaviour of calf thymus histone on different grades of Sephadex was carried out by Johnson and collaborators[35]. Four or five incompletely separated heterogeneous fractions were obtained using Sephadex G-50, G-75 and G-100; the molecular weights of the constituents were assumed to lie in the range 2000–40,000. High molecular weight biliproteins have been separated on a long column (2 × 150 cm) of Sephadex G-75 in 0.1 *M* citrate-phosphate buffer, pH 7. Phycocyanin migrated faster than the phycoerythrins in one blue and two red zones. The blue zone contained C-phycocyanin and traces of allophycocyanin, the second zone C-phycoerythrin, and the third zone an unknown pigment previously called B(C)-phycoerythrin[36].

Under selected conditions valuable information on the state of aggregation of proteins can be obtained. Crestfield *et al.*[37] were puzzled by the varying yields of ribonuclease A obtainable after

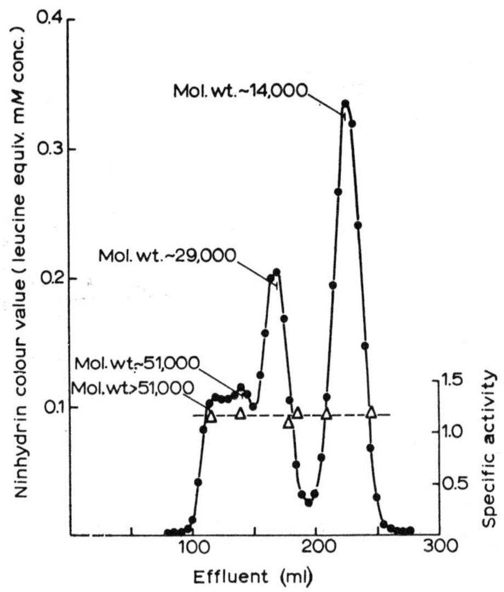

Fig. 83. Separation of ribonuclease A and its aggregates on a 2 × 143 cm-column of Sephadex G-75 with 0.2 *M* sodium phosphate buffer, pH 6.47, as eluent. Specific activity against cyclic cytidylic acid is recorded as μequiv. of proton/minute/μmole of leucine colour equivalents. Crestfield *et al.*[37].

chromatography of different samples on the resin IRC-50. One preparation which had been freeze dried from 50% acetic acid was fully active on a weight basis but gave only a 50% recovery from the column. The solution to this problem was found by examining the behaviour of the samples on Sephadex columns. Sephadex G-75 separates aggregates of two, four, or more molecules of ribonuclease A from one another and from the monomer (Fig. 83). Ribonuclease polymers not detected by chromatography on IRC-50 were transformed to the monomer again by heating at 65° for 15 min at pH 6.5. A later investigation of a reversible polymerization of α-chymotrypsin on Sephadex G-100 was reported by Winzer and Scheraga[38].

Protein chemists engaged in structural work, and faced with the problem of fractionating complex peptide mixtures, quickly realised the value of a new tool to lighten the burden of their labours. Bennich[39] investigated the optimal conditions for fractionation of tryptic hydrolysates of α-casein on Sephadex G-25, G-50 and G-75. Four peaks were obtained on G-25 and seven on G-75 when the mixture was run in 1 *M* acetic acid.

By a combination of counter-current distribution and gel filtration (0.2 *M* acetic acid as eluent) Konigsberg and collaborators[40,41] were able to resolve all the peptides obtained by the tryptic digestion of the α- and β-chains of human haemoglobin. Similar applications in the peptide field are: separation of the ferriporphyrin-c peptide[42] from other products of peptic digests of cytochrome c on Sephadex G-50; separation of three peptide chains from α-chymotrypsin after performic acid oxidation[43]; separation of α-MSH from lysine vasopressin[44] in the solvent system pyridine–acetic acid–water (15:55:30); purification of oxytocin[44] on a 2 × 240 cm column of Sephadex G-25; fractionation of glycopeptides from ovalbumin[45,46]; isolation of antigenic fragments of human serum albumin[47]; and fractionation of small proteins and peptides[48] in the solvent system phenol–acetic acid–water (1:1:1 w/w).

Purification of protein conjugates with fluorescent dyes has been reported[49,50]. In one instance gel filtration gave complete removal of unreacted dye as well as fractionation of protein into a highly labeled fraction and one less completely labeled[50]. Sephadex G-25 was used in a study of beef pancreas carboxypeptidase to demonstrate enzyme-substrate complexes and the binding of zinc to the apoenzyme[51]. Other similar investigations have been concerned

References p. 275

with the measurement of the interaction between small ions and proteins[52], the binding of iron[53] and the binding of drugs[54].

A disadvantage of the highly porous gels made by the block polymerization technique has been the inconveniently low flow-rates. In order to overcome this difficulty AB Pharmacia developed a bead polymerization process offering spherical particles of considerably lower resistance in a packed column. Grades G-100 and G-200 more porous than G-75 were made available. The relative selectivity of the gels G-75, G-100 and G-200 is illustrated by the fractionation of plasma proteins carried out by Flodin[33]. One peak was obtained with G-75, two peaks with G-100, and three with

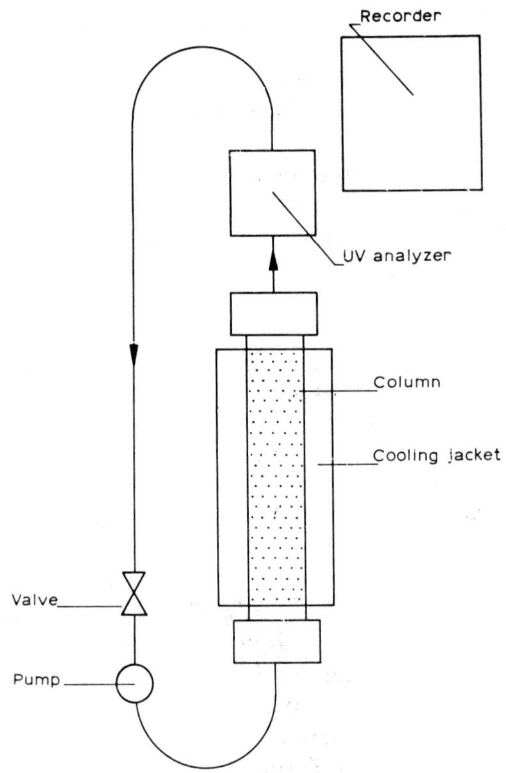

Fig. 84. Schematic diagram of apparatus for recycling chromatography. Porath and Bennich[55].

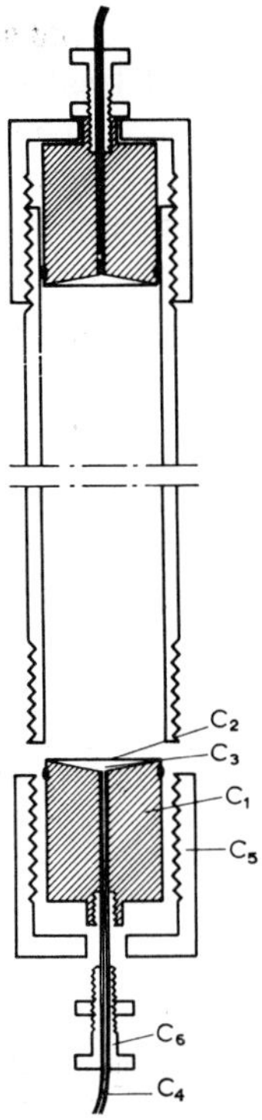

Fig. 85. Diagram of column for recycling chromatography: C_1, adjustable plunger; C_2, polystyrene membrane; C_3, free liquid space; C_4, capillary tubing; C_5 and C_6, screws. Porath and Benich[55].

G-200. The three zones from the G-200 column contained as major components, high molecular weight lipoproteins, γ-globulin, and serum albumin + transferrin.

Convenient flow rates and available head room impose a limit to the resolution possible with gels operated in conventional chromatographic apparatus. An ingenious method which circumvents this limitation and exploits the gel filtration principle to the full, was introduced by Porath and Bennich[55] and termed *recycling chromatography*. A schematic diagram of the whole apparatus consisting of column, U.V. flow analyzer, selector valve and pump, and the design of the column are shown in Figs. 84 and 85. Care must be taken to remove air from the gel suspension and from the membranes. The gel bed is stabilized and equilibrated by pumping the buffer upwards at a flow rate of about twice that to be used later in the actual run. The function of the selector valve is to facilitate sample application and withdrawal through appropriate capillary tubes. The flexibility of the process is demonstrated by the elution record shown in Fig. 86. Recycling filtration of a human cerulo-

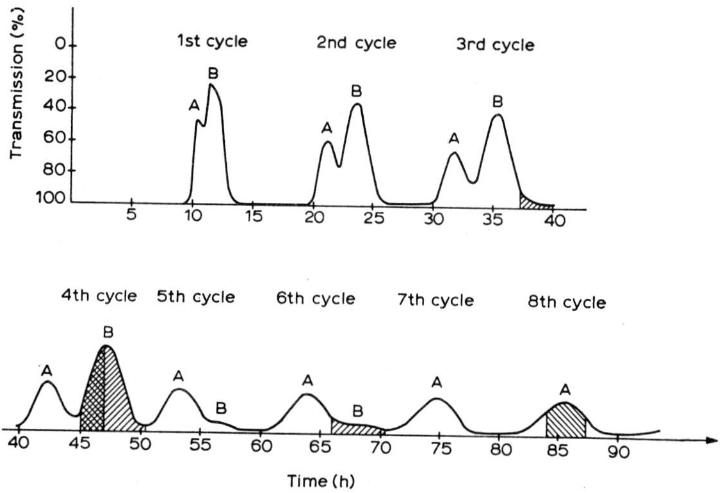

Fig. 86. Recycling filtration of human ceruloplasmin on Sephadex G-100, 3.2 × 93 cm-column, at 10°. Sample, 5.2 ml of a 3% solution; flow rate, 20 ml/h; absorption at 254 mμ, 3-mm cell. Hatched areas indicate bleedings of the column. Porath and Bennich[55].

plasmin specimen (5.2 ml) was performed on a column of Sephadex G-100 of dimensions 3.2 × 93 cm. Three cycles were performed before re-mixing of the components occurred and in order to extend the separation, trailing material of component B was bled from the column. Component B was almost completely removed from the system after four cycles.

Thin-layer gel filtration

In the recycling filtration example given above the bed height corresponding to eight cycles was 7.4 m. By contrast small scale gel filtration in thin layers of Sephadex has been carried out successfully, thus providing a rapid, micro method for estimating molecular weights of small proteins. Recent accounts of this technique have been given by Johansson and Rymo[56,57], Determann[58], Downey and Andrews[59], and Morris[60].

Initially Johansson and Rymo[56] experimented with Sephadex

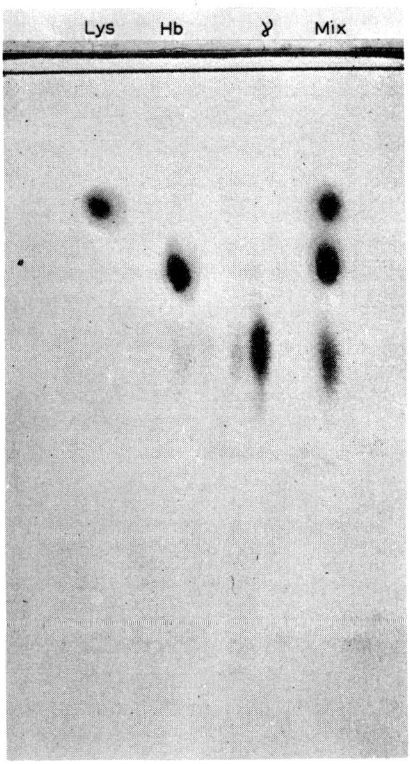

Fig. 87. Thin-layer chromatogram of lysozyme (Lys), haemoglobin (Hb) and γ-globulin (γ) on Sephadex G-200–0.5 *M* NaCl. Morris[61].

References p. 275

G-25 (fine), G-75 (fine) and G-75 (<400 mesh). G-25 (10 g) was mixed with buffer (50 ml), and after equilibration for 1 h, applied to 4 plates of dimensions 8 × 20 cm or 2 plates of dimensions 20 × 20 cm. A Desaga applicator was used with the slit adjusted to 0.5 mm. Sephadex G-75 (5 g) was spread as a suspension in buffer (70 ml) after equilibration for 7 h. Filter paper wicks formed the bridge between the plate and the eluent. After a running time of 5–8 h, the plates were dried at 80° and sprayed with ninhydrin or dyed.

Later a procedure for Sephadex G-100 and G-200 ('superfine', <400 mesh) was described[57]. Gels were prepared in 0.02 *M* sodium phosphate, pH 7.0, containing 0.2 *M* NaCl; the volumes of buffer required for G-100 and G-200 were 19 ml and 25 ml respectively. An equilibration period of at least 24 h was required. Samples of 1–5 μl were applied to the plates and suitable flow rates were found to be 1–2 cm/h. Before drying, the plates were covered with a piece of filter paper (avoiding trapped bubbles of air) and stained in a saturated solution of Lissamine Green in 5% acetic acid for 1 h. The filter paper, which loosened rapidly from the plate in the dye bath, was washed several times in 5% acetic acid.

Morris[60] used CO-haemoglobin as a reference marker and determined R_{Hb} values of 12 test proteins. These values were found to be related linearly to the logarithms of the molecular weights of the proteins examined. A separation of lysozyme, haemoglobin and γ-globulin[61] is shown in Fig. 87.

Comments

In the absence of salts many proteins are adsorbed on Sephadex and cannot be removed by extensive washing with distilled water. However, recovery of adsorbed proteins can usually be effected by elution with a dilute salt solution (>0.02 *M*). Zone precipitation of proteins on Sephadex has been discussed by Porath[62] who suggests the technique might be useful for finding suitable conditions for crystal formation of proteins. Desalting of an aqueous solution of particular proteins is possible with water as eluent but the use of a volatile buffer may be advantageous. Adsorption of proteins is assumed to be due to the presence of charged groups in the gel[63,64], which are neutralized in salt solutions of molarity greater than 0.02 *M*. When boric acid buffers are employed a marked ion exchange effect may be expected since the boric acid binds to the gel.

Sephadex also shows slight adsorption of aromatic and heterocyclic substances and, depending on the ionic strength and the pH of a gel column, strong adsorption of some basic substances may occur[65]. Disturbances in ionic strength and pH at certain points in a column can give rise to artifacts in the form of false peptide or protein peaks. Thus Lindqvist[66] observed several conductivity peaks when a column equilibrated with a buffer containing polyvalent ions was eluted with pure water.

The best measure that can be taken to avoid contamination of Sephadex by bacteria or fungi is to saturate the buffer with chloroform before storage. Void volumes may be determined using Indian ink, or Blue Dextran, a product of molecular weight about 2×10^6 developed for this purpose by AB Pharmacia.

The three main reasons for non-reproducible performance of Sephadex columns are incomplete equilibration, excessively high flow rates, and imperfect packing of the columns.

Agarose

The neutral polysaccharide agarose is conveniently prepared from agar by the method of Hjertén[67]. Agar (32 g) is added cautiously with stirring to a boiling solution (800 ml) of 0.02 *M* sodium citrate and when all is dissolved cetylpyridinium chloride (12 g) in the same buffer (400 ml) is added slowly. This precipitates agaropectin. The stoppered flask is placed in an oven at 65° for 30 min and the mixture is then transferred to centrifuge tubes also warmed to 65°. Centrifugation at 13,000 *g* for 30 min at room temperature gives a sediment sharply demarcated from the agarose supernatant which usually forms a stiff gel by the end of the run. The agarose gel is broken up, washed, liquified, and treated at 65° with Fuller's earth which is added in small portions until the foam almost entirely disappears. After freezing and thawing the salts can be squeezed out and the agarose dehydrated by freeze drying or by precipitation with alcohol. Agarose spheres are prepared by forming an emulsion with the aid of organic solvents[68] or by spraying a hot solution under pressure into cooled ether[69].

As a medium for gel filtration Hjertén[70] recommends a concentration of 5–15% for the separation of proteins and a lower concentra-

References p. 275

tion of 1–4% for the separation of sub-cellular particles. Flow rates for a column of internal diameter 1.5 cm should be in the range 1–4 ml/h. The R_F values for R-phycoerythrin (mol.wt. 290,000), haemoglobin, and cytochrome *c* as a function of the concentration of agarose are shown in Fig. 88. Resolution of proteins on granulated agar appears to be largely dependent on the differences in diffusion constants and to a lesser extent on molecular weight differences[26]. Steere and Ackers[71] have discussed the use of the gels for the determination of the size of particles.

Polyacrylamide

The preparation of polyacrylamide is described on p. 323. Gels are best granulated by compression through metal nets of different mesh sizes; a press is required for gels of concentration higher than 10%. The relation between R_F value and total concentration of 5%

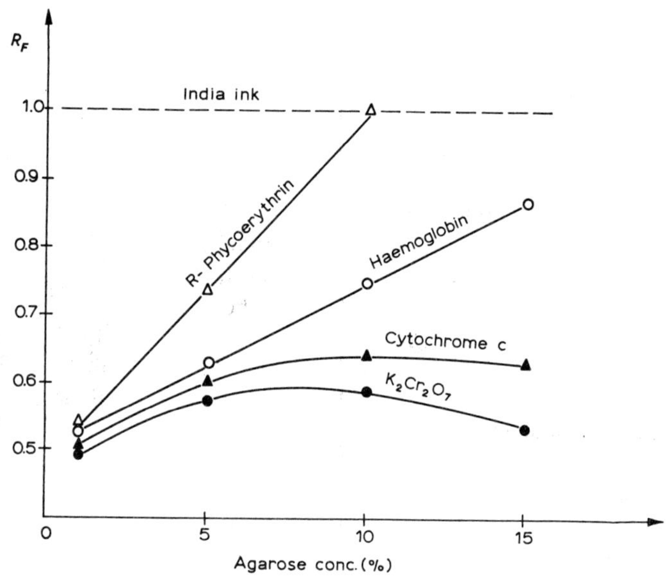

Fig. 88. The R_F values of R-phycoerythrin (mol. wt. 290,000), haemoglobin (mol. wt. 68,000), cytochrome *c* (mol. wt. 13,000) and potassium dichromate (mol. wt. 294) as a function of the concentration of agarose. Reference substance, Indian ink. Hjertén[70].

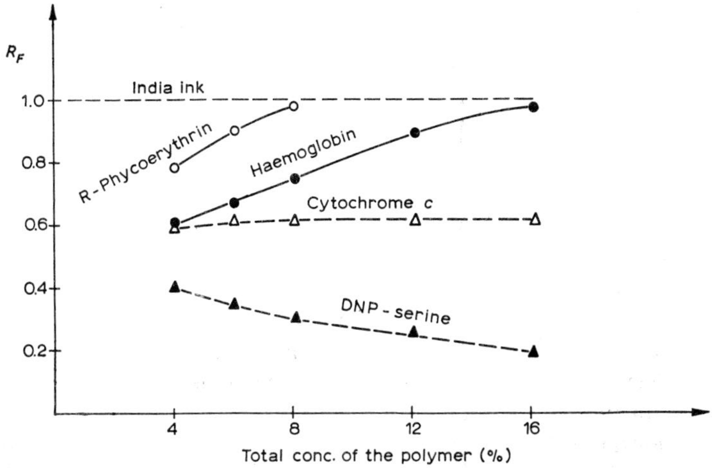

Fig. 89. Relation between R_F value and total concentration of cross-linked polyacrylamide for R-phycoerythrin (mol.wt. 290,000), human haemoglobin (mol. wt. 68,000), cytochrome *c* (mol. wt. 13,000 and DNP-serine (mol. wt. 271). Cross-linking, 5%; 0.05 *M* sodium phosphate buffer, pH 7.3; temperature +4°, Hjertén[73].

cross-linked polyacrylamide for molecules of different size is shown in Fig. 89. Zone distribution due to 'wall effects' is a common experience with this type of gel, but this can be usually avoided by coating the walls of the column with methyl cellulose (Methocel, visc. 7000 cps; Dow Chemical Co.). The treatment is the following, Formaldehyde (50 ml) and formic acid (10 ml) are added to a 0.4% aqueous solution of methylcellulose (140 ml). The tube is exposed to the solution for 5 min, allowed to drain for 5–10 min., and finally dried in an oven at 120° for 40 min. The operation is repeated once more[72]. Human transferrin (mol.wt. 90,000) has been isolated[73] free from proteins of higher molecular weight on a 2 × 61 cm column consisting of 100-mesh grains of gel (*conc.* 6%; cross-linking 5%) in 0.05 *M* sodium phosphate buffer, pH 7.3.

Polyacrylamide gels should be washed thoroughly to remove all traces of monomer. This is important where effluent protein zones are being monitored by nitrogen estimations. It is claimed that gels prepared by the copolymerization of diethylaminoethyl methacrylate (DEAE methacrylate), acrylamide, and N,N-methylenebisacrylamide are less likely to shed trace amounts of substances which may interfere with biological test systems[74].

Theory of gel filtration

The basic equation describing the operational parameters in the form given by Gellotte[65] is

$$V_e = V_o + K_D V_i$$

where V_e is the elution volume, V_o is the void volume, V_i the volume of the water imbibed by the gel particles, and K_D the 'distribution coefficient'. K_D bears a formal resemblance to the similar expression used in normal chromatography which is the ratio between the concentration of a solute in the mobile and stationary phases. In gel filtration the stationary phase may be considered to be the water imbibed by the gel particles and K_D the fraction of the imbibed water available, for distribution of the solvent. The actual concentration of the solute interior to the gel pores may be assumed to be practically identical with that of the external liquid phase. Knowing the weight of dry gel *a*, and the water regain W_r, K_D can be determined from experimental data using the relation $V_i = \mathrm{a}W_r$, hence

$$K_D = \frac{V_e - V_o}{V_i} = \frac{V_e - V_o}{\mathrm{a}W_r}$$

A parameter K_{av} similar but no identical to K_D is defined as the fraction of the volume of the gel available for the solute; K_D includes a correction for the volume occupied by the dextran molecules.

The gel filtration process has been explained in terms of two effects. One is a simple exclusion of molecules from part of the gel, the other is a decreased diffusion rate of molecules in the gel. There is experimental evidence to support both mechanisms. In a recent theoretical treatment Laurent and Killander[75] tacitly assumed that the volume in the gel phase available for a particular solute determines its position in the gel chromatogram. It is argued that since the elution volume of a solute does not vary appreciably with the flow rate of the eluent the diffusion effect cannot be operational.

To test this hypothesis for Sephadex G-200, the molecular radii and K_{av} values for each substance were obtained experimentally and compared with theoretical values derived from diffusion data and Stoke's formula (molecular radii), and Ogston's formula for the available volume in the gel particles. The close correspondence of the experimental data with the theoretical values supports the hypothesis and a mechanism based on steric exclusion of solutes from the gel phase.

In a different approach Ackers[76] compared equilibrium distribution coefficients with column distribution coefficients and found good correspondence for Sephadex G-75 and G-100 gels but not for G-200 gels in which the equilibrium molecular exclusion effect was less sensitive to molecular size. A mechanism based on steric and frictional hindrance to molecular diffusion within the gel was then examined experimentally and a method of calibration derived which involved one parameter, the effective gel pore radius. The latter is obtained from the effluent peak position of a solute of known Stoke's radius. A calibrated column will therefore provide

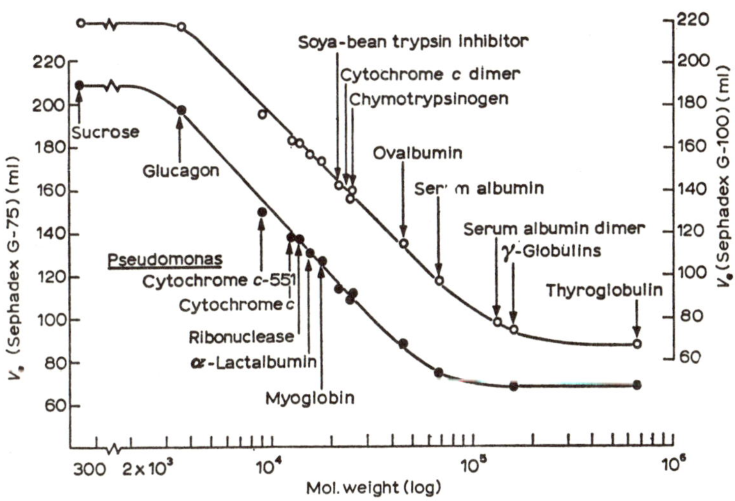

Fig. 90. Plots of elution volumes, V^e, against log (molecular weight) for proteins on Sephadex G-75 (●) and G-100 (O) columns [2.4 cm × 50 cm; equilibrated with 0.05 *M*-Tris hydrochloride buffer, pH 7.5, containing KCl (0.1 *M*)]. Andrews[78].

a value for the Stoke's radius of a protein from which the diffusion coefficient may be derived[77]. The diffusion coefficient may then be combined with the sedimentation coefficient or intrinsic viscosity to determine an upper limit for the hydrated molecular weight. In this treatment K_g is interpreted as the ratio of restricted to free area available for diffusion.

Andrews[78] investigated the correlation between elution volumes and molecular weight and found that plots of V_e *vs.* log (molecular weight) were linear over the molecular weight ranges 3000–35,000 and 5000–60,000 for Sephadex G-75 and G-100 respectively (Fig. 90). Calibrated columns of Sephadex G-100 were used to estimate the molecular weights of a number of enzymes and the figures found were in excellent agreement with literature values. In this type of work 10 μg of protein with enzymic activity is sufficient for an

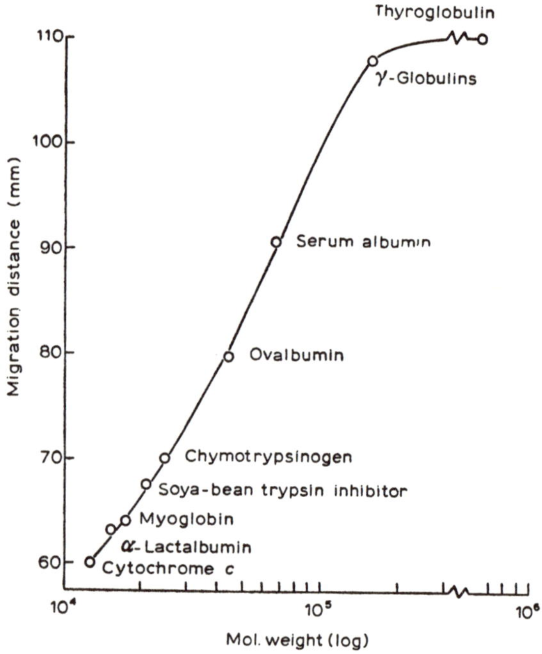

Fig. 91. Plot of migration distance against log (molecular weight) for thin-layer gel-filtration of proteins on Sephadex G-100 equilibrated with 0.05 *M*-Tris hydrochloride buffer, pH 7.5 containing KCl (0.1 *M*). Andrews[78].

estimation. Good correlation was also found between molecular weight and migration rate in thin layers of Sephadex G-100, although the gel columns offered better resolution (Fig. 91). Further relevant studies are those of Porath[79], Whitaker[80], and Sanfelippo and Surak[81].

REFERENCES

1 H. W. Siegelman and E. M. Firer, *Anal. Biochem.*, 3 (1962) 435.
2 R. Signer, H. Hänni, W. Koestler, W. Rottenberg and P. von Tavel, *Helv. Chim. Acta*, 29 (1946) 1984.
3 L. C. Craig and T. P. King, *J. Am. Chem. Soc.*, 77 (1955) 6620.
4 L. C. Craig, T. P. King and A. Stracher, *J. Am. Chem. Soc.*, 79 (1957) 3729.
5 L. C. Craig and W. Konigsberg, *J. Phys. Chem.*, 65 (1961) 166.
6 R. L. M. Synge, *Biochem. J.*, 65 (1957) 266.
7 W. J. Elford, *J. Pathol. Bacteriol.*, 34 (1931) 505.
8 W. J. Elford and J. D. Ferry, *Britt. J. Exptl. Pathol.*, 16 (1935) 1.
9 D. L. Mould and R. L. M. Synge, *Biochem. J.*, 58 (1954) 571.
10 G. S. Adair, *Biochem. J.*, 62 (1956) 26P.
11 C. W. Carr, D. Anderson and I. Miller, *Science*, 125 (1957) 1245.
12 R. E. Stauffer, in: A. Weissberger (Ed.), *Technique of Organic Chemistry, Separation and Purification, Part* 1, Interscience Publ., New York, 1956, p. 65.
13 G. K. Ackers and R. L. Steere, *Biochim. Biophys. Acta*, 59 (1962) 137.
14 L. C. Craig, E. J. Harfenist and A. C. Paladini, *Biochemistry*, 3 (1964) 764.
15 D. J. Lyman, B. H. Lou and E. W. Crawford, *Biochemistry*, 3 (1964) 985; D. J. Lyman, *Trans. Am. Soc. Artificial Internal Organs*, 10 (1964) 17.
16 L. C. Craig and A. O. Pulley, *Biochemistry*, 1 (1962) 89.
17 H. Theorell and A. Åkeson, *Arkiv Kemi (Mineral. Geol.)*, 16A (1942) 8.
18 C. H. Haselbach and A. R. Piquet, *Helv. Chim. Acta*, 35 (1952) 2131.
19 R. L. M. Synge, *Biochem. J.*, 49 (1951) 642.
20 F. Sanger and H. Tuppy, *Biochem. J.*, 49 (1951) 463.
21 S. Katz and P. H. Maurer, *J. Am. Chem. Soc.*, 79 (1957) 2170.
22 S. Katz and F. Ellinger, *Biochemistry*, 2 (1963) 406.
23 R. L. M. Synge and A. Tiselius, *Biochem. J.*, 46 (1950) xli.
24 G. H. Lathe and C. R. J. Ruthven, *Biochem. J.*, 62 (1956) 665.
25 S. M. Partridge, *Nature*, 169 (1952) 496.
26 A. Polson, *Biochim. Biophys. Acta*, 50 (1961) 565.
27 S. Hjertén, *Arch. Biochem. Biophys.*, Suppl. 1 (1962) 147.
28 T. C. Laurent, K. Hellsing and B. Gellotte, *Acta Chem. Scand.*, 18 (1964) 274.
29 D. J. Lea and A. H. Sehon, *Can. J. Chem.*, 40 (1962) 159.
30 J. Porath and P. Flodin, *Nature*, 183 (1959) 1657.
31 J. Porath and P. Flodin, *Protides Biol. Fluids*, 10 (1963) 290.
32 J. Porath, *Advan. Protein Chem.*, 17 (1962).
33 P. Flodin, *Dextran Gels and their Applications in Gel Filtration, Dissertation*, Uppsala, 1962, p. 73.

34 J. PORATH, *Biochim. Biophys. Acta*, 39 (1960) 193.
35 L. D. JOHNSON, A. DRIEDGER AND A. M. MARKO, *Can. J. Biochem.*, 42 (1964) 795.
36 W. NULTSCH, *Biochim. Biophys. Acta*, 59 (1962) 213.
37 A. M. CRESTFIELD, W. H. STEIN AND S. MOORE, *Arch. Biochem. Biophys.*, Suppl. 1 (1962) 217.
38 D. J. WINZER AND H. A. SCHERAGA, *Biochemistry*, 2 (1963) 1263.
39 H. BENNICH, *Biochim. Biophys. Acta*, 51 (1961) 265.
40 R. J. HILL AND W. KONIGSBERG, *J. Biol. Chem.*, 235 (1960) PC 2122.
41 G. GUIDOTTI, R. J. HILL AND W. KONIGSBERG, *J. Biol. Chem.*, 237 (1962) 2184.
42 J. PORATH, *Biochim. Biophys. Acta*, 39 (1960) 193.
43 D. VAN HOANG, M. ROVERY AND P. DESNUELLE, *Biochim. Biophys. Acta*, 58 (1962) 613.
44 J. PORATH AND A. V. SCHALLY, *Endocrinology*, 70 (1962) 738.
45 G. S. MARKS, R. D. MARSHALL, A. NEUBERGER AND H. PAPKOFF, *Biochim. Biophys. Acta*, 63 (1962) 340.
46 Y. C. LEE AND R. MONTGOMERY, *Arch. Biochem. Biophys.*, 97 (1962) 9.
47 E. M. PRESS AND R. R. PORTER, *Biochem. J.*, 83 (1962) 172.
48 M. BAGDASARIAN, N. A. MATHESON, R. L. M. SYNGE AND M. A. YOUNGSON, *Biochem. J.*, 91 (1964) 91.
49 J. E. FOTHERGILL AND R. C. NAIRN, *Nature*, 192 (1961) 1073.
50 W. GEORGE AND K. W. WALTON, *Nature*, 192 (1961) 1188.
51 J. E. COLEMAN AND B. L. VALLEE, *J. Biol. Chem.*, 237 (1962) 3430.
52 J. P. HUMMEL AND W. J. DREYER, *Biochim. Biophys. Acta*, 63 (1962) 530.
53 A. A. BARBER, C. DEMPSTER AND N. G. ANDERSON, *Clin. Chim. Acta*, 8 (1963) 143.
54 T. L. HARDY AND K. R. L. MANSFORD, *Biochem. J.*, 83 (1962) 34-35P.
55 J. PORATH AND H. BENNICH, *Arch. Biochem. Biophys.*, Suppl. 1 (1962) 152.
56 B. G. JOHANSSON AND L. RYMO, *Acta Chem. Scand.*, 16 (1962) 2066.
57 B. G. JOHANSSON AND L. RYMO, *Acta Chem. Scand.*, 18 (1964) 217.
58 H. DETERMAN, *Experientia*, 18 (1962) 430.
59 W. K. DOWNEY AND P. ANDREWS, *Biochem. J.*, 92 (1964) 6P.
60 C. J. O. R. MORRIS, *Biochem. J.*, 92 (1964) 6P.
61 C. J. O. R. MORRIS, *J. Chromatog.*, 16 (1964) 167.
62 J. PORATH, *Nature*, 196 (1962) 47.
63 F. MIRANDA, H. ROCHAT AND S. LISSITZKY, *J. Chromatog.*, 7 (1962) 142.
64 A. N. GLAZER AND D. WELLNER, *Nature*, 194 (1962) 862.
65 B. GELLOTTE, *J. Chromatog.*, 3 (1960) 330.
66 B. LINDQVIST, *Acta Chem. Scand.*, 16 (1962) 1794.
67 S. HJERTÉN, *Biochim. Biophys. Acta*, 62 (1962) 445.
68 S. HJERTÉN, *Biochim. Biophys. Acta*, 79 (1964) 393.
69 S. BENGTSSON AND L. PHILIPSON, *Biochim. Biophys. Acta*, 79 (1964) 399.
70 S. HJERTEN, *Arch. Biochem. Biophys.*, 99 (1962) 466.
71 R. L. STEERE AND G. K. ACKERS, *Nature*, 196 (1962) 475.
72 S. HJERTÉN AND R. MOSBACH, *Anal. Biochem.*, 3 (1962) 109.
73 S. HJERTÉN, *Arch. Biochem. Biophys.* Suppl. 1 (1962) 147.
74 C. C. CURTAIN, *Anal. Biochem.*, 6 (1963) 512.

[75] T. C. Laurent and J. Killander, *J. Chromatog.*, 14 (1964) 317.
[76] G. K. Ackers, *Biochemistry*, 3 (1964) 723.
[77] E. J. Cohn and J. T. Edsall, *Proteins, Amino Acids and Peptides*, Reinhold, New York, 1943, pp. 402 and 422.
[78] P. Andrews, *Biochem. J.*, 91 (1964) 222.
[79] J. Porath, *Proceedings, International Symposium on Pharmaceutical Chemistry, Florence*, Butterworth, London, 1962.
[80] J. R. Whitaker, *Anal. Chem.*, 35 (1963) 1950.
[81] P. M. Sanfelippo and J. G. Surak, *J. Chromatog.*, 13 (1964) 148.

CHAPTER 9

COLUMN CHROMATOGRAPHY OF PROTEINS

Since the beginnings of biochemistry the two most common procedures for the purification of proteins have been precipitation with salts or specific reagents and adsorption to on various solids. Although these methods are still widely used today, especially in batch operations, a detailed account is beyond the scope of this volume. A useful aid when working with ammonium sulphate as a precipitant is the nomogram constructed by Dixon[4] (see Fig. 92). Column chromatography, however, is a more satisfactory procedure in analytical work and some recent developments are briefly discussed.

ADSORPTION CHROMATOGRAPHY ON CALCIUM PHOSPHATE

Calcium phosphate has a long history of use in protein purifications but we owe our knowledge of its properties in column chromatography to the systematic studies of Tiselius *et al.*[1,2]. Precipitated calcium phosphate, although crystalline in structure as shown by X-ray patterns, nevertheless is a gel of high water content. The most stable form, especially at high pH values is hydroxyl-apatite [$Ca_5(PO_4)_3OH$] and its preparation is as follows[3].

Two litres of each of 0.5 *M* aqueous solutions of calcium chloride and disodium hydrogen phosphate are allowed to run into a beaker simultaneously with stirring, at a rate of about 120 drops per min. The precipitated brushite is washed with 4 l of distilled water four times by decantation, suspended in 4 l of water and boiled with 100 ml of 40% sodium hydroxide (by weight) for 1 h. The precipitate is washed again with water four times by decantation, 0.01 *M* sodium phosphate buffer, pH 6.8, is added and the mixture heated to incipient boiling. After decantation more buffer is added and the solution boiled for 5 min. The precipitate is then boiled with fresh 0.01 *M* buffer for 15 min and twice with 0.001 *M* phosphate buffer for 15 min. The hydroxyl-apatite is finally stirred in 0.001 *M* phosphate buffer at pH 6.8.

A suspension of the adsorbent is poured into a chromatograph

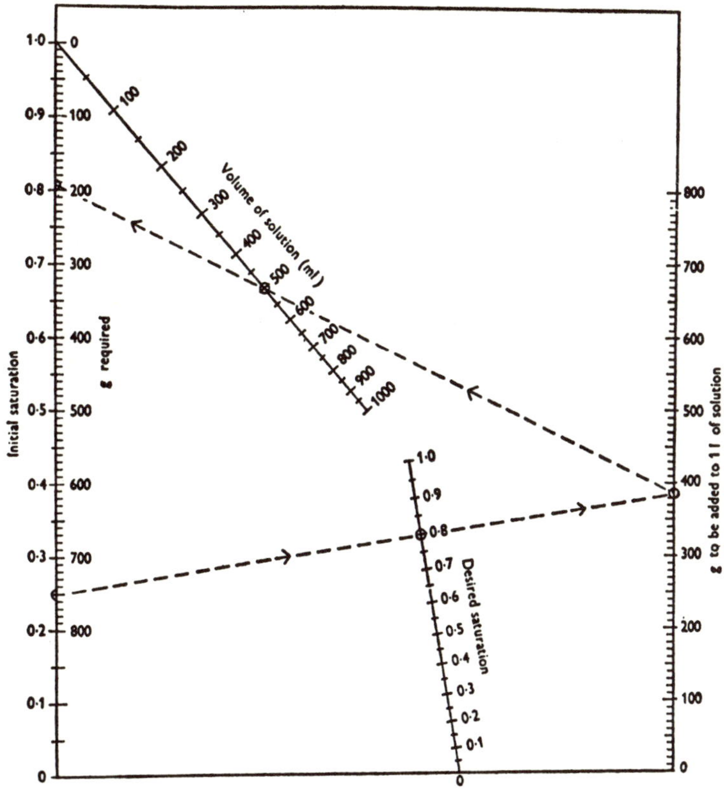

Fig. 92. Nomogram for ammonium, sulphate solutions (Ref. [4], p. 302). A straight line through the initial saturation and the desired saturation gives the amount of solid $(NH_4)_2SO_4$ to be added to 1 litre of the solution. A line from this point passing through the volume of the solution gives the amount required. Reproduced by kind permission of Dr. M. Dixon, and the Biochemical Society.

tube closed at the lower end with a plug of glass wool. The rate of flow can be adjusted by application of gentle pressure to the column when the adsorbent will be sufficiently compressed to remain in the column when the tube is inverted. Adsorbent remaining on the walls of the tube can then be washed out with buffer solution. If necessary the top layer of the column (1 cm) is stirred with a glass rod and allowed to settle in order to form a level surface. A column of

References p. 302

dimensions 1.5 × 15 cm is suitable for elution chromatography of a 50-mg protein sample.

Elution curves obtained from this type of column require careful interpretation as the mechanism of the procedure is different from that in partition chromatography or in ion-exchange chromatography. During adsorption chromatography considerable tailing will occur unless the composition of the eluting solution is such that the R_F of a given solute approaches 1.0. Successful fractionation of a mixture of proteins will depend on the use of buffer concentrations being sufficiently far apart in ionic strength to provide clean cut elution of the components of the mixture from the column. In many cases serial elution of proteins has been accomplished without difficulty, but it must be borne in mind that peaks obtained from a mixture of unknowns may contain several proteins for each elution range. On the other hand, a single protein possessing a strongly curved isotherm might be expected to provide a discrete peak with each increase in ionic strength over the more extended elution range. It is easy to confuse adsorption chromatography with other forms of chromatography. Perhaps the best criterion for the former is a lack of proportionality between the column length and the absolute position of a peak on the elution curve. Where true partition or ion-exchange effects are involved the elution volumes for a given solute will be directly proportional to the column lengths.

It is the experience of many workers that step-wise elution from calcium phosphate columns is a more useful procedure than that employing a continuous concentration gradient (gradient elution). The reason for this is that marked displacement of one protein by another occurs in this type of system and, in fact, displacement development is a good alternative to elution. Proteins that can be chromatographed are in the molecular weight range 10,000 to several million; low molecular weight material such as amino acids and peptides are not adsorbed. The affinity of hydroxylapatite for a particular protein might be expected to change slightly with the nature of the bound cation as it is known that the crystal lattice exchanges calcium ions for sodium and potassium ions to different extents[5]. Wolf and Sly[6] have reported the effects of buffer cations on the elution patterns obtained when soybean proteins are chromatographed on hydroxylapatite. The behaviour of a protein on thin layers of the adsorbent might indicate the feasibility of a separation on a column[7].

Applications

The examples cited here illustrate the use of calcium phosphate in the form of hydroxyl-apatite. Accounts of results obtained from the use of other forms of calcium phosphate are to be found in recent reviews[8,9].

The columns lend themselves particularly well to the separation of the high molecular weight coloured proteins of algae. Tiselius[2] has described the fractionation of protein mixtures from the sea-weed *Ceramium rubrum*. The red protein, phycoerythrin is eluted at 0.005–0.05 *M* phosphate buffer, pH 6.8, and the blue protein, phycocyanin, is eluted at 0.05–0.07 *M* concentration. The red protein can be resolved further by rechromatography into three zones which are eluted in 0.01 *M*, 0.01–0.02 *M* and 0.05 *M* phosphate buffer. Gradient elution over the range at pH 6.8 gives a single extended zone. The molecular weights of phycoerythrin and phycocyanin are 290,000 and 270,000 respectively, and their iso-electric points are at pH 4.25 and 4.85. Haxo *et al.*[10] have studied the chromoproteins from red and blue-green algae. In this case the column was first washed with 1% sodium chloride solution and the proteins were eluted with buffers at pH 6.0–6.5 in the concentration range 0.0025–0.1 *M*.

When bovine plasma albumin was applied to a column 1 × 13 cm in 0.02 *M* buffer at pH 6.8 and eluted step-wise, three separate zones were obtained and the recovery of protein was 100%. Rechromatography of these fractions after dialysis showed that they appeared again at the same three different phosphate concentrations (Fig. 93). Gradient elution of human γ-globulin gave a very broad peak in the range 0.005–0.13 *M* phosphate, pH 7.0, in the presence of 0.2 *M* sodium chloride. Rechromatography of portions of the zone gave small peaks in the expected places. Other proteins studied were ovalbumin, lysozyme, protamine sulphate and hemocyanin (*Helix*).

Examples of enzyme purifications on calcium phosphate columns are liver catalase[11] (mol.wt. 249,000), β-galactosidase[12] from an ascomycete, choline esterase[13] from serum, a nitrate reductase[14] from wheat leaves, the flavoprotein diaphorase and cytochrome reductase from heart muscle[15]. An interesting separation of lact-albumin from other whey proteins has been described[16]. 250 mg of the proteins soluble at 50% saturation with ammonium sulphate

References p. 302

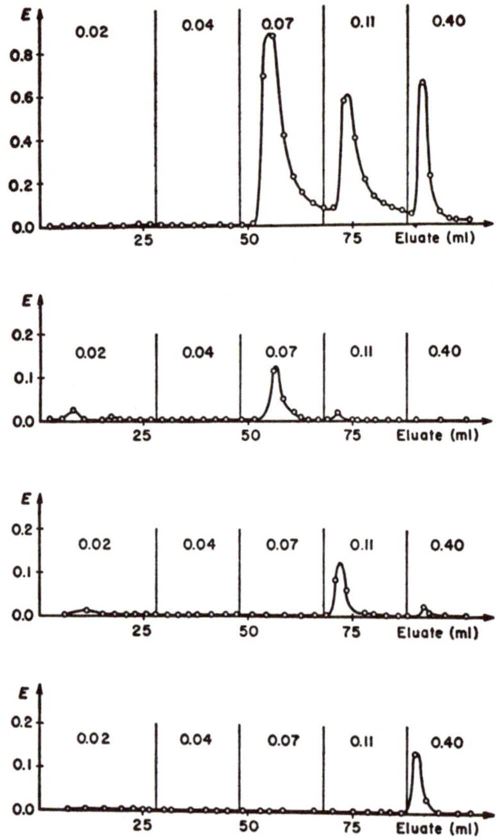

Fig. 93. Chromatographic behaviour of bovine serum albumin in step-wise elution (top diagram) and rechromatography of the fractions obtained (three lower diagrams). Tiselius *et al.*[3].

were applied to a 2.8 × 8 cm column of calcium phosphate in 0.03 *M* sodium phosphate, pH 6.9, which was 0.1 *M* with respect to sodium chloride. Lactalbumin, serum albumin and an iron-containing globulin were successively eluted from the column with 0.03 *M*, 0.12 *M* and 0.5 *M* phosphate buffer containing 0.1 *M* sodium chloride.

CHROMATOGRAPHY OF PROTEINS ON ION-EXCHANGE RESINS

Whereas the principles governing ion-exchange chromatography of small molecules like amino acids are fairly well understood, the behaviour of proteins on resins presents certain features for which there is still no satisfactory explanation. One difficulty is to reconcile the fact of true reversible distribution with the concept of a multi-point attachment of the protein to the resin. The problem has been discussed in a recent review by Moore and Stein[8] who point out that the effective polyvalency of a protein molecule might be diminished through steric effects. Many cationic sites within the interior of small dense proteins are possibly sufficiently screened to become unavailable for bonding with the resin.

A detailed study of the effect of pH and cation concentration on the adsorption of the basic protein cytochrome *c* on the resin Amberlite IRC-50 was made by Boardman and Partridge[17]. The resin Amberlite IRC-50 is a polycarboxylic acid of structure similar

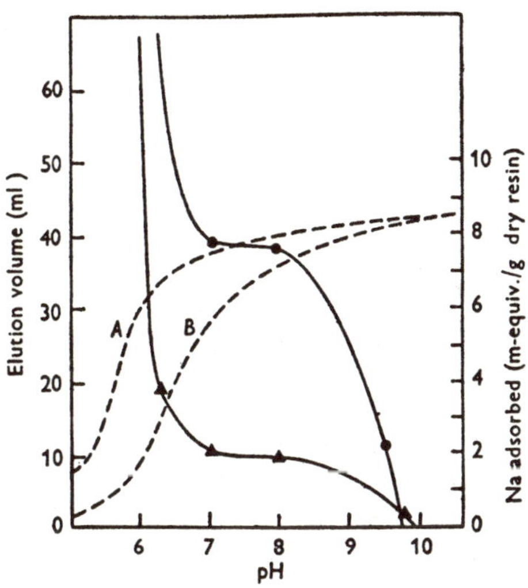

Fig. 94. Effect of pH on the elution of cytochrome *c* by sodium phosphate buffers. Column, IRC-50, 0.9 × 16.0 cm; load, 6.0 mg; temperature, 25°; Na^+ concentration, ● = 0.25 g ions/l, ▲ = 0.34 g ions/l. The broken lines show the amount of Na^+ taken up by the resin; A, from 1.0 *M* NaCl; B, from 0.1 *M* NaCl. Boardman and Partridge[17].

to that shown on p. 63. The weakly acidic carboxyl groups are ionized at pH 3 and the resin is not completely saturated with Na^+ below pH 8 or 9 as will be seen from Fig.94. Cytochrome *c* has an isoelectric point of 10.1 and a molecular weight of 13,000. Fig. 94 shows the effect of pH on the elution of the protein at two different Na^+ concentrations. This may be compared with the curve for the effect of pH on the adsorption of lysine (isoelectric point, pH 9.7), Fig. 95. Whereas both cytochrome *c* and lysine are desorbed at high

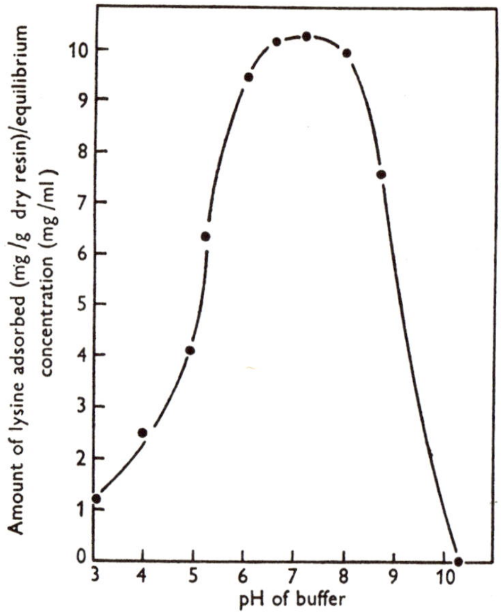

Fig. 95. Effect of pH on the adsorption of lysine by IRC-50 (from Boardman and Partridge[17]).

pH as the isoelectric points are approached, the two substances behave quite differently at low pH values. Although the carboxyl groups of the resin are no longer charged at low pH cytochrome *c* is not desorbed under acid conditions.

It is the view of Boardman and Partridge that secondary short-range forces (hydrogen bonding and van der Waals forces) are responsible for bonding of the protein at low pH, most probably due to removal of a layer of electrostatically bound water. The fact that two proteins of similar molecular weight and net charge, lysozyme and cytochrome *c*, can be fractionated on IRC-50 is explained by different secondary forces operating in proteins dissimilar in configuration.

Amberlite IRC-50

Although a good chromatographic grade of IRC-50 is available it is advisable to screen the commercial product through a 200 mesh sieve or use the procedure of Hamilton (see p. 74), to prepare a batch resin of particle size in the range 60–70 μ. Before use the resin should be converted to the sodium form by adding 40% sodiun hydroxide solution to a suspension of the resin in water. When the pH of the suspension reaches about 11 the resin is filtered and washed thoroughly with water until the washings are about pH 10. The resin is then converted back to the acid form by passing 3 *N* hydrochloric acid through the filter and finally washing with water. Equilibration of the resin to a suitable working pH is a rather slow process. The best way is to suspend the resin in the buffer solution (usually 0.2 *M* sodium phosphate) and stir in 40% sodium hydroxide to maintain the pH between 5 and 6. After each addition of alkali time must be given for complete neutralization of the carboxyl groups; this is particularly important as the desired pH is approached. The resin is then washed with 10–20 litres of buffer until the pH of the filtrate is identical with that of the inflowing solution. Columns are poured in sections and for analytical purposes are usually 1 cm in diameter and 15–30 cm long. Some workers have used a roll of silver gauze inserted in the stem of the tube and touching the sintered plate to promote even flow of the effluent from the column. The pH of the effluent is checked before the sample, containing 5–10 mg of protein, is applied to the column. A suitable reservoir is connected to the top by means of flexible plastic tubing and the head of liquid adjusted to give a flow rate of 1–1.5 ml/h. The hold-up volume of a 30 cm column is 6–7 ml. After a run the column is washed with many volumes of buffer. If the column becomes contaminated the resin is heated on a steam bath with 4 *N* hydrochloric acid for 4 h and then washed with 1 *N* sodium hydroxide. The effluent fractions may be analyzed by any of the methods described in Chapter 11.

Applications

It was Paleus and Neilands[18] who introduced IRC-50 as a useful tool for the purification of proteins. Taking advantage of the high isoelectric point of cytochrome *c* useful purification was possible on a column at pH 9.0. This was confirmed in later studies by Margoliash[19]. Extensive studies by Hirs[20,21], and Tallan *et al.*[22] led to the purification of a number of proteins by the procedure outlined above. Thus ribonuclease, lysozyme and chymotrypsinogen were purified at pH 6.47, 7.18 and 6.02 respectively. The isoelectric points of these proteins are at about pH 8, 10 and 9, and the molecular weights are 13,700, 14,000 and 22,500. Fig. 96 shows the frac-

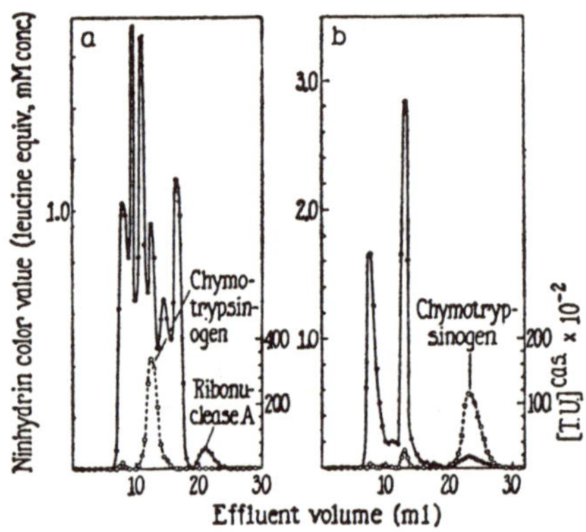

Fig. 96. Chromatography of 0.25 *N* sulphuric acid extracts of comminuted beef pancreas on a 0.9 × 30 cm column of IRC-50 (XE-64). a, curve obtained at pH 6.47 with an undialyzed extract; b, curve obtained after dialysis of the extract and chromatography at pH 6.02. ●, ninhydrin value; ○, proteolytic activity after activation with trypsin (from Hirs[21]).

tionation of 0.25 *N* sulphuric acid extracts of comminuted beef pancreas on two columns of IRC-50. Recovery of protein and enzyme activity was practically quantitative and upon rechromatography the enzymes were recovered at the same effluent volumes. As much as 20 g of pancreatic extract was fractionated on preparative columns of dimensions 7.5 × 60 cm.

A 10- to 15-fold concentration of ribonuclease with respect to the buffer salts is described by Hirs[20]. A solution of the enzyme in phosphate buffer, pH 5.35–5.40, is passed through a short column (2 cm in diameter) of IRC-50 which has been equilibrated with 0.2 *M* sodium phosphate buffer at pH 5.38. Under these conditions the enzyme is adsorbed to the extent of 30 mg/ml of resin bed. Desorption is effected by suspending the resin in an equal volume of phosphate buffer, pH 7.96 (0.2 *M* in phosphate, 0.38 *N* in sodium ion), and adding 5 *N* sodium hydroxide to bring the pH to 8 during 30 min. Most of the enzyme is then eluted and the resin can be filtered and washed with a little buffer.

The barium form of IRC-50 has been used by Crampton *et al.*[23,24] to fractionate the highly basic histones obtained from cell nuclei of various tissues. The less basic half of the total histones were eluted with barium acetate of concentration gradient 0.1–1.0 *M*, at pH 6 or 7. Other interesting separations have been testicular hyaluronidase[26], phosphodiesterase from rattlesnake venom[25], phospholipase A[27] and bacterial pectinase[28].

The work of Boardman and Partridge[17] on the CO haemoglobins (isoelectric points at about pH 7) has shown that neutral proteins may also be fractionated on the resin. A curious effect found by Huisman and Prins[29] is that the order of elution of four CO haemoglobins from IRC-50 at pH 6.5 is different from that which might be expected from their electrophoretic mobilities at pH 6.75. From the resin the rates of elution are $F>A>B>C$; the electrophoretic mobilities are in the order $A<F<B<C$. It would appear that overall structure is a factor governing the separations on the resin. Another neutral protein purified on IRC-50 is the interstitial cell stimulating hormone (β-ICSH)[30] of molecular weight 30,000 and isoelectric point at pH 7.3.

Bovine thrombin[31] and the growth hormones[32] obtained from human and monkey pituitary glands have acid isoelectric points yet they have been purified on the resin.

A new departure in protein chromatography is the use of strong hydrogen bonding or dissociating agents such as urea and dimethylformamide in the eluting solutions. Some proteins and large peptides are insoluble in strong salt solutions or possess isoelectric points within the pH-ranges employed in chromatography. Separation of proteins with similar isoelectric points is usually enhanced by operating chromatograms near the isoelectric points of the protein[33]. In these circumstances the incorporation of solubilizing

References p. 302

agents like urea will prove beneficial provided it can be demonstrated that no denaturation of the particular protein takes place. It is not generally realized that dimethylformamide is a formylating agent[34] and that strong urea solutions may contain significant amounts of reactive cyanate ions. The irreversible inactivation of proteins brought about by prolonged exposure to urea has been discussed by Stark *et al.*[35].

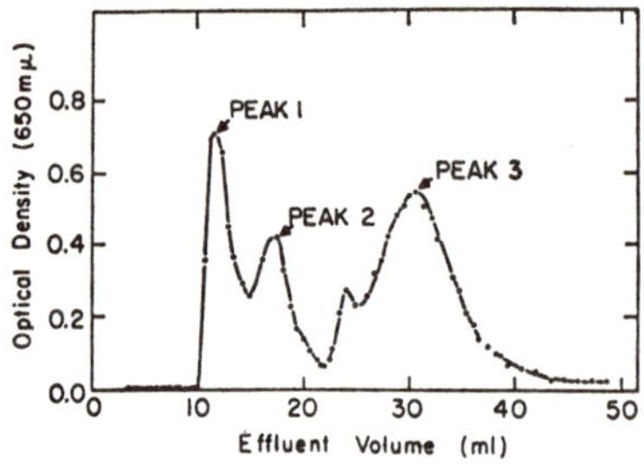

Fig. 97. Chromatography of amorphous porcine insulin which contains a significant amount of deamido-insulin. Column, IRC-50, 0.9 × 20 cm. Eluent, 0.13 *M* phosphate–7 *M* urea at pH 6.0. Peak 1, deamido-insulin; Peak 2, product of transformation of insulin by urea; Peak, 3 main insulin peak. Cole[36].

The principles of ion-exchange chromatography appear to operate just as well in strong urea solutions as in ordinary buffer solutions. The separation of the A and B chains of insulin on a column of Dowex 50-X2 at pH 1.9 is described on p. 130. Cole[36] has reported the chromatography of different species of insulin on IRC-50 at pH 6.0 in 0.13 *M* phosphate–7 *M* urea and the partial separation of glucagon and insulin under the same conditions on a column 1 × 30 cm[37]. One of the samples of porcine insulin studied by Harfenist and Craig[38] and known to contain a large proportion of deamido-insulin was submitted to chromatography in phosphate–urea buffer by Cole. The result shown in Fig. 97 indicates a likely identity of peak 1 with deamido-insulin. In studies of the chromatographic behaviour of *Myrothecium* cellulose a urea gradient was employed[39]. Urea in effluents from the columns was estimated by the colórimetric method of Fawcett and Scott[40].

Dowex 2 and Dowex 50

These resins and the corresponding Amberlites IRA-400 and IR-120 are more often employed in a mixed bed for desalting protein solutions than for true chromatographic operations[41,42]. With a finely pulverized form it is possible to effect useful separations although the low-capacity is a serious drawback. Boman and Westlund[43] used an 8–10% cross-linked Dowex 2, 200–400 mesh for a systematic study and a number of interesting facts were elucidated. Before use the resin was washed with 1 *N* hydrochloric acid. After equilibration at pH 7.3 serum proteins were adsorbed at low ionic strength (0.02–0.04 *M*) and eluted by step-wise increases in molarity. Better control of pH was possible with the cation buffer Tris-HCl than with anionic buffers such as acetate. The capacity of Dowex 2 for serum albumin at pH 7.3 is 0.5 mg/ml resin so it was necessary to operate the column near the saturation level. As with calcium phosphate a single protein might be eluted in a number of small peaks representing the differences between the amounts of protein necessary to saturate the resin at the particular concentrations. Boman found that the order of elution of serum proteins from Dowex 2 was not only different from the electrophoretic one but also different from that obtained with calcium phosphate columns and cellulose

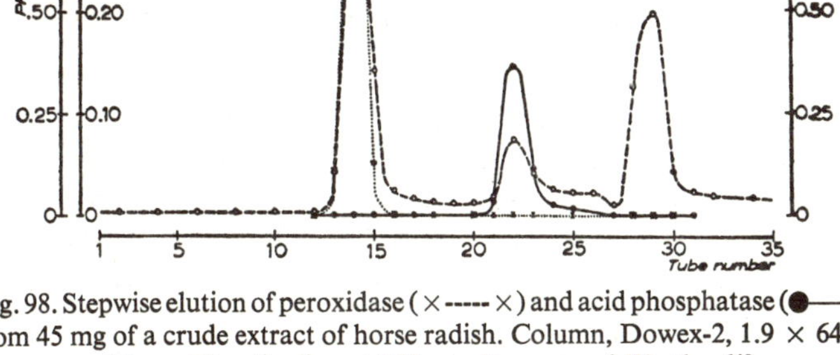

Fig. 98. Stepwise elution of peroxidase (×----- ×) and acid phosphatase (●——●) from 45 mg of a crude extract of horse radish. Column, Dowex-2, 1.9 × 64 cm. (○-----○) extinction at 280 mμ. Boman and Westlund[43].

ion exchangers. Here again, as with IRC-50, non-ionic adsorption affinities might be assumed to play a role in the total adsorption effect. Fig. 98 shows the separation of peroxidase and acid phosphatase from horse radish obtained by step-wise elution. For preparative purposes it is recommended to run the column with the displacement technique and then re-run using step-wise or gradient elution.

A five-fold purification of a phosphatase with a 75% recovery of activity was obtained by Boman[44,45] on a column of Dowex-50. Frieden *et al.*[46] reported a 15-fold purification of relaxin using the same resin but the recovery of activity was very low (30–50%). Most of the protein in the sample travelled through the column unretarded.

Fractionation of egg yolk granule proteins (5.4 g) has been carried out on a 2.3 × 55 cm column of Dowex 1-X2 with sodium acetate as eluent[47].

CELLULOSE ION EXCHANGERS

A most important advance in the subject of protein chromatography was made in 1956 when Peterson and Sober[48] described new types of ion exchangers based on cellulose. These materials, which are made by modifying cellulose chemically, have a very high capacity for proteins and can be endowed with any desired ion-exchange characteristic. Organic chemistry can provide a wide range of chemicals suitable for combination with cellulose and there is no reason why derivatives based on sulphydryl groups, lipid groups, antigens etc. should not be explored. The preparation of seven different derivatives of cellulose will be described.

Diethylaminoethyl (DEAE-) cellulose[48] *(anion exchanger)*

A solution of 40 g of sodium hydroxide in 170 ml of water is stirred into 60 g of Solka-Floc, care being taken to distribute the liquid as evenly as possible through the particles. The mixture is allowed to stand in an ice-bath for 30 min with occasional stirring, and a solution of 35 g of recrystallized 2-chlorotriethylamine hydrochloride in 45 ml of water is then added in several portions (fume-cupboard). After thorough blending, the mixture is immersed in an oil-bath at 80–85° for 35 min and stirred occasionally. The product is cooled in an ice-bath and 250 ml of 2 *M* sodium chloride is thoroughly mixed in. The thick suspension is filtered through a 5 inch coarse sintered-glass funnel and the residue washed with 1 *N* sodium hydroxide until the emerging filtrate is practically colourless. The material is resuspended in about 350 ml of 1 *N* hydrochloric acid to make a strongly acid suspension which is then filtered immediately. The residue is washed successively with 350 ml portions of 1 *N* sodium hydroxide, 1 *N* hydrochloric acid and 1 *N* sodium hydroxide, without intermediate washing with water. The cellulose is finally suspended in 250 ml of 1 *N* sodium hydroxide and diluted with water to 3 litres. After standing overnight, the cloudy supernatant liquid is decanted, and the sediment washed 5 or 6 times by decantation to remove more fine particles. Coarse, aggregated material is also removed at this stage. The sediment is filtered to remove as much water as possible and then washed with three 500-ml portions of ethanol. The product can be stored moist with alcohol or *in vacuo* after drying. When cellulose finer than 325 mesh is used a yield of 45 g of adsorbent is obtained; coarse cellulose gives greater yields. Approximately 1.4% of nitrogen (1.0 mequivs. per g) is combined with the cellulose.

Some workers have experienced difficulty in obtaining a product that will filter reasonably well and permit useful flow rates in a column. It is advisable then to treat the cellulose with a smaller proportion of 2-chlorotriethylamine. In the writer's experience products obtained by treatment of Solka-Floc (Solka Floc SW-40B, Johnsen, Jorgensen and Wettre, Ltd., London) with 50–70% of the amount of base hydrochloride quoted above filtered readily and gave fast running columns. Adsorbents prepared in this way contained 0.40–0.60 mequivs. of basic nitrogen/g.

2-Chlorotriethylamine hydrochloride is prepared as follows. 2-Diethylaminoethanol (100 g) in 200 ml of chloroform are mixed in a 1-litre flask and cooled in ice. Thionyl chloride (140 ml) is added slowly with cooling and shaking during 40 min. When the addition is complete the mixture is allowed to stand at room temperature for 3–4 h. Chloroform and excess thionyl chloride are then distilled off in apparatus protected from atmospheric moisture under a hood. The distillation is stopped when the volume of liquid remaining in the distillation flask is about half of the starting volume. Ether is added to the cooled liquid to precipitate the crystalline

References p. 302

hydrochloride. On standing overnight at 0° the mixture sets to a solid mass. This is transferred to a Buchner funnel, the mother liquor sucked off, and the crystals washed with ether. After one recrystallization from methanol–ether the crystals melt at 206°.

Little difficulty should be experienced in packing a column of cellulose if it is supported at the base by a disc of porous plastic (Porous Plastics, Ltd., Dagenham Works, Essex). Any tendency to channeling can be corrected by compression of the bed using air or nitrogen at 1 psi, or by stirring the bed to about 2 cm below the loaded portion. Before use it is advisable to precycle two or three times with 0.2 *N* HCl and 0.2 *N* NaOH using boiled out water in the preparation of the reagents.

ECTEOLA-cellulose[48] *(weak anion exchanger)*

A solution of 60 g of sodium hydroxide in 150 ml of water is stirred into 60 g of Solka-Floc in the same way as for DEAE-cellulose and treated with a mixture of 35 ml of triethanolamine and 60 ml of epichlorhydrin added in several portions. After thorough mixing the mixture is immersed in an oil-bath at 75–80° for 30 min and stirred occasionally. The gummy mass is cooled in ice and worked into a smooth suspension with 250 ml of 2 *M* sodium chloride solution. The product is filtered, washed and dried in the manner described for DEAE-cellulose. About 60 g of adsorbent containing 0.25–0.30 mequivs. of basic nitrogen per g is obtained.

Carboxymethyl (*CM-*) *cellulose*[48] *(cation exchanger)*

A solution of 90 g of sodium hydroxide in 200 ml water is stirred into 60 g of Whatman cellulose powder (325 mesh) and cooled in an ice-bath as above. A solution of 30 g of chloroacetic acid in 40 ml of water is added in 3 or 4 portions with good mixing and the mixture heated in an oil-bath at 70° for 20 min. After cooling in ice the yellow product is treated with 500 ml of 10% acetic acid in several portions. The strongly alkaline suspension is diluted to two litres with water and allowed to settle. The sediment is then washed several times by decantation to remove colour, alkali and fine particles. Subsequent filtration of the cellulose is aided by adding sufficient glacial acetic acid (about 10 ml) to make the suspension just acid to Congo red. The product is finally washed free of acid and stored wet with ethanol or dried. About 45 g of white powder are obtained containing 0.7 mequivs. of acidic groups per g.

Material containing a high number of ionizable groups tend to become too gelatinous for use in columns. Better products (about 0.45–0.50 mequivs. per g) are obtained when about half the quantity of chloroacetic acid quoted above is used.

Phosphato (P-) cellulose[48]

A solution of 75 g of sodium hydroxide in 185 ml of water is stirred into 50 g of Whatman cellulose and allowed to stand in ice for 30 min. Water (330 ml) is added during 15 min with slow mechanical stirring, followed by a solution of 40 ml of phosphorus oxychloride in 80 ml of ether added dropwise during 50 min. During the reaction the temperature is maintained at 25–30° and the final pH is about 6. Water is then added and the cellulose washed repeatedly by decantation. White powder (38 g) is obtained containing 1.9% phosphorus (0.6 mmole phosphate per g).

Sulphomethyl cellulose[49]

Methylene dichloride (85 g) dissolved in 200 ml of ethanol is added to 126 g of sodium sulphite (Na_2SO_3) in 300 ml of water and the mixture heated to 120° in an autoclave for 6 h with stirring. The sulphonated product ($ClCH_2SO_3Na$) is extracted from the dried residue after cooling by boiling with methanol (yield about 70 g). A solution of 200 g of sodium hydroxide in 300 ml of water is stirred into 100 g of cellulose powder and the mixture allowed to swell for a few hours. The cellulose is then treated with 10 g of $ClCH_2SO_3Na$ in 60 ml of water. The mixture is then dried at 90–95° to remove most of the water, cooled, suspended in 1 litre of 95% ethanol and treated with 1 litre of 1 *N* hydrochloric acid. The cellulose is washed with 1 litre of 0.5 *N* hydrochloric acid and then with water until the filtrate is neutral.

References p. 302

Triethylaminoethyl (TEAE-) cellulose[49]

This material has been prepared by heating 80 g of DEAE-cellulose with 350 ml of 10% ethyl bromide in ethanol for 4 h. The product was suitable for chromatography after it had been washed with 1 litre of 1% w/v sodium hydroxide solution and finally with water and acetone.

p-Nitrobenzyl-cellulose[50]

Solka-floc (4 g) is mixed with *p*-nitrobenzyl chloride (12 g) and 40% sodium hydroxide (30 ml) and stirred vigorously at 95°. Cooling is required during the first part of the exothermic reaction. After 4 h the mixture is poured into a large excess of cold water, filtered, and the residue washed with water, ethanol and acetone. The material contains about 1.5 benzyl groups per glucose unit.

Diazobenzyl-cellulose[50]

p-Aminobenzyl-cellulose is first prepared from the nitro compound by heating a suspension in ethanol (50 ml) with sodium hydrosulphite ($Na_2S_2O_4 \cdot 2H_2O$; 5 g) for 30 min. The light yellow product (5 g) is washed, suspended in 2 *N* HCl (10 ml), mixed with 20 ml of water and cooled in an ice-bath. 0.5 *M* $NaNO_2$ solution is added slowly with stirring until a test with KI-starch paper remains positive for 15 min after addition of the last portion of nitrite. After a further 15 min the material is filtered and washed with weakly acid ice water.

Applications

Cellulose ion exchangers are now being used in a variety of biochemical investigations and a few examples will illustrate the wide range of applications. The effluent diagram of freshly prepared human serum shown in Fig. 99 is taken from the studies of Sober *et al.*[51]. 20 ml of dialyzed serum (156 mg N) were applied to a 2.5 × 40 cm column of DEAE-SF (28 g of adsorbent) and eluted at a rate of 11 ml/h according to the following schedule.

I. 0.005 *M* sodium phosphate, pH 7.0.
II. Gradient to 0.02 *M* sodium phosphate, pH 6.0.
III. Gradient to 0.05 *M* sodium dihydrogen phosphate.
IV. Gradient to 0.02 *M* sodium chloride–0.05 *M* sodium dihydrogen phosphate.
V. Gradient to 0.05 *M* sodium chloride–0.05 *M* sodium dihydrogen phosphate.
VI. Gradient to 0.1 *M* sodium chloride–0.05 *M* sodium dihydrogen phosphate.

This elution schedule makes use of a substantial salt concentration gradient combined with a slow reduction of pH from 7 to about 5. Owing to the buffering action of the adsorbent and the proteins a peak eluted from the column during application of a particular gradient is never in contact with solution of that composition. Other elution schedules are possible and guidance may be taken from the following facts established by Sober *et al.*[51]. Affinity for a specific protein can be reduced in at least three ways:

(1) a decrease in the pH of the solution, reducing the anionic character of the protein;

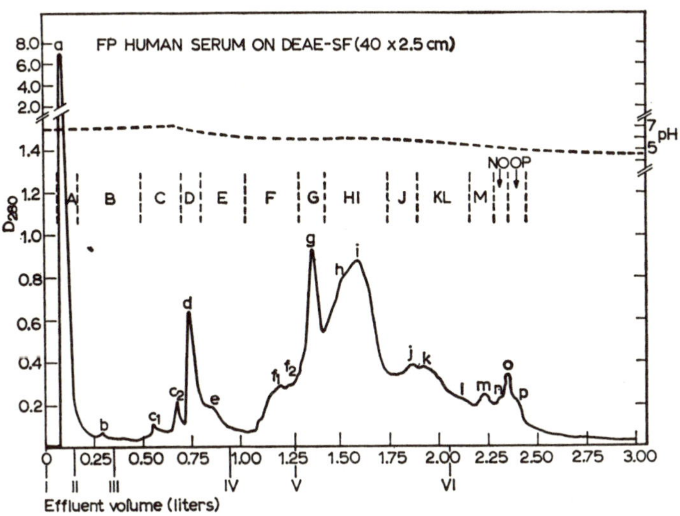

Fig. 99. Chromatography of freshly prepared human serum on a column of DEAE-cellulose (from Sober *et al.*[51]). See text for conditions.

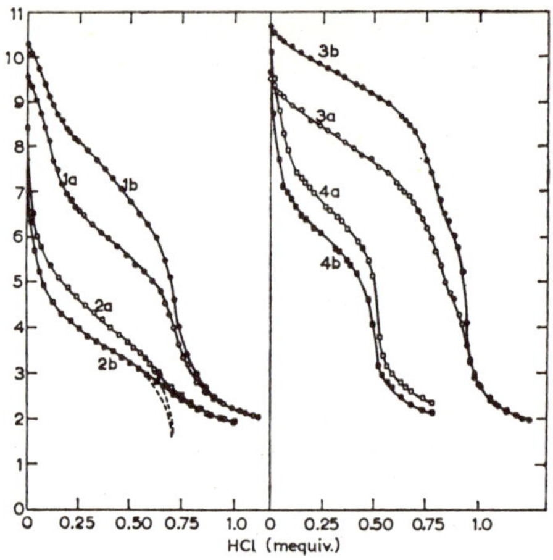

Fig. 100. Titration curves of four cellulose ion exchangers. 1, ECTEOLA-SF; 2, CM-W; 3, DEAE-SF; 4, P-W. (a) Curves obtained in water alone, (b) those obtained in 0.5 *M* NaCl. The small shoulder on the acid side of the DEAE-SF was apparently due to bicarbonate formation during the washing procedure. Peterson and Sober[48].

(2) an increase in the pH of the solution, suppressing the ionization of the basic groups of the adsorbent and thus reducing its anion-binding capacity;

(3) an increase in the ionic strength of the buffer.

A valuable property of these exchangers is their remarkable adsorption capacity for proteins. Good resolving power was still maintained when the chromatogram shown in Fig. 99 was run with 5 times the load (775 mg N). The buffering power of the adsorbents is indicated by the titration curves reproduced in Fig. 100. The addition of sodium chloride at the start of the titration resulted in p*K* values which were closer to those exhibited by similar basic or acidic groups in free solution.

Fig. 101 shows an interesting comparison of paper and boundary electrophoresis patterns obtained from fractions taken from a chromatogram similar to that in Fig. 99. The letters correspond to the zones in Fig. 99. As far as fraction K the components are eluted in the expected reverse order of their respective electrophoretic mobilities. The titration curves of DEAE-cellulose reported by O'Donnell and Thompson indicate a marked effect of temperature

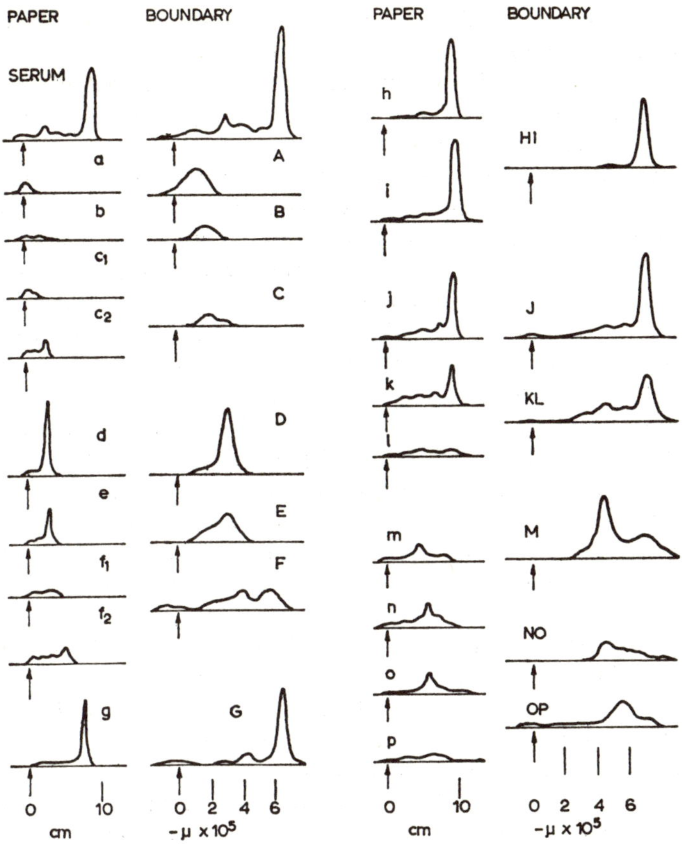

Fig. 101. Comparison of paper and boundary electrophoresis patterns of chromatographic fractions from Fig. 99 (from Sober *et al.*[51]).

on the ionization of the basic groups[52]. A decrease of temperature lowers the elution volume of a protein.

Cathepsin D, the principal proteolytic enzyme of ox spleen has been found to exist in at least ten different forms (Press *et al.*[53]). Four different groups, α, β, γ, δ, were separated initially on DEAE-cellulose at pH 8.4 using sodium borate–phosphate buffer in the concentration range 5.25–21 m*M*. Each group was then resolved further on carboxymethylcellulose columns by gradient elution with 0.01–0.2 *M* sodium acetate buffer, pH 5.5. Rechromatography on carboxymethylcellulose seems to have established the existence of two α enzymes, three β enzymes, two γ enzymes and three or more δ enzymes. Electrophoresis studies showed that in general the most rapidly eluted enzymes from DEAE-cellulose had the lowest mobilities at pH 9.1.

The pepsin family of enzymes has been investigated using DEAE-cellulose[54]. Crude pepsin (6 g) was applied to a 5 × 35 cm column in 0.1 *M* acetate, pH 4.0, and eluted in the gradient 0–0.2 *M* sodium chloride in the same buffer. Two proteolytic enzymes, parapepsin I and parapepsin II, were obtained as small peaks running ahead of the main pepsin peak. They differed from pepsin in their N-terminal residues, phosphorus content, and in their general stability and activity. When crude pepsinogen was chromatographed on DEAE-cellulose in 0.02 *M* phosphate, pH 6.9, and a salt gradient applied (0–0.45 *M* NaCl) a small peak was eluted after the main pepsinogen peak. The small component is considered to be the zymogen of parapepsin II[55]. Parapepsinogen II is more basic than parapepsin II. Pepsin itself is very much more basic than its zymogen and requires 1% sodium hydroxide for elution from DEAE-cellulose[56].

L-Myosin has been purified[57] on DEAE-cellulose in 0.25 *M* potassium chloride, 5 m*M* Tris-HCl, pH 7.5. Although the specific ATPase activity was not constant along the eluted peak the total activity was higher than that of the starting material. Ability to combine with actin to form actomyosin was retained.

Another example of the purification of an enzyme from very crude sources is illustrated by the isolation of the latent polyphenolase activity from whole juice of broad bean leaves[58]. The enzyme was easily located in the effluent from a column of DEAE-cellulose, 1 × 30 cm, to which had been applied whole juice from 50 g of fresh leaves. The DEAE-cellulose contained 0.50 mequivs. of basic nitrogen/g and was initially equilibrated in 5 m*M* potassium phosphate buffer, pH 7. After 20 ml had passed through the column 0.5 *M* potassium hydroxide was applied through a 100 ml mixing chamber to establish a concentration gradient and a pH gradient to 8.5. Rechromatography on a similar column gave a single and fairly symmetrical peak which represented a 500-fold increase in a catecholase activity in 90% overall yield. Adsorption probably plays a role in this fractionation.

CM-cellulose has been particularly useful in the purification of a number of protein hormones. A thyroid-stimulating hormone preparation from the anterior lobes of bovine pituitaries was adsorbed at pH 6 in 0.01 *M* sodium phosphate buffer and the TSH activity eluted with 1 *M* sodium chloride[59]. It is interesting to note that quantitative recovery was possible from CM-cellulose but the yield from IRC-50 was only 50%. Commercial trypsin has been purified on a 0.7 × 9 cm column of CM-cellulose with 0.05–0.1 *M* sodium citrate–0.013 *M* $CaCl_2$ at pH 6.02 as eluent[60].

Wheat gluten proteins have been fractionated on carboxymethyl cellulose[61] by discontinuous elution with aqueous hydrochloric acid (0.006–0.030 *N*) over the pH range 3.40–1.50. The proteins isolated corresponded to the major electrophoretic components of gluten and were eluted in the order of increasing electrophoretic mobilities. Simmonds and Winzer[62] chromatographed wheat flour proteins on CM-cellulose at the higher pH of 4.1 and were able to keep the proteins soluble by using buffers containing 1.0 mole of dimethylformamide per litre. Six components were eluted using a sodium chloride concentration gradient followed by a pH gradient (4.1–12).

Sulphomethyl cellulose in its Mg^{2+} form was used by Malmström to purify yeast enolase in gram quantities[63]. A highly active enzyme was eluted from the adsorbent with 1 *M* potassium chloride. ECTEOLA has been used for the fractionation of viruses from contaminating nucleic acids usually in a sodium chloride concentration gradient. Tobacco mosaic virus[64], ECHO-13 virus[65], and T_2-bacteriophage[66] from lysed cultures of *E. coli* are reported to have been purified in this way. Antibodies free of inactive protein may be prepared in reasonable yield utilizing a column of cellulose-linked antigen[50,67]. Protein antigens are coupled to a diazocellulose derived from aminocellulose. A new chromatographic material, 'organomercurial polysaccharide', has been developed[68,69] for the specific separation and isolation of SH-proteins.

Chromatography on cellulose ion exchanges followed by gel filtration or *vice versa* are procedures common to many schemes for the fractionation of biological materials. The two separational principles operate simultaneously in columns of CM- or DEAE-Sephadex, and this may offer some advantage. Some recent examples exploiting this new medium are the purification of a streptococcal proteinase[70], the separation of various isozymes of lactatedehydro-

References p. 302

genase[71,72], the preparation of purified collagenase[73], and the separation of the peptide chains of insulin[74].

Comments

The rate of ion-exchange by ECTEOLA-cellulose in the range pH 5–8 is very slow as a certain amount of hydroxyl ions is tenaciously retained during equilibration of the columns with neutral buffers[75,76].

Björk[77,78] has drawn attention to the risk of chromatographic artifacts arising from the interaction of anion exchangers and alkaline buffers with atmospheric carbon dioxide. Many workers have at some time come across artificial zoning on anion exchangers; an example in the literature[79] where highly purified bovine carbonic anhydrase was split up into two active peaks (probably identical) provides a salutary warning against rash assumptions regarding heterogeneity.

GRADIENT ELUTION

Gradient elution chromatography begins with a weakly displacing eluent and finishes with a sufficiently strong eluent to remove the most strongly held substances from the column. The gradient illustrated in Fig. 102, curve c (often termed a convex gradient) will be found satisfactory for many purposes, particularly for initial exploratory experiments. Linear and concave gradients (Fig. 102, curves a and b) are more useful systems as they provide the best elution characteristics for solutes whose adsorption isotherms are non-linear. The eluting solvent is introduced gradually at first and then more rapidly so that strongly retained solutes are eluted with considerably less tailing of the zones. A further advantage is the greatly increased sensitivity. The very broad bands characteristic of conventional elution chromatography tend to get lost within the drift of the base line and escape detection.

Theoretical elution curves can be calculated from the formula[80]:

$$C = C_2 - (C_2 - C_1)(1 - V)^{\frac{A_2}{A_1}}$$

where C is the concentration of eluting solute in the solution present-

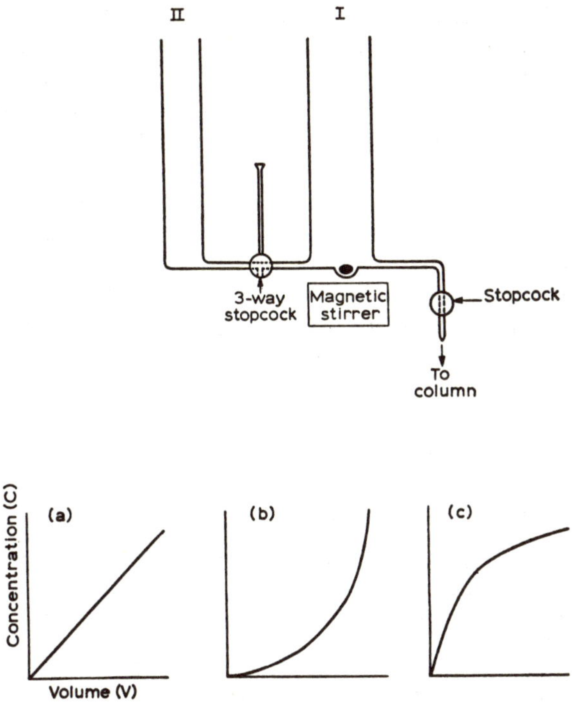

Fig. 102. Schematic diagram of apparatus for gradient elution. Curves in lower part of figure; (*a*) linear gradient, (*b*) concave gradient, (*c*) convex gradient. See text for further explanation.

ed to the column (Fig. 102), C_1 the initial concentration of eluting solute in vessel I, C_2 the initial concentration of solute in vessel II, V the fraction of the total volume already passed, A_1 the cross-sectional area of vessel I, and A_2 the cross-sectional area of vessel II. If $A_1 = A_2$, C will increase as a linear function of volume; if A_2 is greater than A_2 a concave gradient will result. With vessel I stoppered the volume of solution in I will remain constant until the solution has completely drained from vessel II. This is the situation in the ordinary fixed-volume mixing device which provides a convex-gradient.

The theory of gradient elution chromatography has been treated by several authors since the method was introduced by Alm *et al.*[81]

in 1952. An article by Svensson[82] should be consulted for a general review; other recent contributions dealing with different aspects are by Bader and Morgan[83], Katz[84], and Snyder[85]; design of apparatus is discussed by Peterson and Sober[86], Nelson[87], and Whitaker and Mittelstadt[88]. A theory of the development of pH gradients was presented by Reiner and Reiner[89].

Equipment is available commercially for recording salt concentrations and pH gradients. Apparatus for continuous recording of absorbance of column effluents at selected wavelengths in the ultraviolet range is also available although a good quality spectrophotometer can be readily modified for satisfactory monitoring. Simple arrangements are those of Crestfield[90] and McClendon[91]; more ambitious systems are described by Anderson[92].

REFERENCES

1 S. M. Swingle and A. Tiselius, *Biochem. J.*, 48 (1951) 171.
2 A. Tiselius, *Arkiv Kemi*, 7 (1954) 443.
3 A. Tiselius, S. Hjertén and Ö. Levin, *Arch. Biochem. Biophys.*, 65 (1956) 132.
4 M. Dixon, *Biochem. J.*, 54 (1953) 457.
5 W. F. Neuman, T. Y. Toribara and B. J. Mulryan, *Arch. Biochem. Biophys.*, 98 (1962) 384.
6 W. J. Wolf and D. A. Sly, *J. Chromatog.*, 15 (1964) 247.
7 A. F. Hofmann, *Biochim. Biophys. Acta*, 60 (1962) 458.
8 S. Moore and W. H. Stein, *Advan. Protein Chem.*, 11 (1956) 191.
9 C. A. Zittle, *Advan. Enzymol.*, 14 (1953) 319.
10 F. Haxo, C. O'hEocha and P. Norris, *Arch. Biochem. Biophys.*, 54 (1955) 162.
11 V. E. Price and R. E. Greenfield, *J. Biol. Chem.*, 209 (1954) 363.
12 B. van Hofsten, *Nature*, 177 (1956) 844.
13 B. G. Malmström, Ö. Levin and H. G. Boman, *Acta Chem. Scand.*, 10 (1956) 1077.
14 W. Anacker and V. Stoy, *Biochem. Z.*, 330 (1958) 141.
15 V. Massey, *Biochim. Biophys. Acta*, 37 (1960) 310.
16 B. Johansson, *Nature*, 181 (1958) 996.
17 N. K. Boardman and S. M. Partridge, *Biochem. J.*, 59 (1955) 543.
18 S. Paleus and J. B. Nielands, *Acta Chem. Scand.*, 4 (1950) 1024.
19 E. Margoliash, *Biochem. J.*, 56 (1954) 529, 535.
20 C. H. W. Hirs, S. Moore and W. H. Stein, *J. Biol. Chem.*, 200 (1953) 493.
21 C. H. W. Hirs, *J. Biol. Chem.*, 205 (1953) 93.
22 H. H. Tallan, S. Moore and W. H. Stein, *J. Biol. Chem.*, 200 (1953) 507.
23 C. F. Crampton, S. Moore and W. H. Stein, *J. Biol. Chem.*, 215 (1955) 787.
24 C. F. Crampton, S. Moore and W. H. Stein, *J. Biol. Chem.*, 225 (1957) 363.
25 P. S. Rasmussen, *Biochim. Biophys. Acta*, 14 (1954) 567.
26 M. P. de Garilhe and M. Laskowski, *Biochim.Biophys. Acta*, 18 (1955) 370.

[27] W. P. Neuman and E. Habermann, *Biochem. Z.*, 327 (1955) 170.
[28] P. W. Tallboys, *Nature*, 166 (1950) 1077.
[29] T. H. S. Huisman and H. K. Prins, *J. Lab. Clin. Med.*, 46 (1955) 255.
[30] P. G. Squire and C. H. Li, *Science*, 127 (1958) 32.
[31] P. S. Rasmussen, *Biochim. Biophys. Acta*, 16 (1955) 157.
[32] C. H. Li and H. Papkoff, *Science*, 122 (1955) 920.
[33] R. L. Hill, J. R. Kimmel and E. L. Smith, *Ann. Rev. Biochem.*, 28 (1959) 97.
[34] E. Campaigne and W. L. Archer, *J. Am. Chem. Soc.*, 75 (1953) 989.
[35] G. R. Stark, W. H. Stein and S. Moore, *J. Biol. Chem.*, 235 (1960) 3177.
[36] R. D. Cole, *J. Biol. Chem.*, 235 (1960) 2294, 3484.
[37] R. D. Cole, *J. Biol. Chem.*, 235 (1960) 2300.
[38] E. J. Harfenist and L. C. Craig, *J. Am. Chem. Soc.*, 74 (1952) 3083.
[39] D. R. Whitaker, K. R. Hanson and P. K. Datta, *Can. J. Biochem. Physiol.*, 41 (1963) 671.
[40] J. K. Fawcett and J. E. Scott, *J. Clin. Pathol.*, 13 (1960) 156.
[41] E. O. P. Thompson, *Nature*, 169 (1952) 495.
[42] S. M. Partridge, *Nature*, 169 (1952) 496.
[43] H. G. Boman and L. E. Westlund, *Arch. Biochem. Biophys.*, 64 (1956) 217.
[44] H. G. Boman, *Nature*, 173 (1954) 447.
[45] H. G. Boman, *Biochim. Biophys. Acta*, 16 (1955) 245.
[46] E. H. Frieden, M. W. Noall and F. Alonso de Florida, *J. Biol. Chem.*, 222 (1956) 611.
[47] R. W. Burley and W. H. Cook, *Can. J. Biochem. Physiol.*, 39 (1961) 1295.
[48] E. A. Peterson and H. A. Sober, *J. Am. Chem. Soc.*, 78 (1956) 751.
[49] J. Porath, *Arkiv Kemi*, 11 (1957) 87.
[50] D. H. Campbell, E. Luescher and L. S. Lerman, *Proc. Natl. Acad. Sci. U.S.*, 37 (1951) 575.
[51] H. A. Sober, F. J. Gutter, M. M. Wyckoff and E. A. Peterson, *J. Am. Chem. Soc.*, 78 (1956) 756.
[52] I. J. O'Donnell and E. O. P. Thompson, *Australian J. Biol. Sci.*, 13 (1960) 69.
[53] E. M. Press, R. R. Porter and J. Cebra, *Biochem. J.*, 74 (1960) 501.
[54] A. P. Ryle and R. R. Porter, *Biochem. J.*, 73 (1959) 75.
[55] A. P. Ryle, *Biochem. J.*, 75 (1960) 145.
[56] I. E. Liener, *Biochim. Biophys. Acta*, 37 (1960) 522.
[57] S. V. Perry, *Biochem. J.*, 74 (1960) 94.
[58] J. L. Bailey, *Biochem. J.*, 79 (1961) 514.
[59] P. G. Condliffe and R. W. Bates, *J. Biol. Chem.*, 223 (1956) 843.
[60] S. Maroux, M. Rovery and P. Desnuelle, *Biochim. Biophys. Acta*, 56 (1962) 202.
[61] J. H. Woychik, R. J. Dimler and F. R. Senti, *Arch. Biochem. Biophys.*, 91, (1960) 235.
[62] D. H. Simmonds and D. J. Winzer, *Nature*, 189 (1961) 306.
[63] B. G. Malmström, *Arch. Biochem. Biophys.*, 70 (1957) 58.
[64] B. Commoner, J. A. Lippincott, G. B. Shearer, E. Richman and J. H. Wu, *Nature*, 178 (1956) 767.
[65] B. H. Hoyer, E. T. Bolton, R. A. Ormsbee, G. LeBouvier, O. B. Ritter and C. L. Larson, *Science*, 127 (1958) 859.

[66] A. TAUSSIG AND E. H. CREASER, *Biochim. Biophys. Acta*, 24 (1957) 448.
[67] N. R. MOUDGAL AND R. R. PORTER, *Biochim. Biophys. Acta*, 71 (1963) 185.
[68] L. ELDJARN AND E. JELLUM, *Acta Chem. Scand.*, 17 (1963) 2610.
[69] E. JELLUM AMD L. ELDJARN, *Biochem. J.*, 92 (1964) 7P.
[70] T-Y. LIU, N. P. NEUMANN, S. D. ELLIOTT, S. MOORE AND W. H. STEIN, *J. Biol. Chem.*, 238 (1963) 251.
[71] R. RICHTERICH, P. SCHAFROTH AND H. AEBI, *Clin. Chim. Acta*, 8 (1963) 178.
[72] E. D. WACHSMUTH AND G. PFLEIDERER, *Biochem. Z.*, 336 (1963) 545.
[73] S. KELLER AND I. MANDL, *Arch. Biochem. Biophys.*, 101 (1963) 81.
[74] S. FITTKAU, *Naturwiss.*, 15 (1963) 522.
[75] H. A. VEDER, *J. Chromatog.*, 10 (1963) 507.
[76] H. A. VEDER AND C. N. PASCHA, *J. Chromatog.*, 13 (1964) 408.
[77] W. BJÖRK AND H. G. BOMAN, *Biochim. Biophys. Acta*, 34 (1959) 503.
[78] W. BJÖRK, *J. Chromatog.*, 2 (1959) 536.
[79] W. BJÖRK AND I. SVENSSON, *J. Chromatog.*, 8 (1962) 479.
[80] R. M. BOCK AND N. LING, *Anal. Chem.*, 26 (1954) 1534.
[81] R. S. ALM, R. J. P. WILLIAMS AND A. TISELIUS, *Acta Chem. Scand.*, 6 (1952) 826.
[82] H. SVENSSON, in: P. ALEXANDER AND R. J. BLOCK (Eds.), *A Laboratory Manual of Analytical Methods of Protein Chemistry, Vol.* 1, Pergamon Press, London, 1960.
[83] H. BADER AND H. E. MORGAN, *Biochim, Biophys. Acta*, 57 (1962) 562.
[84] S. KATZ, *Anal. Biochem.*, 5 (1963) 7.
[85] L. R. SNYDER, *J. Chromatog.*, 13 (1964) 415.
[86] E. A. PETERSON AND H. A. SOBER, *Anal. Chem.*, 31 (1959) 857.
[87] G. J. NELSON, *Anal. Biochem.*, 5 (1963) 116.
[88] D. R. WHITAKER AND K. A. MITTELSTADT, *Can. J. Biochem. Physiol.*, 42 (1964) 149.
[89] J. M. REINER AND B. REINER, *Anal. Biochem.*, 4 (1962) 1.
[90] A. M. CRESTFIELD, *Anal. Chem.*, 35 (1963) 1762.
[91] J. H. McCLENDON, *Anal. Biochem.*, 3 (1962) 94.
[92] N. G. ANDERSON, *Anal. Biochem.*, 4 (1962) 269.

CHAPTER 10

ZONE ELECTROPHORESIS OF PROTEINS

Recent years have witnessed a flood of communications dealing with the electromigration of proteins, peptides and amino acids in stabilizing media as a result of keen interest in separational methods. Boundary separations provide only partial separation of components and an anticonvectional medium was necessary to stabilize the boundaries and facilitate the subsequent separation of the components in zones. For a number of years interest has centred largely on paper electrophoresis for the separation of proteins and where a rapid preliminary examination of a protein mixture is required this technique can provide useful information. As a medium for preparative work and for fine analytical separations it has been almost completely superseded by other inert materials.

PAPER ELECTROPHORESIS

The literature on paper electrophoresis has been complicated a little by the varied terminology used by different workers. The term 'electrophoresis' was used by Michaelis[1] in 1909 to describe the migration of colloid ions in an electric field and in their review of 1945 Martin and Synge[2] defined ionophoresis as the electromigration of low molecular weight ions in stabilizing media. Separations based on charge differences of both small and large molecules are done in similar or identical apparatus and for this reason the majority of workers tend to use the term 'electrophoresis' with a suitable qualifying adjective to indicate the supporting medium. The apt expression 'zone electrophoresis' was suggested by Tiselius and his collaborators to distinguish between these methods and moving boundary electrophoresis. It would seem unnecessary to introduce special terms to indicate the influence of secondary effects due to chromatography, adsorption or evaporation.

The main physical factors effecting migration rates in paper are the type and concentration of buffer, control of evaporation, electro-endosmosis, and the electric field. These factors will be considered separately.

References p. 337

Effect of pH and ionic strength of the buffer

The technique of paper electrophoresis in its simplest form consists in the application of a spot or band of solute to a moistened paper strip and maintenance of a potential difference across the ends. Concentration changes in the applied spot or band during electrophoresis will influence migration and an interesting analysis of the conditions which may arise was made by Edward[3].

Consider two strips of paper I and II uniformly moistened with buffer electrolyte B^+X^- on which is placed a very small volume of solute A^+X^- containing the ion A^+ which is the sample under study (Fig. 103). It is assumed that an equal volume of background solution is displaced from the area of the spot and uniformly distributed over the rest of the strip by capillary suction. In strip Ia the solution AX has displaced completely the background solution from the area of spot PQ. In strip IIa the solution of AX has mixed with a much larger volume of the background solution, so that the area of the spot PQ contains a solution of BX of concentration similar to the background, and a very small concentration of AX.

The passage of current through the solutions represented in Ia causes the ions A^+X^- to migrate in opposite directions, resulting in a shift in the zone AX shown in Ib and Ic. The migration of A^+ beyond

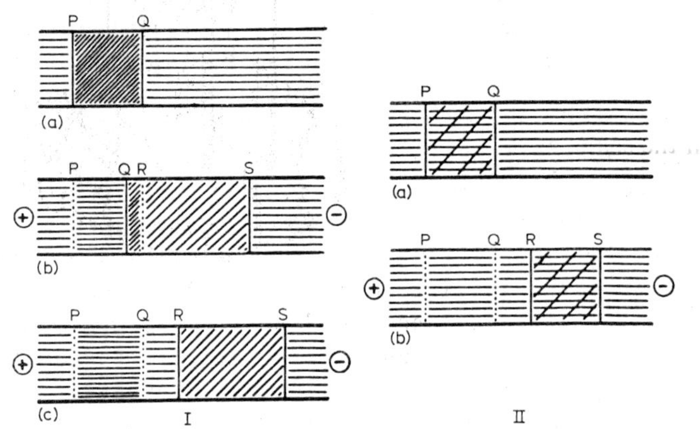

Fig. 103. Changes in positions and concentrations of AX (diagonal shading) and BX (horizontal shading) on passing a current through the solutions (relative concentrations shown by depth of shading). (From Edward[3].)

the boundary Q leads to the formation of a zone QS (Ib) of AX, in which the concentration of $A^+(C'_A)$ may differ from the concentration (C_A) in the original zone PQ (Ia), and in RQ of Ib. Eventually the rear boundary R of the zone of AX passes Q (Ic), and the concentration of A^+ throughout the zone RS is C'_A. From the Kohlrausch 'regulating function' (discussed in de Wael and Wegelin[4]) it may be shown that

$$C'_A(U_A + U_X)/U_A = C_B(U_B + U_X)/U_B \quad (1)$$

where C_B is the concentration of B^+ in the original background solution, and U_A, U_B and U_X are the mobilities of A^+, B^+, and X^- respectively, and are defined by

$$u = \lambda/F$$

where λ is the equivalent conductance of an ion and F the Faraday equivalent in coulombs.

The following possibilities may be distinguished:

(a) $U_A = U_B$ and $C_A = C_B$. It follows from equation (1) that $C'_A = C_A$. The concentration of the zone of A^+ is unchanged as it travels from PQ (Ia) to RS (Ic), and the zone will remain unchanged in size and shape.

(b) $U_A = U_B$ but $C_A > C_B$. This is the situation represented in Ia–Ic. Then $C'_A < C_A$. This can happen only if the front boundary S of the zone A^+ travels more rapidly than the rear boundary R, as in Ib, so that the zone becomes elongated in the direction of the current. After R has passed through Q, as in Ic, front and rear boundaries travel at the same speed and there is no further change in the shape of the zone from this effect.

(c) $U_A = U_B$ but $C_A < C_B$. The zone contracts in the direction of the current.

(d) $C_A = C_B$ but $U_A > U_B$. The zone contracts.

(e) $C_A = C_B$ but $U_A < U_B$. The zone elongates.

Zones tend to be most compact when the background concentration is high. As Edward points out, it is known from experience that if the ionic strength on the spot is much greater than in the background solution a long trailing zone results or, in extreme cases there is no movement of the spot at all. Diffusion of an ion is proportional to the square of the concentration and when $U_A = U_B$ blurring due to diffusion is unaffected by the passage of the current. When how-

References p. 337

ever the ion behind a moving boundary has a lower mobility than the ion ahead of it, the boundary is sharpened during the passage of current. On the other hand blurring is accelerated when the ion behind the boundary has a higher mobility than the ion ahead. Thus the conditions in case (d) above may explain the comet-shaped zone sometimes observed in the migration of non-adsorbed ions. In IIa–IIb the potential gradient is approximately equal along the length of the strip and the zone should move without change in concentration or shape and the effects predicted above will be diminished.

The pH of the background buffer will determine the net charge of the migrating ion. A particle with a net positive charge migrates towards the cathode and a particle with a net negative charge migrates towards the anode. The size and shape of the particle will influence the mobility as may be seen from the expression where Q

$$u = \frac{Qd}{4\pi r^2 \eta}$$

is the charge on the particle, d the distance of the ionic double layer, r the radius of the particle and η the viscosity of the solution. A dissymmetrical molecule moves less easily in paper than a globular one. Thus the tendency of a protein to denaturation is manifest in paper electrophoresis by elongation of the zone. Mobility is increased considerably by a reduction in the radius of the molecule and consequently during electrophoresis of serum proteins for example, any low molecular weight peptides or amino acids present will migrate off the paper.

The isoelectric point of a pure protein is the pH at which the protein has zero mobility in an electric field. Its value for a particular protein will depend on the nature of the solvent and the concentration of electrolytes. Isoionic protein is obtained by removal of all non-colloidal ions from a solution of the protein in water and the pH of the solution thus obtained has been termed the *isoionic point*. There are, however, a few examples in the literature where the isoionic point is defined as the pH of the protein solution which does not change on the addition of more isoionic protein (see Ref. [22], p. 275).

The ionic strength of a buffer is given by the expression:

$$\mu = \tfrac{1}{2}\sum_{1}^{n} C_i Z^2_i$$

where C_i is the concentration in moles per litre, and Z_i the valence of each type of ion i in a solution containing n ions. As the ionic strength increases more current is carried by the buffer ions resulting in a depressing effect on the mobility of a protein and a sharpening of the boundary.

Disturbances due to evaporation, and changes in ionic strength created by the application of the spot render it impossible to determine absolute mobilities in a paper medium. In a comparison of mobilities of serum proteins in paper and in free electrophoresis Kunkel and Tiselius[5] point out that the difference in values can be explained by the more tortuous route followed by the protein molecules in paper. Whereas in free electrophoresis the protein migrates through a distance,

$$d = \frac{uti}{qk}$$

(where u = mobility, t = time, i = current density in amperes, q = cross-section in cm^2, k = conductivity in ohms^{-1}cm^{-1}), in paper

$$d' = \frac{uti}{q_a k}$$

where $d' = d \times (l'/l)$ and q_a is the cross-sectional area of the paper strip. The correction factor l'/l depends on the type and condition of the paper employed, and once this is determined for different papers the mobility values become practically identical. The approximate cross-sectional area q_a of buffer is obtained by weighing the filter paper between glass plates before and after wetting with buffer. The path of ionic migration in the paper l' is obtained from $l' = Rq_a k$ where R is the resistance of the paper strip.

When an electric field is applied to moist paper the immobile paper phase being negatively charged causes an *endosmotic flow* of solution towards the cathode. It is possible to change the polarity of the paper by chemical means[6] but it is more usual to tolerate the flow or neutralize it to some extent with an opposing gravitational flow. Endosmotic flow is roughly proportional to the potential gradient and is less in buffers of high ionic strength. It can be measur-

References p. 337

ed with substances which are electrically neutral or have very low mobilities. Where mobility values have to be corrected a substance is chosen having a similar molecular volume to the molecule under study. Dextrans have been used to run with proteins and glucose, sucrose, hydrogen peroxide, *o*-nitroaniline[6] or 2,4-dinitrophenylethanolamine[7] with simple ions like amino acids or peptides.

Evaporation and electric field

The most difficult problem associated with zone electrophoresis is control over the evaporation of the buffer solution. The problem has been studied thoroughly and it is even possible to classify electrophoresis methods on the basis of the degree of evaporation control. A paper strip with its very small cross-section of buffer is more susceptible than other systems to the effect of evaporation. Simple exposure to the air induces salt concentration gradients in the strip which will result in a non-uniform field when current is applied. It has been shown that salts will accumulate at the centre of the strip as well as along the edges.

The passage of an electric current accentuates evaporation owing to the conversion of electrical energy to heat energy. The Joule heat evolved is given by the expression

$$H = \frac{RI^2}{A} = \frac{VI}{4.8}$$

where H = no. calories per sec., A = mechanical heat equivalent, I = current and R = resistance of strip. As the temperature of the strip rises the resistance falls slightly and the current tends to rise. Migration velocity is a function of the current so the use of a constant current DC power supply offers an advantage over the constant voltage type. Another effect of the current is ionophoresis of the buffer electrolytes. Differences in the mobilities of the ions making up the buffer may lead to pH changes within the strip and in the bridge section of the apparatus.

Apparatus

Apparatus for low voltage paper electrophoresis may be constructed very simply from glass or plastic and the form described by Grass-

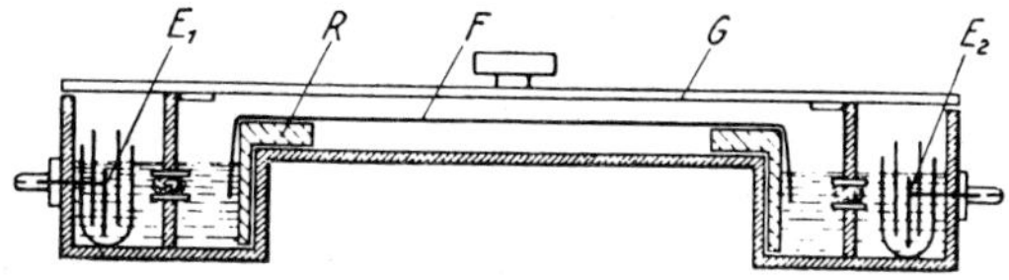

Fig. 104. Electrophoresis chamber according to Grassmann *et al.*[8]. G, glass cover; E_1 and E_2, electrodes; F, filter paper strip; R, plastic rack.

mann *et al.*[8] in 1951 is suitable for most purposes (Fig. 104). In this apparatus F is the filter paper, R a plastic rack, E_1, and E_2 the electrodes and G a glass cover. The electrodes are at the centre of a baffle system which prevents electrolysis products diffusing too rapidly into the bridge chamber. A gap below the glass lid allows gaseous products to escape and the two compartments are connected by a small tube plugged with glass wool. The levels of buffer solution at each end of the strip must be equal to avoid syphoning.

Application of a sample is best carried out on paper that has been moistened with buffer solution and partially dried between filter papers. A smooth-tipped pipette or capillary tube is useful for this, care being taken not to scratch the surface of the paper. The line of application should be clear of both edges of the paper by 7–8 mm. It is very easy to overload a paper with a protein solution; the best concentrations to use are between 2 and 10% and the sample volume need not exceed 0.005 ml per cm.

It is usual to apply a potential of 5–10 V per cm across the strip; high potential gradients generate more heat and reproducible results are more difficult to obtain. Before experimenting with an unknown protein solution it is advisable to check the apparatus and conditions with a known material such as serum albumin. If a small amount of bromophenol blue solution is added to the serum albumin the migration behaviour can be observed. A selection of suitable buffers is listed in Table 49.

Detection and estimation of proteins

The usual method of detection of the protein zones is by staining with dyes. Where the dye-binding capacity is known for a particular protein a fairly good quantitative assessment is possible. Before staining, however, the proteins must be rendered insoluble by chemical means or by heat denaturation in an oven at 110° for 20 min.

TABLE 49

BUFFERS FOR PAPER ELECTROPHORESIS

Buffer	*pH*	*μ*	*Composition for 1 litre*
Veronal	8.6	0.1	3.68 g diethylbarbituric acid 20.6 g sodium salt
Veronal-acetate	8.6	0.1	9.8 g sodium diethylbarbiturate 3.9 g sodium acetate 60 ml 0.1 *N* HCl
Phosphate	7.8	0.12	0.29 g $NaH_2PO_4.H_2O$ 3.25 g Na_2HPO_4
Phosphate	4.6	0.15	20.4 g KH_2PO_4
Phosphate	6.5	0.1	3.11 g KH_2PO_4 1.49 g Na_2HPO_4
Ammonium acetate	6.8		260 ml 2 *M* acetic acid 500 ml 1 *M* NH_4OH
Borate	9.0		0.62 g boric acid 7.63 g sodium borate
Phthalate	5.9		5.10 g potassium acid phthalate 0.86 g sodium hydroxide
Glycine	9.6	0.1	3.0 g glycine 0.829 g sodium hydroxide 4.68 g sodium chloride
Tris*	9.0		60.5 g Tris 6.0 g EDTA 4.6 g Boric acid

* Aronsson and Grönwall[9].

The optimum staining conditions are best determined in trial experiments.

(1) *Amido Black 10B (Naphthalene Black 12B.200)*. In the original method of Grassmann and Hannig[40] the strip is dipped in a saturated solution of the dye in methanol containing 10% acetic acid for 10 min. The excess dye is washed out with methanol containing

10% acetic acid or with many changes of 1% aqueous acetic acid. The coloured zones may be eluted from the strip with 10% aqueous acetic acid, 50% methanol or ethanol–0.5 *N* sodium hydroxide (1:1). Solutions of the dye in the latter solvent should be stabilized with EDTA (500 mg/l)[10]. Maximal absorption for this dye is at 620 mμ.

(2) *Lissamine Green (Light Green SF150 I.C.I.)*. A solution of 5 mg of dye in 100 ml 50% w/v trichloroacetic acid is generally used. After 12 h in the solution the excess dye is removed with 2% v/v aqueous acetic acid. The zones are eluted with 40% (v/v) acetic acid or with alkali. Light absorption is measured at 600 mμ.

(3) *Bromophenol Blue*. A litre of solution contains 100 mg of dye, 50 ml acetic acid and 50 g zinc sulphate. The strip is heated to 120° for 30 min before staining or treated with an aqueous solution containing 1% mercuric chloride, 0.05% bromophenol blue and 2% acetic acid. Excess dye is removed in 0.5% acetic acid and the zones are eluted with 0.5% v/v sodium carbonate or with 50% (v/v) methanol. Light absorption is measured at 585 mμ.

For quantitative work the zones can be located by the absorption in ultraviolet light prior to staining and the appropriate areas can then be cut out for elution. Many excellent texts are available which deal with the direct photometry of paper strips especially with reference to routine clinical and diagnostic investigations.

Dyes for staining lipoproteins are Sudan Black, Oil Red O and Oil Blue N. They are used as saturated solutions in 60% (v/v) ethanol and applied during a period of 4–12 h. Glycoproteins are conveniently detected with a boiling solution of diphenylamine in acetic acid[11].

ZONE ELECTROPHORESIS IN OTHER SUPPORTING MEDIA

As a general method for the fractionation of protein mixtures and for the study of homogeneity, electrophoresis in media such as granular starch, powdered cellulose, or resin beads offers considerable advantages over paper electrophoresis. Large-scale separations are possible usually yielding amounts of the resolved materials sufficient for further analytical or enzymic study. The evaporation problem is not as acute as with paper electrophoresis and some of the media are suitable for the determination of mobilities and isoelectric points.

References p. 337

Starch block

Electrophoresis in a starch block or trough is the simplest of all the methods using a supporting medium. The apparatus used in most laboratories is much the same as that originally devised by Kunkel and Tiselius[5] for paper chromatography and an example is shown in Fig. 105.

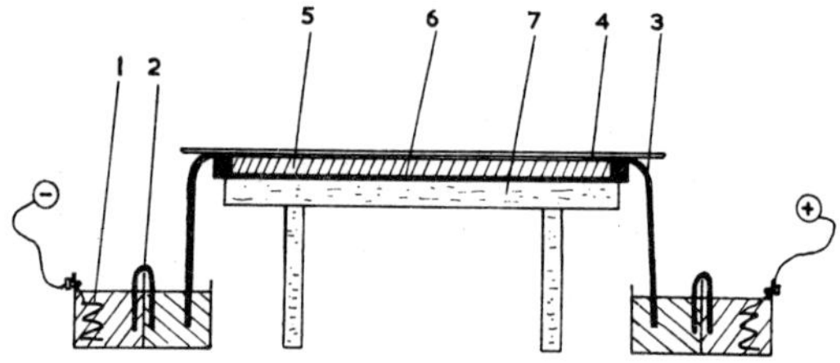

Fig. 105. Apparatus for electrophoresis in a starch block. 1. platinum or silver–silver chloride electrode; 2. filter paper; 3. filter paper bridge; 4. glass plate; 5. starch block; 6. plastic or glass trough; 7. adjustable table.

Ordinary potato starch is washed with distilled water several times by decantation and equilibrated thoroughly with the buffer solution to be used in the experiment. A thick slurry of the starch in buffer is poured into a trough with movable sides made of glass or plastic. Exploratory runs are conveniently carried out in a block of dimensions 40 × 5 × 0.7 cm; thicker blocks 50 cm wide will take up to 1 g of protein applied as a 2 cm band along the width. When the starch grains have settled excess buffer is drained off by application of thick filter paper to both ends. The optimum moisture content of the block is such that it will allow a neat slit to be cut for insertion of the sample. The latter is applied by means of a fine pipette and the slit closed up by manipulation with a spatula. An alternative method is to remove a segment of the block at the required starting point by means of two parallel knife blades, blot with filter paper, and mix thoroughly with the sample. When the starch is of the right consistency it is packed between the blades again and returned to the block.

In the procedure of Kunkel[12] the block is wrapped round with polythene sheeting, an operation requiring a good deal of experience. Other workers have achieved success by simply lowering a glass plate gently on to the surface of the block or even leaving the block uncovered. If the glass cover is used the trough must possess rigid sides; in the other methods the starch is self-supporting and requires no supports. The block is mounted across the two electrode vessels and contact between these and the block is made with thick filter paper, cotton cloth or thin sponge rubber sheet. If pools of liquid form on the surface of the block during assembly more moisture must be drained from the ends. Excessive endosmotic flow can cause pool formation at the cathode end. This can be remedied either by increasing the height of the block relative to the level of electrolyte or by covering the wicks with a tight sleeve of cellophane (see p. 43). In the method of Bloemendal[13] the trough, covered with a glass plate, is clamped vertically so that the endosmotic flow is largely neutralized by the gravitational flow. In this procedure the upper end is made the cathode, and no cooling is necessary with a potential gradient of 5 V/cm.

At the end of the run excess moisture is removed from the ends and the block is cut into sections. Each segment is eluted with buffer solution through a small sintered-glass funnel or, in order to avoid dilution, the segment is centrifuged and washed by centrifugation. Incomplete recovery of the sample is a recognized drawback to the procedure. Low molecular weight proteins penetrate the starch grains and will not migrate freely. Various substances, *e.g.* carbohydrate, are leached from the starch grains and must be taken into account during analysis of the eluted zones.

Starch gel

A major development in electrophoretic techniques has followed the discovery made by Smithies[14] in 1955 of the remarkable resolving power of starch gels for serum proteins. An authoritative. and detailed review of the subject has been published by Smithies[15]

Two types of apparatus have been described. One is similar to the universal apparatus of Fig. 105, the other is a vertical arrangement (Fig. 106). A feature of Smithies' technique has been the use of reversible silver–silver chloride electrodes. These are made very simply from 25 mm, 28 gauge silver ribbon by electrolysis in dilute hydrochloric acid. The polarity of the current is reversed three or four times until the ribbons are evenly covered with silver chloride. They are washed and stored under distilled water. Two compart-

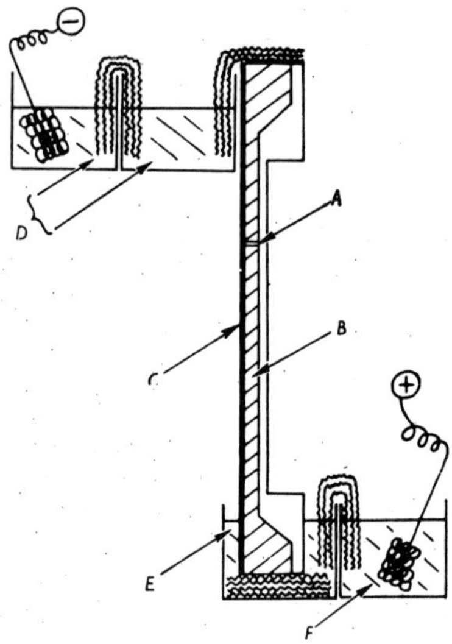

Fig. 106. Experimental arrangement for vertical starch-gel electrophoresis. A, position of sample slots; B, gel; C, petroleum jelly seal; D, trays containing bridge solution (0.3 *M* boric acid, 0.06 *M* NaOH); E, tray containing bridge solution with approx. 10 thicknesses of filter paper at the bottom; F, tray containing concentrated NaCl solution for positive silver–silver chloride electrode. Negative electrode is in outer tray D. (From Smithies[39].)

ments to each electrode vessel are essential. One is required for the electrode reaction and contains 1 *N* sodium chloride solution or a gel made up with 5–6% sodium chloride; the other forms the bridge between gel and electrode and contains an appropriate buffer solution, usually containing 0.30 *M* boric acid and 0.06 *M* sodium hydroxide. Electrical continuity is made by means of thick filter paper.

A suitable starch is available commercially ('Starch-hydrolyzed', Connaught Laboratories, Toronto), but the preparation of small quantities is not difficult. Samples of potato-starch (30 g) are suspended in 60 ml of acid–acetone (1 vol. of concentrated hydrochloric acid to 100 ml acetone) in small flasks. Danish or Idaho starch have been recommended for the best results but English starch is quite suitable. The samples are heated in a waterbath at 37° and one is removed at half-hour intervals during a total period of 2½ h. On removal of each sample the reaction is stopped by addition of aqueous 1 *M* sodium acetate solution. Gels are then prepared from the hydrolyzed samples (14 g/100 ml buffer) and selection is based on correct mechanical properties. A firm gel should be obtained with good mechanical strength and which breaks cleanly under tension[16].

A suspension of 13–15 g of hydrolyzed starch in 100 ml of buffer is warmed over a naked Bunsen flame in a 1 litre longnecked flask. At the 'gel point' of approximately 65° the suspension becomes very viscous and a strong circular movement is necessary during the subsequent heating which produces a more fluid gel. As soon as the gel reaches the mobile stage a vacuum is applied (water pump) for a few seconds to remove air bubbles. An asbestos glove can be worn for greater comfort during this operation. Care must be taken not to boil away too much water or the gel will become stiff again. The hot gel is poured quickly into the trough which has been lightly smeared with liquid paraffin. Plastic sheeting (PVC) treated with liquid paraffin is layered gently on to the surface of the gel avoiding enclosure of air bubbles and a heavy glass sheet is laid on top while the gel is setting. The following are the buffer systems which have been tried so far.

(1) 0.03 *M* boric acid–0.012 *M* sodium hydroxide per litre, pH 8.48. General purpose buffer.

(2) Sodium potassium phosphate, pH 7.8, di-sodium hydrogen phosphate, 42 g/l adjusted to pH 7.8 with saturated potassium dihydrogen phosphate. A 1:50 dilution is used[17].

(3) 0.076 *M* Tris–0.005 *M* citric acid; 0.30 *M* boric acid–0.06 *M* sodium hydroxide buffer solution in electrode compartments and in filter-paper bridges. In this system a visible boundary passes through the gel[18].

(4) Gel buffer, 90% 0.0033 *M* citric acid–0.016 *M* Tris and 10% 0.02 *M* lithium hydroxide–0.076 *M* boric acid giving a pH at 20° of 8.1. Buffer in electrode compartments 0.1 *M* lithium hydroxide, 0.38 *M* boric acid, pH 8.5. This discontinuous system gives a non-linear voltage gradient which changes as an area of high voltage gradient spreads through the gel from cathode to anode[19].

(5) 0.017 *M* Aluminium lactate of ionic strength 0.1, adjusted to pH 3.2 with lactic acid[20]. The buffer is prepared from activated aluminium foil[21].

Smithies has described a procedure for selecting the optimum buffer concentration for serum separations for a given batch of hydrolyzed starch. Different concentrations of starch are tested

References p. 337

with borate buffer, 0.021 *M* boric acid–0.0084 *M* sodium hydroxide, until one is obtained in which the migration distance of transferrin (β-globulin) is roughly 2.5 times that of Sα_2-globulin (Fig. 107). The total buffer concentration is then varied until the front and back of the albumin zone are equally sharp.

The sample is best applied as an intimate mixture with granular starch into narrow slots (1–2 mm) cut in the gel. Insertion of liquid samples directly into the slots is not recommended as it has been shown that strong electrodecantation concentrates the fast moving component at the bottom of the gel thus giving uneven distribution in the vertical plane. The vertical gel was devised to overcome electrodecantation effects and although the apparatus is a little more complicated the improved separations amply justify the extra time in construction. In this method the gel is poured around a former which is removed when the gel is set leaving accurately positioned

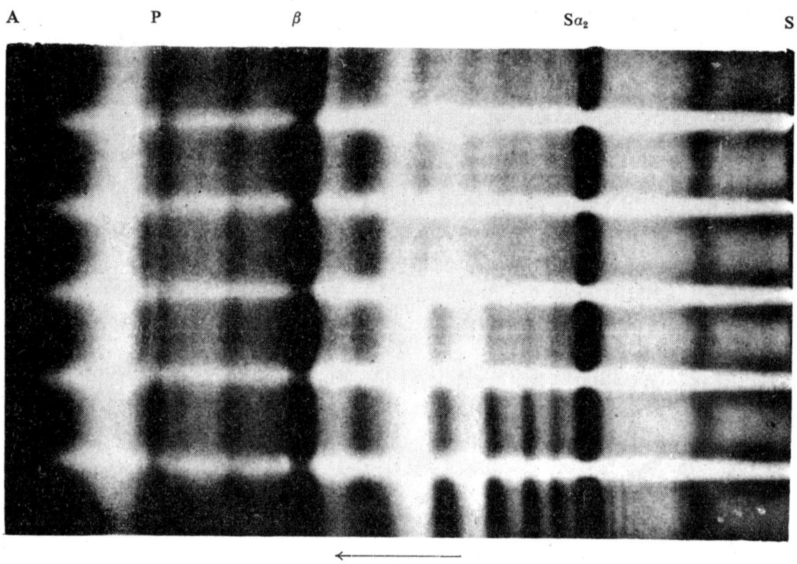

Fig. 107. Vertical starch-gel electrophoresis (approx. 19 h at 5 V/cm) of serum samples from six healthy individuals. S, sample slots; Sα_2, S$_{\alpha_2}$-globulin (probably identical with heat-labile glycoprotein and α_2-macroglobulin); β, main β-globulin considered to be the iron-binding protein, transferrin; P, post-albumins; A, albumin. Smithies[22].

slots. Liquid samples can be pipetted into these slots and sealed with paraffin jelly. Bussard[25] has described a novel technique for covering gels. A solution of vinyl chloride in dichloroethane is carefully run on to the gel and allowed to evaporate in the air when a protective covering of plastic rapidly forms on the surface of the gel. Furthermore, the thin film reduces evaporation during a run to about 10% compared to an open block. If cool air is blown across the surface 0.1 W/cm^2 of Joule heat can be dissipated.

After a run a thin sheet of glass is slid under the gel and the latter is sliced longitudinally in one movement with the aid of a convenient length of fine steel wire. The protein zones are stained with Amido Black 10B or Nigrosin in methanol–water–acetic acid (50:50:10 by vol) and the gel washed exhaustively with the same solvent. It has been found useful to prestain serum lipoproteins before electrophoresis. A suitable dye is acetylated Sudan Black B[23] which should be dissolved in propylene glycol[24].

Removal of proteins from the gel is not easy. Freezing and thawing the gel produces a sponge-like material from which it is possible to suck off a solution containing protein but higher yields are obtained by centrifugation at moderate speeds. It is reported that irreversible adsorption of some proteins can result from prolonged storage of the gel in the frozen state. Gordon[26] has described an ingenious method whereby proteins from segments of a gel are removed by means of a second electrophoresis and trapped in a small volume of solution on one side of a cellophane membrane barrier. The technique was successful when applied to serum proteins with the exception of γ-globulin.

Gordon[27] later modified the elution apparatus to accommodate four pieces of gel for simultaneous elution. A diagram of the cell is shown in Fig. 108. Current is applied to the upper surfaces of the pieces of gel by means of a hanging cellophane membrane (*B*) filled with 0.02 *M* phosphate buffer, pH 8.3. The thickness of the pieces of gel is such that will allow an air gap to remain under each Perspex bar (*D*) without any loss of contact between gel and membrane. Overheating and pH changes are prevented by means of a pumping system which circulates buffer continuously between the cathode and anode chambers. Air bubbles trapped under the bottom frame must be removed after the apparatus is filled with buffer. Operation for 3 h at 200 mA gave a 95% recovery of albumin.

References p. 337

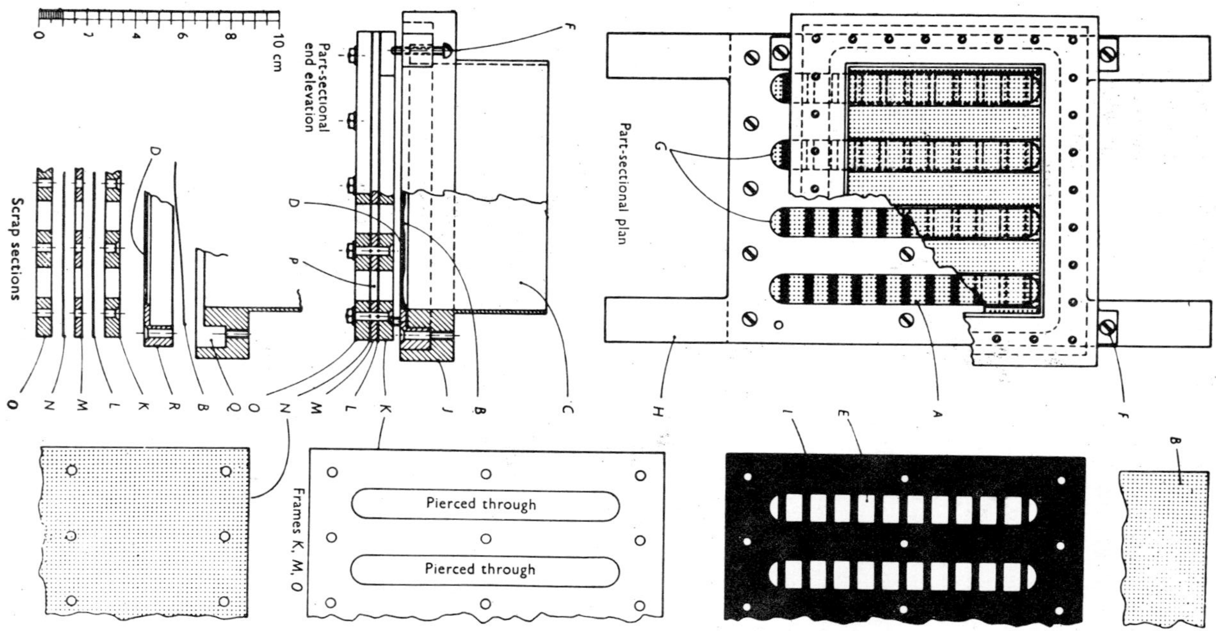

Fig. 108. Diagram of elution cell *A*, gel compartments; *B*, cellophane membrane which makes contact with upper surfaces of pieces of gel; *C*, open box; *D*, supporting bars; *E*, holes in polythene sheet; *F*, adjusting screws; *G*, open ends of compartments which contain solution into which protein is eluted; *H*, bars on which cell rests; *J*, frame forming upper part of the base of the open box (the frame is fabricated from sheet Perspex; details of fabrication are not shown); *K*, frame for holding pieces of gel; *L*, polythene sheet cut to form bridges; *M*, intermediate frame, similar to *K*; *N*, cellophane membrane; *O*, bottom frame, similar to *K*; *P*, compartment which contains buffer solution into which protein is eluted; *Q*, slot in frame into which base of open box is screwed, thus sealing edges of cellophane membrane; *R*, base of open box with support bars for membrane. Gordon[27].

Mention is made in Chapters 4 and 9 of certain chromatographic procedures where strong urea solutions proved beneficial and it is interesting to note that starch gel electrophoresis has also been carried out in the presence of a high concentration of urea. A convincing demonstration of the products of reductive cleavage of an α_2-macroglobulin was recently reported by Poulik[28] who used a gel prepared in the following way. Hydrolyzed starch (60–65 g) was mixed with 250 g of urea and added in small portions to 300 ml buffer with vigorous agitation. The viscous solution was then heated at 70° for 10 min, the trapped air was removed, and the gel poured into an electrophoresis tray. The principal components of casein have been reinvestigated following the demonstration[29] of their apparent heterogeneity by electrophoresis in the presence of urea, and an effective evaluation of the purity of preparations of β- and γ-casein is possible[30].

A more promising approach to the preparative aspect of starch gel electrophoresis is possibly that of Bernfeld and Nisselbaum[31] who employed lower concentrations of starch. In this method a slurry made from freshly prepared corn amylase containing a total of 3 g carbohydrate is added to a boiling mixture of 30 ml of water and 100 ml Tris–citrate buffer pH 8.8, ionic strength 0.06 (18.3 g Tris, 2.1 g citric acid monohydrate per litre). Water is then added to bring the volume to 190 ml, 6 g of Hyflo Super Cel are added followed by a suspension of 6 g corn-starch in 10 ml of water added slowly with stirring. After boiling for 10 min the mixture is poured into a trough and allowed to cool overnight. Good separations are obtained in a gel of dimensions 30 × 5 × 1.3 cm with up to 150 mg of serum proteins in a 2-ml sample (0.03 ionic strength). A run takes about 18 h at 6.7 V/cm and the zones can be seen under ultraviolet light. Recoveries of proteins are very good but it is necessary to remove the amylase by centrifugation at 20,000 g. Large-scale separations have been carried out with 50 ml of serum proteins in a trough of the same length and width but 30 cm high. Cellophane membranes separated the gel from the electrode compartments in which the electrodes were surrounded by an ion-exchange resin in the chloride form. At the other end of the scale techniques have been developed in which starch gels are supported on glass slides[32] or cover slips[33] (24 × 50 × 0.1 mm). With the latter method a 0.2 μl sample of human serum gave 18 bands.

References p. 337

There have been numerous applications of starch gel electrophoresis to the field of enzyme fractionation. Instead of staining the proteins with a dye the enzyme zones are located by treatment of the gel with appropriate substrates. For example, polyphenolase-active proteins in leaf extracts may be readily located after electrophoresis by application to the gel of a solution of β-(3,4-dihydroxy phenyl) alanine (DOPA). Hunter and Markert[34] have described the detection of esterases after electrophoresis of a sample obtained from a tissue homogenate. Electrophoretic patterns of the enzyme lactate dehydrogenase (LDH) are composed of several bands believed to be different molecular forms of the enzyme (isozymes)[35,37]. The patterns have been examined using differential isozyme inhibitors[36] and tissue specificity has been established.

Agar gel

Electromigration of colloids in agar gel was studied as long ago as 1907 by Field and Teagre[38] but serious application of the method to problems of protein structure dates back to 1946 when Consden *et al.*[39] analyzed mixtures of peptides from wool hydrolysates. Gordon *et al.*[40] later developed the method still further for protein separations in a 1% gel. A special feature of their apparatus is a simple means of continuous buffer irrigation and efficient cooling beneath the thin glass plate supporting the agar block. When buffer solutions of ionic strength less than 0.05 are used rapid electro-endosmosis takes place between the jelly and the glass and a further disadvantage is the difficulty of removing last traces of agar from the eluted proteins. By means of the arrangement shown in Fig. 109 the proteins are made to migrate out of the gel. The section of agar containing the protein is placed on a sheet of wet cellophane lying in the trough and the edges of the cellophane are raised to leave a space 1 cm wide along the edge of the gel. The proteins migrate into

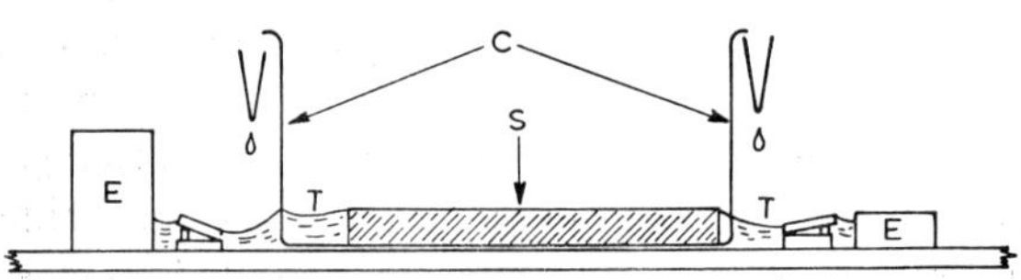

Fig. 109. Apparatus for electrophoresis in agar gel. C, cellophane; E, electrodes; T, buffer solution; S, slab of gel. Gordon *et al.*[40].

the buffer filling the space between the cellophane barrier and the gel. Column electrophoresis in agarose (see p. 269) has been described by Hjertén[41]; adsorption of proteins to this medium is extremely low and migration velocities are similar to those in free solution. Agarose is a good anticonvection medium at a concentration as low as 0.16%.

A combination of electrophoresis in agar gel with serological precipitation as developed by Grabar and Williams[42] is another splendid example of the enhanced analytical sensitivity obtained by a simple extension of established techniques. Following a normal electrophoretic run antiserum is added to the gel in a groove cut along its length and allowed to diffuse laterally to meet the fractionated protein zones. Factors which favour agar in this type of work are the low adsorption of proteins in spite of the coarse texture, the comparatively large free volume approaching that in free elctrophoresis, and the general ease of manipulation. Being a strongly acidic substance its use in alkaline buffers brings about considerable endosmotic flow which is perhaps the major disadvantage. As with starch gel little is known about the materials that can be leached out to combine with or otherwise contaminate proteins. Nevertheless the technique is of great value in the fields of immunology and clinical biochemistry and provides an additional criterion of purity and homogeneity for protein studies. An example of its use in a somewhat different field of endeavour is provided by studies in progress on the characterization of plant proteins.

Polyacrylamide gel

Procedures for electrophoresis in polyacrylamide were worked out independently by Ornstein and Davis[43], and Raymond and Weintraub[44]. In contrast to starch gels, polyacrylamide gels are transparent, thermostable and non-ionic; the pore size can be varied over a wide range, and the absence of charged groups reduces endosmotic flow to a negligible amount. As an analytical tool they surpass the starch gels for high resolving power and sensitivity in performance; unfortunately their low electrical resistance sets a limit to the scale of operation and best results are obtained in very small columns.

The following stock solutions are required for the preparation of the gels:

References p. 337

A. Acrylamide, 36 g; Eastman Organic Chemical 5521. N,N′-Methylenebis (acrylamide), 600 mg; Eastman 8383. 0.03–0.1 *M* Tris, borate, or Tris-borate buffer, pH 8–9, to 100 ml.
B. 3% β-Dimethylaminopropionitrile in same buffer as A.
C. 1% Ammonium persulphate in buffer.

Solutions A and B are diluted with an appropriate amount of buffer (*see below*), deaerated, and mixed with an equal volume of C. Gels form in a few minutes to about 2 h. The pore size of the gel depends on the concentration of polymer and good general purpose gels of 1.66% cross-linking are obtained with the following combinations (parts by volume).

TABLE 50

COMBINATION OF STOCK SOLUTIONS GIVING GENERAL PURPOSE GELS

Concentration (%)	*A*	*B*	*Buffer*
5	2	1	4.2
6	2	1	3
7	2	1	2.2
8	2	1	1.5
9	2	1	1

A procedure will be described which, in the matter of technique, is basically that of Ornstein and Davis[43]. Gels are formed in small glass tubes 5–6 cm in length and 5–7 mm internal diameter. Leak-proof rubber caps are fitted at the lower end of the tubes and water is layered over the monomer solution in order to obtain a smooth, flat gel surface. If necessary the polymerization may be delayed by incorporation of an inhibitor such as potassium ferricyanide (1 mg to every 100 ml of final solution). The tubes are pushed into holes, fitted with grommets, in the base of a circular plastic dish which forms one electrode chamber (cathode). A hanging drop of buffer is placed at the bottom of each tube and the vessel with attached tubes lowered to immerse the ends of the tubes in the buffer of the anode chamber. With a potential difference of 150–200 V the current for each tube should be 5–8 mA. At the end of the run (20–30 min) the gels are teazled out of the tubes under water using a needle, and stained with a suitable dye dissolved in 7% acetic acid. Excess dye is removed either by soaking the gel in a bath containing equal volumes of methanol and 20% acetic acid, or by electrophoretic destaining. Fig. 110 A shows the effect of gel pore size on the migration rates of water soluble proteins of spinach beet chloroplast[45]; the striking difference in behaviour of these proteins in borate and Tris buffer is illustrated in Fig. 110 B.

Many workers use an analytical technique similar to the one above, differences in charge and molecular size or shape providing the basis for the separations. The original method of Ornstein and Davis, which they termed 'disc' electrophoresis, made use of a gel composed of three layers: (*1*) a large-pore anticonvectional gel containing the sample; (2) a large-pore spacer gel in which electrophoretic concentration takes place; (3) a small-pore gel in which electrophoretic separation is accomplished. Both the sample and spacer gels were photo-polymerized in the presence of a small quantity of riboflavin. Discontinuities in pH were incorporated as well as conditions set up by the Kohlrausch 'regulating function'[4].

Layering the proteins in an appropriate buffer was suggested by Barka[46]; Reisfeld[47] added sucrose to the sample so that the density gradient diminished diffusion. Broome[48] used a gel overloaded with a 1-cm column of a buffer–sucrose–Sephadex G-200 mixture. Sephadex G-200 was suspended in 20% (w/v) sucrose and a working mixture prepared by diluting two volumes of the sedimented Sephadex with one of Tris-chloride buffer (pH 6.7, 180 m*M* with respect to Cl′). Two vols of the mixture weie added to one vol. of the sample.

Vertical gel apparatus has been described by Raymond and Nakamichi[49] and Radhakrishnamurthy *et al.*[50]. A small scale preparative technique allowing migration elution of proteins from the gel column has been reported[51,52] and permanent record and scanning of disc electrophoretic patterns has also been discussed[53]. If it is intended to scan stained gels with the aid of a densitometer they should be prepared in matched colorimeter tubes.

The transparency and inert nature of polyacrylamide gel make it an ideal medium for the application of histochemical techniques. Slow diffusion of proteins within the gel permit enzyme staining reactions without previous fixation. Acid phosphatases[46] and thiamine pyrophosphatase[54] have been examined in this way. Zeldin and Ward[55] used the gels in a study of internal changes of protein pattern with morphogenesis and differentiation in a slime mould.

References p. 337

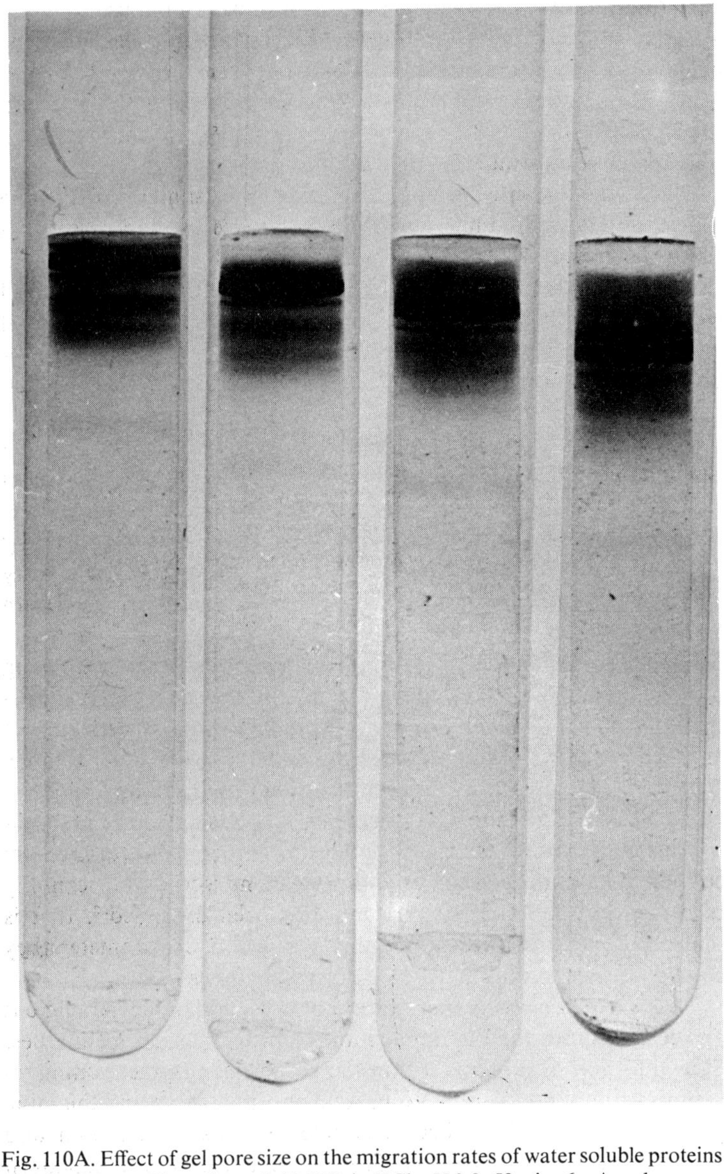

Fig. 110A. Effect of gel pore size on the migration rates of water soluble proteins of spinach beet chloroplasts; 0.03 *M* Tris-HCl, pH 8.9; 50 min; 6 mA: gel concentration (*left to right*): 9, 8, 6 and 5%.

Fig. 110B. Migration of same proteins in 0.03 *M* Tris-HCl, pH 8.9, 50 min (*left*), and in 0.03 *M* sodium borate, pH 8.9, 25 min (*right*).

Every few years the problem of microheterogeneity and polymorphism is re-investigated following the development of methods offering improved resolving power. Fig. 111 shows the resolution of oligomeric and isomeric forms of plasma albumin from the work of Sogami and Foster[56] and it is thought that the various isomeric forms exist in an essentially fixed state. Tombs[57] has demonstrated a possible polymorphism in the seed protein arachin; samples for examination were prepared from single seeds of different strains of peanut. The application of polyacrylamide gel electrophoresis to the problem of heterogeneity in nucleohistones has been ably discussed by Driedger *et al.*[58] and McAllister *et al.*[59]. A fractionation of calf-thymus histone at pH 4.9 is illustrated in Fig. 112. Gels have also been run in acetic acid[60] and in 8 *M* urea[61].

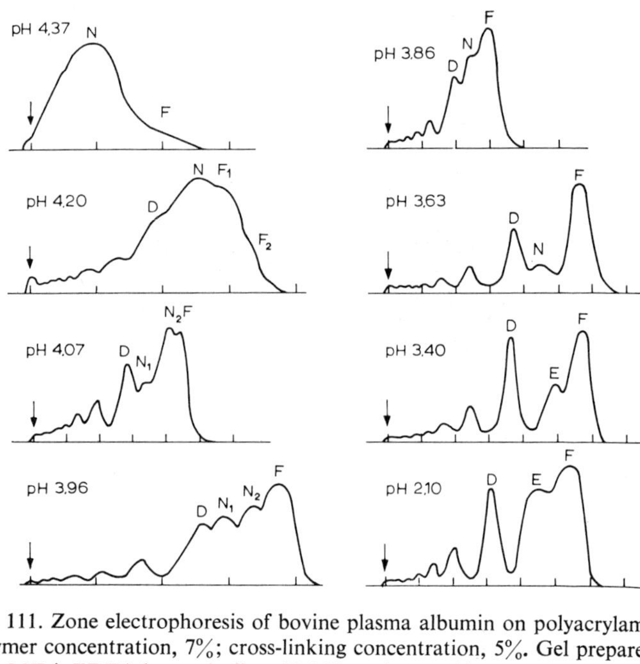

Fig. 111. Zone electrophoresis of bovine plasma albumin on polyacrylamide; polymer concentration, 7%; cross-linking concentration, 5%. Gel prepared in 0.10 *M* Tris EDTA-borate buffer, pH 9.00, and poured into mould, 22 × 3.3 × 0.16 cm. Microdensitometer tracings in 0.10 *M* chloride at various pH values. The monomer yields N (N_1 and N_2) forms and F (F_1 and F_2) forms which migrate ahead of dimer, D. E may correspond to an expanded form of the protein. Sogami and Foster[56].

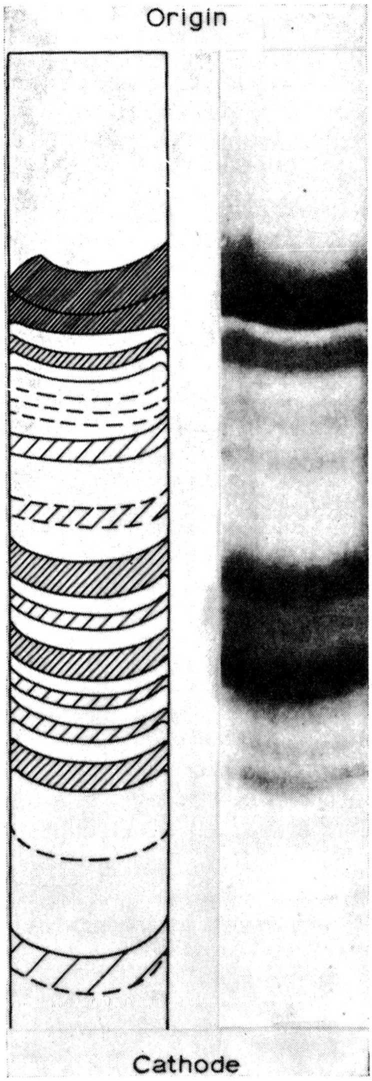

Fig. 112. Electrophoresis of calf thymus histone on polyacrylamide gel in acetate buffer, pH 4.9, ionic strength 0.02. (From Driedger *et al.*[58])

Cellulose acetate strip

Remarkably clean separations on a micro-scale are possible on strips of cellulose acetate[62]. This material is practically a gel being very brittle when dry but quite strong and pliable when wet. It is obtainable as strips 5 cm wide (Oxo Ltd., London) in a high state of purity. Much the same principles operate during electrophoresis with this medium as were discussed in connexion with paper electrophoresis. A slightly lower current density is used of 0.4–0.5 mA/cm width of strip in order to avoid excessive evaporation and the appearance of 'burned' areas. Difficulty is sometimes experienced in wetting the strips thoroughly. A reliable and quick method is to drop the strip on to the surface of some buffer in a small dessicator, evacuate for a few minutes, release the vacuum, and lay the strip between two glass plates for 10–15 min. Blotting of the strip is done between filter papers that have been sprayed lightly with buffer. Too vigorous drying with the appearance of opaque white patches is less likely in this way. An adequate sample is one containing 200–300 μg of protein in about 5 μl of solution; volumes down to 0.1 μl containing 5 μg protein will give a band still detectable on staining with the usual stains. From the results of Fazekas de St. Groth *et al.*[63]. Amido Black does not emerge as a satisfactory dye for the evaluation of protein concentration on the strips. Of a number of dyes tested Coomassie Blue showed the greatest sensitivity; Procion Blue was about three times less sensitive, while Azocarmine, Amido Black and Bromophenol Blue were found 5–10 times less so. Staining with Procion Blue R.S (I.C.I.) was the most superior by criteria of reproducibility and accuracy. Procion Blue is made up in methanol (5 g/l of dye, 20 ml/l of conc. HCl) and staining is allowed to proceed for 5–10 min. Excess dye is removed in a bath of pure methanol.

A valuable property of cellulose acetate is the ease with which it can be rendered transparent in fluids of refractive index about 1.474 such as decalin, and various oils. As a plastic it is soluble in a number of organic solvents, *e.g.* chloroform, acetone and phenol. Extremely pliable gelatinised cellulose acetate is available commercially ('Cellogel', from Shandon Scientific Co., London; Chemitron, Milan). After use this material is rendered transparent by gentle heating after destaining.

Electrophoresis in cellulose acetate is occasionally used in conjunction with paper chromatography[64] or with a second electrophoretic separation in a different medium. From recent work it appears that cellulose acetate is an ideal medium for immunoelectrophoresis[65].

ZONE ELECTROPHORESIS IN VERTICAL COLUMNS

The obvious advantage of a column system of electrophoresis is the ease of recovery of the separated zones by the usual elution methods of column chromatography. In conjunction with a fraction collector the eluate can be analyzed in several ways by taking aliquots. The technical aspects have been developed largely in the Institute of Biochemistry, Uppsala, by Haglund and Tiselius[66], Porath[67], and Flodin and Kupke[68] and a widely used arrangement of the column is based on the apparatus described by Porath[69] in 1957 (Fig. 113).

Columns suitable for analytical work should be 75–150 cm long and 1.5 cm in diameter; very large preparative columns have been operated but a convenient size for general purposes is 50 cm long and 3–4 cm in diameter. The contents of the tube are supported by a sintered glass plate, nylon net or perforated plastic. The writer has found Araldite Type AY103 suitable for the purpose. The tube is mounted firmly on a piece of cellophane on the bench and liquid resin mixed with hardener is poured in to form a plug 7 mm thick. When the resin has set the plug is perforated with a fine drill. If transparent plastic tubing is available one end may be machined to provide an external screwthread on which may be fitted a perforated plastic plug.

A flexible connexion between the top of the column and the long connecting arm is advisable. For the large preparative columns the associated electrode vessels must be of sufficient capacity to prevent too rapid accumulation of electrolysis products. Converted polythene flasks of 5–7 litre capacity as used by Porath are robust and much safer than vessels made of glass. The electrodes are No. 26 gauge platinum wire wound round thick glass rods.

Glass beads, starch and cellulose powder have been used as stabilizing media. Excellent results are obtained with cellulose that has been treated with dry methanolic or ethanolic hydrogen chloride[68]. Ethanolyzed cellulose is prepared as follows. A good quality

References p. 337

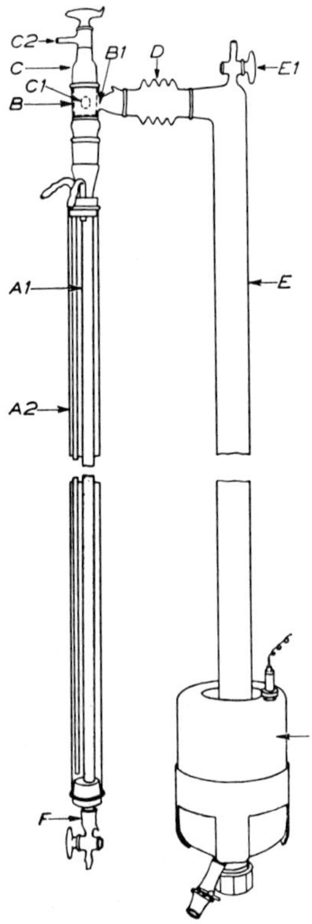

Fig. 113. Column electrophoresis apparatus of Porath[69]. A1, electrophoresis tube; A2, cooling jacket. B and C connect A1 to E via the flexible tube D. G, converted polythene flask. F permits control of liquid flow when mixture is introduced and displaced to the starting position in the column.

cotton or linters is refluxed with absolute ethanol 1 *M* with respect to hydrogen chloride in the proportions 115 g cotton to 4 litres ethanolic hydrogen chloride. The latter is made by adding the calculated amount of acetyl chloride to absolute ethanol. If bad pumping is experienced the mixture can be warmed in a bath at 70°, with occasional shaking, until a sample poured into water settles uniformly without clumping of the particles. After the treatment the material is washed with ethanol until white and then suspended in water several times to remove very fine particles. The powder is allowed to dry after a final wash with ethanol.

A process for preparing cellulose with low adsorption for proteins was described recently by Porath[72]. Munktell Cellulose Powder is extracted with hot water in a continuous infusion extractor[73] for 2–3 days at 90°. The water in the bed is displaced by freshly distilled pyridine and the extraction continued until the liquid leaving the bed is colourless, an operation usually taking 1–2 days. Finally, the cellulose is extracted with ethanol and dried *in vacuo* at a temperature not exceeding 50°. Sorof *et al.*[74] recommend a different procedure involving treatment of the cellulose separately with 0.02 N NaOH and 0.01 N HCl for 1 h at room temperature.

Starch columns are not difficult to pack but the lighter density cellulose powder requires more care if inhomogeneities in the column are to be avoided. Cellulose is packed in the following way. A tube of the same diameter as the column and 150–200 cm in length is connected to the top of the column by means of a ground glass joint. A large funnel and stirrer are mounted at the upper end of the tube, the outlet tap closed, and the whole system is filled with buffer solution. A useful buffer is 0.0115 M $Na_2B_4O_7$-0.0155 M NaH_2PO_4, pH 8.6, ionic strength 0.05. If the column terminates in a plastic plug the latter should be covered with a circle of filter paper. About 100 g of ethanolyzed cellulose is suspended in 1 litre of buffer solution, evacuated to remove air bubbles and poured gently into the funnel. Gentle stirring is maintained while the buffer flows through the column at the rate of 5 ml/min. A sharp, flat boundary should be seen rising steadily in the column. When the desired column height is reached the buffer in the extension tube is sucked off through fine-bore plastic tubing and a circle of filter paper is placed gently on the surface of the cellulose.

The sample is allowed to enter the column under gravity and descend 3–4 cm below the surface of the column. Alternatively, sucrose or glycerol may be added to the sample to increase its density and the mixture layered on the bed surface through the supernatant buffer. The apparatus is then assembled as follows. The outlet below the column is closed and buffer added to the top of the column as far as tap C2. Tube E is then placed in position and its lower end inserted into the electrode vessel G. Suction is applied to E1 until buffer fills the tube and the bridge connexion. Tap E1 is closed to seal the system, and the tap below the column is immersed in the second electrode chamber and removed. The levels of the buffer solutions

References p. 337

in the electrode vessels are then equalized by a syphon. Finally, C is turned to connect the upper part of the column with the rest of the apparatus. It is usual to circulate water at 2° to 10° through the cooling jacket and it will be observed that migration rates at the lower temperature are about one half of the rates at room temperature. If desired, buffer in the anode vessel (beneath the column) can be renewed continuously from a large reservoir, and with the aid of a constant level device. The electroendosmotic backflow of buffer may be reduced considerably by attaching two layers of Visking cellophane to the bottom of the vertical liquid arm[71]. At the end of the electrophoretic run the column is disconnected and mounted immediately over the drop-funnel of a fraction-collector.

Those who have acquired some experience in operating the columns have added the refinement of migration elution, a procedure whereby the zones are removed during electrophoresis by a very slow back-flow of buffer[75]. A very simple way of achieving this has been described by Naughton and Taylor[76] and is illustrated in Fig. 114. The cell consists of a short tube, 5 cm long and of internal dia-

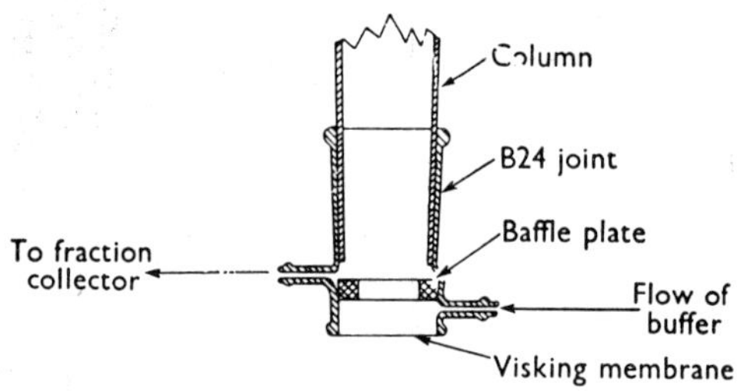

Fig. 114. Cell fixed below column for migration elution of separated zones (from Naughton and Taylor[76]).

meter 2 cm, terminating at its upper end in a B24 ground glass joint, and closed at the lower end by a single layer of Visking cellophane held in place with a rubber band. The lower of the two glass arms is connected to a reservoir of buffer. The protein fractions leave the apparatus via the upper arm and are conveyed through narrow-bore plastic tubing to a fraction collector. A baffle plate, consisting of a single layer of Visking cellophane perforated with pin-holes, is fixed in position between the side arms by means of a split polythene ring. During the removal of albumin from a column of dimensions 2 × 50 cm a buffer flow rate of 16 ml/h was maintained.

The cell of Naughton and Taylor might not work in certain instances because insertion of a cellophane membrane into the path between the electrodes leads to disturbances in the pH and conductivity of the operating buffer. These conditions lead, in turn, to loss of uniformity of pH and conductivity in the packed portion of the column. The problem has been discussed by Hochstrasser *et al.*[77] who suggest circulating buffer around the membranes.

Another type of apparatus designed by Porath[78] for migration elution is illustrated in Fig. 115. A slow stream of buffer through the porous plastic plate (*Vyon I*) is maintained by means of a peristaltic pump connected to *C*. Elution zones appearing at the base of the column leave the system via the long capillary tube before they have a chance to diffuse into the lower electrode chamber.

CARRIER-FREE ELECTROPHORESIS

Free electrophoresis for preparative purposes has been carried out in a density gradient or in thin films. The main advantage of using a density gradient for prevention of convection effects rather than a solid support is the elimination of irreversible adsorption of proteins to the supporting medium. Quantitative recovery of all the protein components in a sample may then be realised. The technique and theory has been described in detail by Svensson[79], and Svensson and Valmer[80]. Recent applications are to be found in studies on proteases[81], and phospholipase C from micro-organisms[82].

In 1916 Raleigh[83] put forward a theory that under certain conditions a thin film of fluid contained between two infinite plane-horizontal surfaces can remain at rest indefinitely, with the density increasing upwards. The principle has been exploited for the suppression of convection effects in electrophoresis, first by Barollier *et al.*[84] and later by Hannig[85], Blakebrough and Brookes[86], and Wiek[87]. Separations are accomplished in thin films of buffer (1/10–1/8 ins) sandwiched between specially made glass plates which must be of equal quality. The electric field is applied at 90° to the direction of buffer flow and the separated components leave the end of the cell through a manifold which distributes them to a rack of test tubes. There are a number of practical problems to be solved before

References p. 337

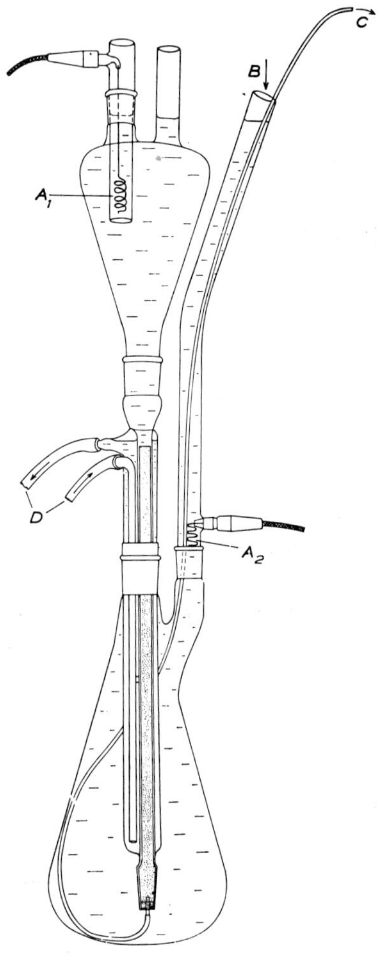

Fig. 115. Apparatus for migration elution according to Porath and Störiko[78]. A_1 and A_2, electrodes; *B*, inlet for electrolyte; *C*, connexion for transport of zones to fraction collector via peristaltic pump; *D*, cooling water.

the method can be fully exploited; one drawback, for example, is spreading of the zones which is much greater than with packed systems.

REFERENCES

1 L. MICHAELIS, *Biochem. Z.*, 16 (1909) 81.
2 A. J. P. MARTIN AND R. L. M. SYNGE, *Advan. Protein Chem.*, 2 (1945) 31.
3 J. T. EDWARD, *Chem. Ind. (London)*, (1958) 276.
4 J. DE WAEL AND E. WEGELIN, *Rec. Trav. Chim.*, 71 (1952) 1035; (F. KOHLRAUSCH, *Ann. Physik u. Chem.*, 62 (1897) 209).
5 H. KUNKEL AND A. TISELIUS, *J. Gen. Physiol.*, 35 (1951) 89.
6 M. A. JERMYN AND R. THOMAS, *Nature*, 172 (1953) 728.
7 A. T. JAMES AND R. L. M. SYNGE, *Biochem. J.*, 50 (1951) 109.
8 W. GRASSMANN, K. HANNIG AND M. KNEDEL, *Deut. Med. Wochschr.*, 76 (1951) 333.
9 T. ARONSSON AND A. GRÖNWALL, *Scand. J. Clin. Lab. Invest.*, 9 (1957) 338.
10 E. B. M. DE JONG, *Rec. Trav. Chim.*, 74 (1955) 1290.
11 B. DREVON AND R. DONAKIAN, *Bull. Soc. Chim. Biol.*, 37 (1955) 1321.
12 H. G. KUNKEL, *Methods Biochem. Anal.*, Vol. 1, p. 141.
13 H. BLOEMENDAL, *J. Chromatog.*, 2 (1959) 121.
14 O. SMITHIES, *Biochem. J.*, 61 (1955) 629.
15 O. SMITHIES, *Advan. Protein Chem.*, 14 (1959) 65.
16 M. D. POULIK AND O. SMITHIES, *Biochem. J.*, 68 (1958) 636.
17 G. C. ASHTON, *Nature*, 180 (1957) 917.
18 M. D. POULIK, *Nature*, 180 (1957) 1477.
19 K. A. FERGUSON AND A. L. C. WALLACE, *Recent Progr. Hormone Res.*, 19 (1963) 1.
20 J. S. D. GRAHAM, *Australian J. Biol. Sci.*, 16 (1963) 342.
21 G. A. H. ELTON AND J. A. D. EWART, *J. Sci. Food Agr.*, 13 (1962) 62.
22 O. SMITHIES, *Biochem. J.*, 71 (1959) 585.
23 W. G. B. CASSELMAN, *Quart. J. Microscop. Sci.*, 95 (1954) 321.
24 L. P. RIBEIRO AND H. J. MCDONALD, *J. Chromatog.*, 10 (1963) 443.
25 A. BUSSARD, *Compt. Rend.*, 239 (1954) 1702.
26 A. H. GORDON, *Biochim. Biophys. Acta*, 42 (1960) 23.
27 A. H. GORDON, *Biochem. J.*, 82 (1962) 531.
28 M. D. POULIK, *Biochim. Biophys. Acta*, 44 (1960) 390.
29 R. G. WAKE AND R. L. BALDWIN, *Biochim. Biophys. Acta*, 47 (1961) 285.
30 M. L. GROVES, T. L. MCMEEKIN, N. J. HIPP AND W. G. GORDON, *Biochim. Biophys. Acta*, 57 (1962) 197.
31 P. BERNFELD AND J. S. NISSELBAUM, *J. Biol. Chem.*, 220 (1956) 851.
32 H. A. RAMSEY, *Anal. Biochem.*, 5 (1963) 83.
33 J. H. DAAMS, *J. Chromatog.*, 10 (1963) 450.
34 R. L. HUNTER AND C. L. MARKERT, *Science*, 125 (1957) 1294.
35 A. E. H. EMERY, *Nature*, 201 (1964) 1044.
36 I. A. BRODY, *Nature*, 201 (1964) 685.

37 I. H. Fine, N. O. Kaplan and D. Kuftinec, *Biochemistry*, 2 (1963) 116.
38 C. W. Field and O. Teagre, *J. Exptl. Med.*, 9 (1907) 222.
39 R. Consden, A. H. Gordon and A. J. P. Martin, *Biochem. J.*, 40 (1946) 33.
40 A. H. Gordon, B. Keil, K. Sebesta, O. Knessel and F. Šorm, *Collection Czech. Chem. Commun.*, 15 (1950) 1.
41 S. Hjertén, *J. Chromatog.*, 12 (1963) 510.
42 P. Grabar and C. Williams, *Biochim. Biophys. Acta*, 10 (1953) 193; *Biochim. Biophys. Acta*, 17 (1955) 67.
43 B. J. Davis and L. Ornstein, in Discussion following *J. Histochem. Cytochem.*, 7 (1959) 291; L. Ornstein and B. J. Davis, *Disc Electrophoresis* (preprint), Distillation Products Industries, Division of Eastman Kodak, 1961.
44 S. Raymond and L. S. Weintraub, *Science*, 130 (1959) 711.
45 J. P. Thornber, S. M. Ridley and J. L. Bailey, unpublished.
46 T. Barka, *J. Histochem. Cytochem.*, 9 (1961) 542.
47 R. A. Reisfeld, *Nature*, 195 (1962) 281.
48 J. Broome, *Nature*, 199 (1963) 179.
49 S. Raymond and M. Nakamichi, *Anal. Biochem.* 3 (1962) 23; 7 (1964) 225.
50 B. Radhakrishnamurthy, K. Chapman and G. S. Berenson, *Biochim. Biophys. Acta*, 75 (1963) 276.
51 D. Racusen and N. Calvanico, *Anal. Biochem.*, 7 (1964) 62.
52 S. Hjertén, *J. Chromatog.*, 11 (1963) 66.
53 D. A. Burns and O. J. Pollak, *J. Chromatog.*, 11 (1963) 559.
54 J. M. Allen, *J. Histochem. Cytochem.*, 11 (1963) 542.
55 M. H. Zeldin and J. M. Ward, *Nature*, 198 (1963) 389.
56 M. Sogami and J. F. Foster, *J. Biol. Chem.*, 237 (1962) 2514.
57 M. P. Tombs, *Nature*, 200 (1963) 1321.
58 A. Driedger, L. D. Johnson and A. M. Marko, *Can. J. Biochem. Physiol.*, 41 (1963) 2507.
59 H. C. McAllister, Y. C. Wan and J. L. Irvin, *Anal. Biochem.*, 5 (1963) 321.
60 R. A. Reisfeld, U. J. Lewis and D. E. Williams, *Nature*, 195 (1962) 281.
61 J. V. Maizel, *Biochem. Biophys. Res. Commun.*, 13 (1963) 483.
62 J. Kohn, *Clin. Chim. Acta*, 2 (1957) 297.
63 S. Fazekas de St. Groth, R. G. Webster and A. Datyner, *Biochim. Biophys. Acta*, 71 (1963) 377.
64 M. A. Naughton, F. Sanger, B. S. Hartley and D. C. Shaw, *Biochem. J.*, 77 (1960) 149.
65 R. Consden and J. Kohn, *Nature*, 183 (1959) 1512.
66 H. Haglund and A. Tiselius, *Acta Chem. Scand.*, 4 (1950) 957.
67 J. Porath, *Acta Chem. Scand.*, 8 (1954) 1813.
68 P. Flodin and D. W. Kupke, *Biochim. Biophys. Acta*, 21 (1956) 368.
69 J. Porath, *Arkiv Kemi*, 11 (1957) 161, 259.
70 P. N. Campbell and N. E. Stone, *Biochem. J.*, 62 (1956) 9P.
71 S. Sorof, E. M. Young, M. M. Spence and P. L. Fetterman, *Biochim. Biophys. Acta*, 38 (1960) 559.
72 J. Porath, *Sci. Tools*, 11 (1964) 21.
73 L. C. Craig and D. Craig, in: A. Weissberger (Ed.), *Technique of Organic Chemistry*, Vol. 3, Interscience Publishers, New York, 1950, p. 238.

[74] S. Sorof, E. M. Young, R. A. McBride and C. L. Binder, *Sci. Tools*, 11 (1964) 27.
[75] J. Porath, E. B. Lindner and S. Jerstedt, *Nature*, 182 (1958) 744.
[76] M. A. Naughton and K. W. Taylor, *Biochem. J.*, 77 (1960) 46.
[77] H. Hochstrasser, L. T. Skeggs, K. E. Lentz and J. R. Kahn, *Anal. Biochem.*, 6 (1963) 13.
[78] J. Porath and K. Störiko, *J. Chromatog.*, 7 (1962) 385.
[79] H. Svensson, in: P. Alexander and R. J. Block, *A Laboratory Manual of Analytical Methods of Protein Chemistry, Vol. 1*, Pergamon Press, London, 1960.
[80] H. Svensson and E. Valmer, *Sci. Tools*, 6 (1959) 13.
[81] A. W. Bernheimer, *Arch. Biochem. Biophys.*, 96 (1962) 226.
[82] A. D. Bangham and R. M. C. Dawson, *Biochim. Biophys. Acta*, 59 (1962) 103.
[83] Lord Raleigh, *Phil. Mag.*, 32 (1916) 529.
[84] J. Barollier, E. Watzke and H. Gibian, *Z. Naturforsch.*, 13b (1958) 754.
[85] K. Hannig, *Z. Anal. Chem.*, 181 (1961) 244.
[86] N. Blakebrough and R. Brookes, *Biotechnology Bioengineering*, 6 (1964) 223.
[87] K. Wiek, *J. Chromatog.*, 13 (1964) 111.

CHAPTER 11

MISCELLANEOUS ANALYTICAL METHODS

ESTIMATION OF PROTEIN

Folin–Ciocalteu reagent

This sensitive method based on the Folin–Ciocalteu phenol reagent[1] is very widely used for the estimation of particular proteins and for monitoring column effluents which may contain many different proteins. The details given below are essentially those of the method of Lowry *et al.*[2] but sodium citrate is used in Reagent B instead of sodium tartrate to give a more stable reagent[3].

Reagents

A. 2% Na_2CO_3 in 0.1 *N* NaOH.

B. 0.5% $CuSO_4.5H_2O$ in 1% sodium citrate.

C. 1 ml of reagent B mixed with 50 ml of reagent A.

D. Folin–Ciocalteu reagent (commercial reagent) diluted with water to give a solution 1 *N* in acid.

0.1 to 0.2 ml of the sample containing 10–100 μg of protein is mixed with 1.0 ml of reagent C and the solution allowed to stand for 10 min at room temperature. 0.10 ml of reagent D is pipetted rapidly into the mixture with thorough mixing and the extinction at 750 mμ read after a time interval of 0.5 to 2 h. As the Folin colour is not strictly proportional to concentration it is advisable to prepare a standard curve for the working concentration range. The portion of reagent C may be increased to 5 ml when very dilute solutions of protein are to be estimated. No interference was observed with the following substances in the given final concentrations: urea (0.5%), guanidine (0.5%), sodium tungstate (0.5%), sodium sulphate (1%), sodium nitrate (1%), perchloric acid (0.5%, neutralized), trichloroacetic acid (0.5%, neutralized), ethyl alcohol (5%), ether (5%), acetone (0.5%), zinc sulphate (0.1%), barium hydroxide (0.5%), and ammonium sulphate (0.15%). Glycine (0.5%) decreases the colour with protein by 50%. Phosphate buffers cause a precipitate to appear at molarities greater than 0.1.

A recent examination of the reaction has shown that any peptide bond will yield some colour, but certain amino acid sequences, not necessarily containing aromatic residues, are more chromogenic than others and largely account for the colour yield of the protein. A preliminary complete hydrolysis reduced the colour-yielding property of albumin by more than two thirds. Cleavage of the disulphide bonds in insulin by oxidation with performic acid gave a loss of about one third of the total colour[4].

Biuret reagent

The following preparation and use of the biuret reagent is that of Gornall *et al.*[5]. It is less sensitive than the Folin reaction but fewer substances interfere with the test.

Reagent

Copper sulphate ($CuSO_4.5H_2O$, 1.50 g) and sodium potassium tartrate ($NaKC_4H_4O_6.4H_2O$, 6.0 g) are dissolved in 500 ml of water. To this solution is added 10% sodium hydroxide (300 ml, carbonate free) with good agitation and 2 g of potassium iodide to prevent autoreduction. The solution is made up to 2 litre and stored in a polythene bottle. To carry out the test, 4.0 ml of the reagent is added to 1.0 ml of a sample containing 1 to 10 mg of protein and the mixture allowed to stand for 30 min at room temperature. The extinction is read at 540 to 650 mμ. The biuret reaction cannot be carried out in the presence of ammonium salts owing to the formation of cuprammonium ions.

Microbiuret reagent[6]

Sodium citrate (173 g) and sodium carbonate 100 g are dissolved in water with warming (not boiling). Copper sulphate (17.3 g) is dissolved in 100 ml of water and added to the first solution. The mixture is made up to 1 litre with distilled water and is known as Benedict's reagent. The reagent (0.2 ml) is added to a sample containing 0.1 to 2.0 mg protein dissolved in 4 ml of 3% sodium hydroxide solution and the extinction at 330 mμ read after 15 min.

References p. 351

Ultraviolet absorption of proteins

The absorption of protein solutions in the ultraviolet is due to tryptophan, tyrosine and to a minor extent phenylalanine. The pH of the solution has a pronounced effect on the absorption maximum and spectrophotometric measurements are usually made in alkaline solutions. Absorption curves for tryptophan, tyrosine and phenylalanine are shown in Figs. 116 and 117. At the points of intersection of the absorption curves, 257 and 294 mμ, the extinction values are proportional to the total tryptophan + tyrosine contents. It is usual to take measurements at 294 mμ since this point is close to the maximum in the tyrosine curve ($\Delta E/\Delta\lambda$ is minimal). In conjunction with

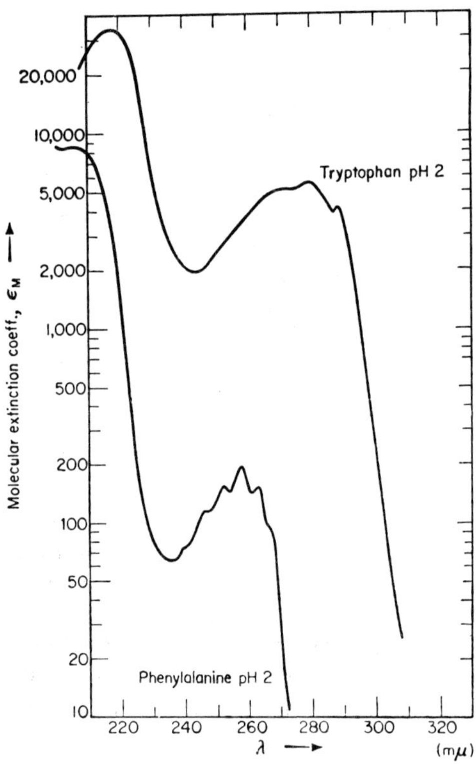

Fig. 116. Ultraviolet absorption spectra of tryptophan and phenylalanine, pH 2, Beaven[35].

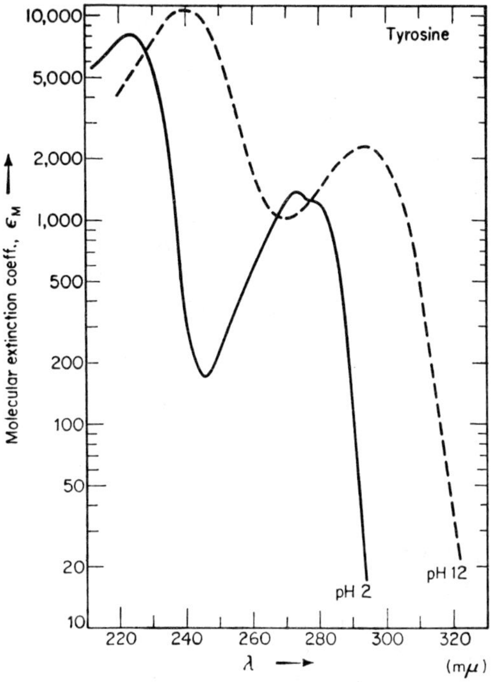

Fig. 117. Ultraviolet absorption spectra of tyrosine, (——) pH 2; (-----) pH 12. Beaven[35].

the extinction at 280 mμ ($\Delta E/\Delta\lambda$ minimal for tryptophan) the concentrations of the two amino acids may be calculated using the method of Goodwin and Morton[7].

x = Total moles/l; y = Tyrosine moles/l; $(x - y)$ = Tryptophan moles/l.
ε_1 = Molar extinction of tyrosine in 0.1 *N* alkali at 280 mμ.
ε_2 = Molar extinction of tryptophan in 0.1 *N* alkali at 280 mμ.

x may be determined from E_{294} knowing the molar extinction at this wavelength.

$$E_{280} = y\varepsilon_1 + (x-y)\varepsilon_2$$

or

$$y = \frac{E_{280} - x\varepsilon_2}{\varepsilon_1 - \varepsilon_2}$$

References p. 351

The following useful formulae were derived by Beaven and Holiday[8].

$$M_{\text{Tyr}} = (0.\ 592K_{294} - 0.263\ K_{280}) \times 10^{-3}$$
$$M_{\text{Try}} = (0.263\ K_{280} - 0.170\ K_{294}) \times 10^{-3}$$

where M_{Tyr} and M_{Try} are the gmoles of tyrosine and tryptophan in one gram of protein, and K_{294} and K_{280} are the extinction coefficients of the protein in 0.1 N alkali at 294 and 280 mμ. When the nitrogen or dry weight content is unknown extinction values can be substituted for the K values to give the molar ratio of tyrosine to tryptophan according to the formula

$$\frac{M_{\text{Tyr}}}{M_{\text{Try}}} = \frac{0.592\ E_{294} - 0.263\ E_{280}}{0.263\ E_{280} - 0.170\ E_{294}}$$

The errors involved in this sort of analysis are discussed in detail by Beaven and Holiday. It will be found that whereas the molar ratios are characteristic of a protein the tyrosine estimate will tend to be high and that of tryptophan low.

Most proteins in 0.1 N NaOH solution show general absorption falling off with increasing wavelength but persisting into the region 330–450 mμ where tyrosine and tryptophan are transparent. Suitable blanks at 294 and 280 mμ are therefore obtained by measuring extinctions at 320 and 360 mμ and extrapolating back to 294 and 280 mμ. When combined in proteins the maximum of the free amino acids is shifted by 1–3 mμ towards the longer wavelengths and pure peptides containing tyrosine and tryptophan should be better standards. A very serious source of error may arise from haze in the solution and if a protein shows a tendency to denature it is advisable to treat with a proteolytic enzyme to get a perfectly clear solution.

Peptides exhibit selective absorption in the 200–220 mμ region and the very strong absorption of proteins at these wavelengths has been favoured by very few workers for use in protein determination. Shortcomings in the instrument used set a limit to the accuracy of measurements and the problem of stray radiation becomes serious. The latter can be checked routinely if desired (see Saidel *et al.*[9]) and the hydrogen lamp should be selected for maximum intensity at the particular wavelength. It might be necessary to substitute a different load resistor to obtain a zero balance. Goldfarb *et al.*[10] have determined the specific extinction coefficient of a number of proteins and peptides at 205 mμ and 210 mμ; Table 51 gives a few values for the concentration expressed in g/l.

TABLE 51

	$E^{0.1\%}_{1\ cm\ 210}$	$E^{0.1\%}_{1\ cm\ 205}$
Insulin	25.0 ± 0.69	38.1 ± 0.53
Egg albumin	21.5 ± 0.27	34.1 ± 0.47
Gelatin	21.1 ± 0.35	33.2 ± 0.57

Tombs *et al.*[11] found the mean specific extinction coefficient for the total proteins of 40 normal and pathological sera at 210 mμ to be 20.5 ± 0.14. At this wavelength a protein concentration of 2 μg/ml gives $E = 0.04$.

Protein determination by the method of Waddell[12] is based on differential measurements at 215 mμ and 225 mμ. The protein solution is diluted with a sodium chloride solution (9 g/l) until the extinction at 215 mμ is less than 1.5. The extinction at 225 mμ is subtracted from that at 215 mμ; the difference multiplied by 144 gives the protein concentration in the sample in μg/ml. With a particular protein under specific conditions accurate measurements of concentration of 5 μg/l are possible. The following relevant points are drawn from the observations of Murphy and Kies[13], Spies *et al.*[14], and of Bendixen[15].

(1) Protein solutions obey Beer's Law at 215 mμ provided the extinction is less than 2.0.

(2) The absorption of proteins in the range 215–225 mμ is practically independent of pH between the values 4 to 8.

(3) Sodium chloride, ammonium sulphate, borate, phosphate and Tris do not interfere.

(4) Acetate, succinate, citrate, phthalate and barbiturate in 0.1 *M* concentration show high absorption at 215 mμ.

The extinction of nucleic acid in the 280 mμ region may be as much as ten times that of protein at the same wavelength and hence a few per cent of nucleic acid can greatly influence the absorption. Warburg and Christian[16] determined the absorption at 260 mμ and 280 mμ of known nucleic acid–protein mixtures and compiled data that enabled them to estimate the amount of protein in an unknown sample from readings at the two wavelengths. Provided the nucleic acid content does not exceed 20% accurate values for proteins are obtained. In order to avoid the use of tables and graphs Kalckar[17] introduced the formula,

$$\text{protein in mg per ml} = 1.45\ E_{280} - 0.74\ E_{260}$$

Determinations based on the use of this expression, the E_{280}/E_{260} ratios or any other convenient ratio, are valid for individual nucleoproteins only. Warburg and Christian pointed out that different proteins and nucleic acids will have different extinction values.

DETERMINATION OF NITROGEN

Micro Kjeldahl method

The most accurate values for the nitrogen content of a protein are obtained from samples free of lipid and carbohydrates. Lipid contamination is comparatively easy to remove by extraction with organic solvents but carbohydrates present a greater problem. It is possible that bacterial proteinases will become available in the near future that will break down proteins completely to amino acids which can then be separated from carbohydrate by absorption on an ion-exchange resin. Material which is all protein will generally have a nitrogen content of 15–16% and a moisture content of 10–12%.

Reagents

Concentrated sulphuric acid, nitrogen-free.

Catalyst mixture. Prepared by grinding to a fine powder 80 g K_2SO_4, 20 g $CuSO_4.5H_2O$ and 0.34 g of sodium selenate (Chibnall *et al.*[18]).

Boric acid (AR), 2% solution in distilled water.

Indicator. Mix equal volumes of 0.2% methyl red and 0.1% methylene blue solutions made up in absolute alcohol.

Standard acid solution, 0.01 *N* HCl.

Sodium hydroxide solution, 40–50%, Analar.

A sample containing 0.5 to 1 mg of nitrogen is weighed into a long-neck 100 ml Kjeldahl flask or transferred as a 1–2 ml aliquot of solution by means of a pipette. Concentrated sulphuric acid

(2 ml) is added carefully and the mixture heated on a digestion rack until charring is just apparent. The flask is removed from the rack, cooled slightly and 1 g of catalyst mixture added. Heating is then resumed keeping the contents of the flask gently refluxing for 8 h. (Brisk boiling for 12 h is required for complete digestion of proteins of high lysine content.) The cooled contents are diluted with 20–30 ml of distilled water.

A convenient distillation apparatus has been described by Markham[19]. Before use the whole apparatus is flushed with steam for a few minutes. The delivery tube is then immersed beneath the surface of 10 ml of boric acid solution with added indicator contained in a 50-ml conical flask. The contents of the Kjeldahl flask are washed quantitatively into the distillation chamber and 10 ml of 30% sodium hydroxide solution added. About 15 to 20 ml of distillate are collected during 4 min, the flask is lowered to free the liquid from the delivery tube and the distillation continued for a further one minute. The ammonia trapped in the boric acid solution is then titrated with standard acid to a neutral grey end-point. 1 ml of 0.01 *N* HCl is equivalent to 0.14 mg N; a distillation carried out on all the reagents should give a blank not exceeding 0.05 ml of standard acid. The factor for the standard acid is obtained from a distillation of a known volume of standard ammonium sulphate solution.

Comments

'Pure' boric acid as distinct from 'Analar' reagent may give some buffering action and end-points will be less sharp[20]. The end-point with the methylene blue-methyl red indicator corresponds to pH 4.9 to 5.3; the solution is greenish grey above this range and a purple grey below.

A very sharp neutral grey end-point at pH 4.6 is obtained with an indicator described by Sher[21]. Bromocresol green (350 mg) is dissolved in 10 ml of 95% ethanol, 1 ml of 0.50 *N* sodium hydroxide solution is added followed by 200 ml of water. A 1% aqueous solution of New Coccine (22.1 ml) and 750 mg of *p*-nitrophenol (in ethanol) are added and the solution diluted to 250 ml with water. 15 ml of this stock solution is diluted with 1% boric acid solution to give a final volume of 2 l.

The Kjeldahl method is made more sensitive by estimation of the ammonia colorimetrically with ninhydrin or by nesslerization.

References p. 351

Estimation of ammonia by ninhydrin (Jacobs[22])

Reagents

Catalyst mixture. Copper sulphate and potassium sulphate (microanalytical reagents), mercuric oxide and selenium, in the proportions 5:15:5:1 by weight.

Hydrogen peroxide. 100 volume, nitrogen-free.

Ninhydrin Reagent. Ninhydrin (2 g) dissolved in 50 ml methyl 'Cellosolve' + 25 ml 4.0 *M* sodium acetate buffer, pH 5.5, + 25 ml water + 0.08 g stannous chloride dihydrate.

For the incineration the sample is mixed with 0.5–1.0 ml of concentrated sulphuric acid and 3 to 5 mg of catalyst mixture and heated for 45 min. If the solution is not clear after 45 min the flask is cooled and 0.2 ml of hydrogen peroxide is added drop by drop. The mixture is then heated for a further 2 h.

The digest (0.5 ml) is dissolved in 25 ml of 0.4 *M* sodium citrate buffer, pH 5.0. A 0.5-ml aliquot is diluted further if necessary with buffer to give a volume of 2 ml and mixed with 2.0 ml ninhydrin reagent. The mixture is heated at 100° for 30 min, cooled and diluted with 6 ml of aqueous ethanol (1:1). Ammonia equivalent to 5.6 μg N gives an extinction of 0.805 ($\pm$ 0.008) for 1-cm path at 570 mμ; the extinction value for a 1 ml sulphuric acid control is 0.045–0.060.

Nesslerization

An incineration procedure adapted for subsequent estimation of the ammonia by the Nessler reagent is described by Lang[23]. The method can be used for samples containing 1–1000 μg N.

Amide nitrogen

The protein is boiled with 2 *N* hydrochloric acid (3 mg/ml) for 2–3 h. Aliquots are neutralized and distilled in a micro-Kjeldahl flask at pH 9–11. Small amounts of ammonia arise during the hydrolysis from deamination of certain amino acids, particularly from serine and threonine, and if desired the correct amide ammonia may be obtained from a plot of the amide-N as a percentage of the total N *versus* the time of hydrolysis and extrapolation to zero time[24].

Chibnall *et al.*[25] have drawn attention to the need to ascertain the free ammonia content of proteins prior to amide determination. This may be done by dissolving 10 mg of the protein in 1 ml of methanolic hydrogen chloride (0.033 *N*) and adding ether (2 ml) to reprecipitate the protein. After removal of the precipitate by centrifugation the ammonia present in the supernatant liquid may be determined by the Conway micro diffusion technique[26].

ESTIMATION OF TRYPTOPHAN IN PROTEINS

Oxidation to kynurenine

The instability of tryptophan in a strongly acid environment is undoubtedly a disadvantage in protein analysis. A fair recovery of the amino acid is possible from a protein or peptide that has a low cystine or cysteine content provided oxygen and heavy metals are rigorously excluded from the acid medium. Formyl kynurenine is a stable oxidation product of tryptophan, which can be deacylated to kynurenine itself according to the following scheme:

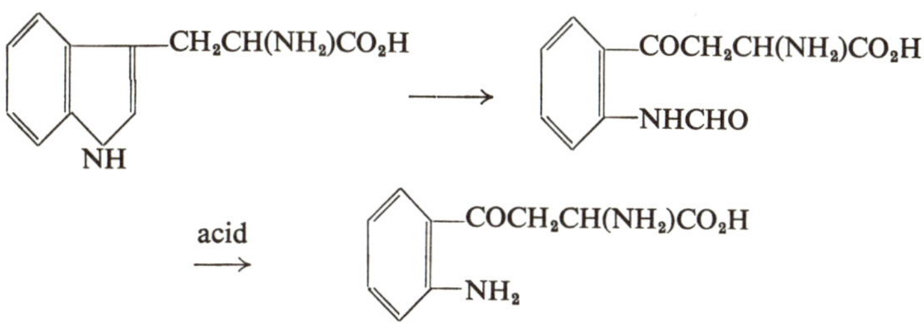

Kynurenine

Various oxidizing agents have been tried in attempts to improve the yield of kynurenine and in fact a direct synthesis of the compound from tryptophan by controlled oxidation with ozone was proposed by Witkop[27] many years ago. Recently, Previero *et al.*[28,29] extended this procedure to the ozonolysis of tryptophan in peptide linkage. The peptide or protein was dissolved in 98–100% formic acid (containing 3% resorcinol in some experiments) and exposed to a slow stream of ozone (4% in oxygen) for 30–40 min at 20° or below. The recovery of kynurenine from acetyl-DL-tryptophanylglycine was 76%. Under these conditions cystine was converted to cysteic acid and methionine to the sulphone; tyrosine was degraded in part to aspartic acid and oxalic acid. It has been reported that a good yield of kynurenine is obtained after oxidation of N-acetyl tryptophanamide with a 10-fold excess of sodium metaperiodate (10^{-2} *M*) at pH 6.8 for 6 h[30].

Alkaline hydrolysis

Hydrolysis of proteins and peptides by alkali is best carried out by the method of Drèze and Reith[31] (see also Drèze[32]): 20 mg of protein, 100 mg of starch (washed with 0.01 *N* HCl) and 600 mg $Ba(OH)_2 \cdot 8H_2O$ are weighed into a tube made of alkali-resistant glass. Water (5 ml) is added and the tube heated in the vertical position for 15 h at 120°. To protect the oven from the effects of a possible fracture the glass tube is enclosed in a jacket made from a piece of iron piping. After the hydrolysis the tube is allowed to cool to room temperature, wrapped tightly in a duster and opened cautiously with a sharp file. 3.5 ml of 6 *N* HCl are added, the mixture made to a volume of 10 ml and a 4-ml portion centrifuged for 30 min 2 ml of the supernatant is then transferred to the top of a 1 × 30 cm column of acid-washed starch poured in 0.1 *N* HCl. After washing in the sample the column is eluted with the same solvent and 80 × 0.5-ml fractions collected and analyzed by the ninhydrin method. The first peak contains most of the amino acids, the aromatic amino acids are retarded slightly and tryptophan emerges as the final peak.

Lugg[33] noticed the stabilizing effect of 5% $SnCl_2.2H_2O$ on the alkaline hydrolysis of tryptophan; Drèze[34] has studied this further and also recommends hydrolysis under nitrogen.

Colorimetric method

Colorimetric estimation of tryptophan with *p*-dimethylaminobenzaldehyde has been widely used since the development of the method by Spies and Chambers[36] in 1949. The main drawbacks to the method are the variations in the colour produced with different proteins[37], the indeterminate reaction times for optimum results, and the unpredictable effects of heavy metal contamination[38,39]. Methods based on the colour reaction obtained with glyoxylic acid and the indole ring were introduced by Shaw and McFarlane[40] and Fischl[41]; the modified procedure of Opieńska–Blauth *et al.*[39] will be described.

Reagents

A. Acetic acid containing 56 mg Fe^{3+}/litre (0.27 g of $FeCl_3 \cdot 6H_2O$

is dissolved in 0.5 ml water and glacial acetic acid added to 1 litre).

B. Concentrated sulphuric acid (sp. gr. 1.84).

Reagent A (2 ml) is mixed with a 1-ml sample containing 2–40 μg of tryptophan, and 2 ml of Reagent B are added. The solution is shaken vigorously, allowed to stand for 15 min, and the extinction at 545 mμ is measured. The reaction is influenced by the amount of water present; chlorides, phosphates, or organic reducing substances have no effect. The colour developed is permanent for at least 10 h.

REFERENCES

1 O. Folin and V. Ciocalteu, *J. Biol. Chem.*, 73 (1927) 627.
2 O. H. Lowry, N. J. Rosebrough, A. L. Farr and R. J. Randall, *J. Biol. Chem.*, 193 (1951) 265.
3 M. Eggstein and F. H. Kreutz, *Klin. Wochschr.*, 33 (1955) 879.
4 Shao-Chia and A. Goldstein, *Biochem. J.*, 75 (1960) 109.
5 A. G. Gornall, C. S. Bardawill and M. M. David, *J. Biol. Chem.*, 177 (1949) 751.
6 J. Goa, *Scand. J. Clin. Lab. Invest.*, 5 (1953) 218.
7 T. W. Goodwin and R. A. Morton, *Biochem. J.*, 40 (1946) 628.
8 G. H. Beaven and E. R. Holiday, *Advan. Protein Chem.*, 7 (1952) 319.
9 L. J. Saidel, A. R. Goldfarb and W. B. Kalt, *Science*, 113 (1951) 683.
10 A. R. Goldfarb, L. J. Saidel and E. Mosovich, *J. Biol. Chem.*, 193 (1951) 397.
11 M. P. Tombs, F. Souter and N. F. Maclagan, *Biochem. J.*, 73 (1959) 167.
12 W. J. Waddell, *J. Lab. Clin. Med.*, 48 (1956) 311.
13 J. B. Murphy and M. W. Kies, *Biochim. Biophys. Acta*, 45 (1960) 382.
14 J. R. Spies, D. C. Chambers and E. J. Coulson, *Arch. Biochem. Biophys.*, 84 (1959) 286.
15 G. Bendixen, *Nord. Med.* 58 (1957) 1487.
16 O. Warburg and W. Christian, *Biochem. Z.*, 310 (1941) 384.
17 H. M. Kalckar, *J. Biol. Chem.*, 167 (1947) 461.
18 A. C. Chibnall, M. W. Rees and E. F. Williams, *Biochem. J.*, 37 (1943) 354.
19 R. Markham, *Biochem. J.*, 36 (1942) 790.
20 S. H. Yuen and A. G. Pollard, *J. Sci. Food Agr.*, 4 (1953) 490.
21 I. H. Sher, *Anal. Chem.*, 27 (1955) 831.
22 S. Jacobs, *Nature*, 183 (1958) 262.
23 C. A. Lang, *Anal. Chem.*, 30 (1959) 1692.
24 S. J. Leach and E. M. J. Parkhill, *Proc. Intern. Wool Text. Research Conf. Australia*, (1955) C 92.
25 A. C. Chibnall, J. L. Mangan and M. W. Rees, *Biochem. J.*, 68 (1958) 111.
26 E. J. Conway, *Microdiffusion Analysis and Volumetric Error*, 4th Ed., Crosby Lockwood, London, 1957.

27 B. Witkop, *Ann.*, 556 (1944) 103.
28 A. Previero, E. Scoffone, P. Pajetta and C. A. Benassi, *Gazz. Chim. Ital.*, 93: 7 (1963) 841, 849, 859.
29 A Previero, M. A. Coletti and L. Galzigna, *Biochem. Biophys. Res. Commun.*, 16 (1964) 195.
30 C. P. Fawcett, *Sixth International Congress of Biochemistry*, New York, Abstracts, II (1964) 53.
31 A. Drèze and W. S. Reith, *Biochem. J.*, 62 (1956) 3P.
32 A. Drèze, *Bull. Soc. Chim. Biol.*, 42 (1960) 407.
33 J. W. H. Lugg, *Biochem. J.*, 32 (1938) 775.
34 A. Dreze, *Bull. Soc. Chim. Biol.*, 38 (1950) 243.
35 G. H. Beaven, in: H. W. Thompson (Ed.), *Advances in Spectroscopy, Vol. 2*, Interscience Publ., New York, 1961, p. 331.
36 J. R. Spies and D. C. Chambers, *Anal. Chem.*, 21 (1949) 1249.
37 P. M. Harrison and T. Hofmann, *Biochem. J.*, 80 (1961) 38P.
38 M. A. Jermyn, *Anal. Biochem.*, 5 (1963) 433.
39 J. Opieńska-Blauth, M. Chareziński and H. Berbeć, *Anal. Biochem.*, 6 (1963) 69.
40 J. L. D. Shaw and W. D. McFarlane, *Can. J. Res.*, 16 (1938) 361.
41 J. Fischl. *J. Biol. Chem.*, 235 (1960) 999.

Chapter 12

SYNTHESIS OF SIMPLE PEPTIDES

INTRODUCTION

Elucidation of the structure of a polypeptide or protein is an important first step towards an understanding of the function of these molecules in biological systems. The problem of structure-function relationships has been tackled experimentally in various ways and some remarkable discoveries have been made, particularly in the field of peptide hormones. The influence of a subtle chemical modification of the structure of a hormone or enzyme on biological activity might be studied, or a controlled degradation might be effected and an attempt made to assess the essential structural requirement for biological function. A different approach is to start a chemical synthesis of a sequence of amino acids known to possess physiological activity and follow the development of this activity with increasing chain length. In this way it has been found, for example, that extension of the peptide chain from either end of the active site of α-MSH increases biological activity until the trideca-peptide stage is reached. Thereafter activity decreases as the chain is lengthened from the carboxyl end[1].

One advantage of the synthetic approach is that the scale of operation can be adjusted to suit the needs of a specific problem and this is important where biological assays are involved consuming large amounts of material; disadvantages are the cost of the starting materials, the time involved, and the need for a good deal of experience. The synthesis of a sizeable peptide (of more than 5–6 residues) requires careful planning and can only be undertaken by experienced workers. However, small peptides are often required as substrates, models or reference substances in a variety of biochemical investigations, and this Chapter deals with methods for their synthesis which have withstood the tests of time and, for the most part, are within the scope of the ordinary biochemical laboratory. Although a useful range of amino acid derivatives and simple peptides is available commercially it is not always possible to obtain a certificate of analysis with these products.

References p. 389

Numerous reviews on the subject of peptide synthesis have appeared in recent years the most substantial of which are those by Fruton[2], Goodman and Kenner[3], Greenstein and Winitz[4], and Hofmann and Katsoyannis[1]. The first two supplement each other and both contain valuable tables of compounds; the article by Hofmann and Katsoyannis deals also with the literature of studies relating primary structure or sequence to physiological activity. For newcomers to peptide chemistry there is an excellent guide published as a lecture by Rydon[5].

GENERAL PRINCIPLES

Owing to the 'zwitter ion' structure of the α-amino acids non-enzymic synthesis of peptides in aqueous media is impracticable. Peptide chemistry has therefore been concerned with methods for masking amino groups and carboxyl groups, the activation of carboxyl groups, and the removal of N- and C-protecting groups. A general scheme for peptide synthesis utilizing these principles is shown below (Fig. 118).

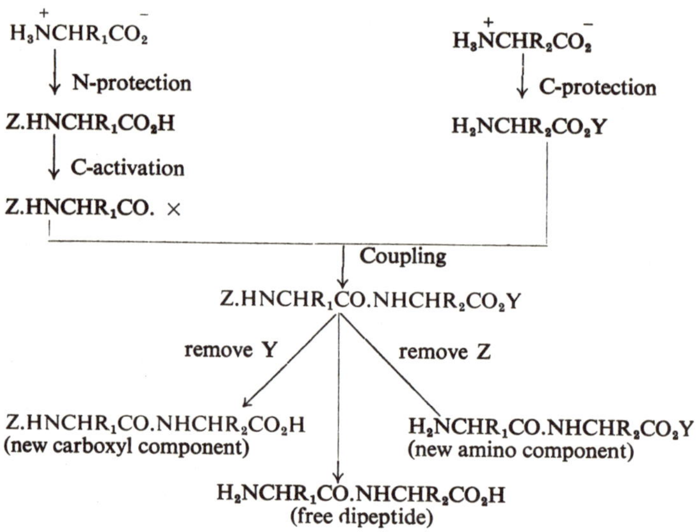

Fig. 118. A scheme for peptide synthesis.

In conventional peptide synthesis Y is removed to release a new carboxyl component for further coupling; a novel procedure in which Z is removed for subsequent coupling of a new amino component is described later (see p. 387).

Rather stringent criteria are applied in an evaluation of a potentially useful protecting group. It is desirable that such groups should be introduced easily and in good yields; it is essential that they should be split off again selectively and under the mildest possible conditions. Rupture of peptide linkages must be avoided, side-reactions must be minimal, optical configuration should be retained. Of the many protective groups that have been proposed only a few satisfy all these requirements. At the coupling stage choice of method is governed by similar considerations. A necessary attribute of a general coupling method is avoidance of substantial racemization.

A summary of the characteristics of N- and C-protecting groups is given in Table 52.

PROTECTION OF THE N-TERMINAL GROUP

Benzyloxycarbonyl ('carbobenzoxy') derivatives

Discovered by Bergmann and Zervas[6] in 1932 the benzyloxycarbonyl procedure is still the most widely used method for N-protection. The derivatives are prepared by reaction of the amino acid with benzyl chloroformate ('carbobenzoxy chloride') under strongly alkaline conditions.

$$PhCH_2OCOCl + H_2NCHRCO_2H \rightarrow PhCH_2OCO.NHCHRCO_2H$$

No appreciable racemization occurs during the reaction, the yields are good, and the derivatives are colourless, crystalline solids, with the exception of those from leucine and isoleucine, which are oils. Sodium hydroxide, sodium or potassium carbonate, sodium bicarbonate or triethylamine have been used as the acid binder, and various solvents, dioxan for example, mixed with water have provided the medium for the reaction[7].

The classical method for the removal of the benzyloxycarbonyl group is that of hydrogenolysis[6] which reduces the substituent to toluene and carbon dioxide.

TABLE 52

CHARACTERISTICS OF N- AND C-PROTECTING GROUPS

Protecting group	*Symbol*[a]	*Terminus*	*Amino acid derivatives*	*Method of removal*
Benzyloxycarbonyl	Z	N	Crystalline excepting Z.Leu and Z.Ileu; prepared in good yield with no loss of optical activity; unstable in strong alkali	(1) Hydrogenolysis: fast reaction, various solvents used; cleaves benzyl esters, slowly reduces nitro groups, unsuitable for sulphur compounds (2) HBr: fast reaction, various solvents used; esters and amides may undergo some hydrolysis (3) Sodium in ammonia: very fast reaction; non-selective
p-Nitrobenzyloxycarbonyl	$Z(NO_2)$	N	All crystalline; give intense UV absorption	(1) Hydrogenolysis: very fast reaction; sulphur compounds amenable[b]; *p*-toluidine by-product (2) HBr: slow reaction[c]
t-Butyloxycarbonyl	BOC	N	Some derivatives difficult to crystallise; no racemization during preparation; acid labile	Diluted mineral acid or anhydrous trifluoroacetic acid; stable to hydrogenolysis and sodium in ammonia
Toluene-*p*-sulphonyl	Tosyl	N	All crystalline; optically active; useful for guanidine group of Arg and ε-NH_2 of Lys	(1) Sodium in ammonia: non-selective, moderate yields, possibility of peptide bond cleavage (2) HBr in phenol/acetic acid

[illegible]	[illegible]	N	Crystalline; some loss in optical activity; alkali labile	Hydrazine or phenylhydrazine: fast reaction, good yields; reagent interacts with esters (except *n*-butyl); incompatible with azide coupling method
Triphenylmethyl	Trityl	N	Crystalline; optically active; extremely acid labile	(1) Dil. mineral acid or 50% aqueous acetic acid; very fast reaction (2) Hydrogenolysis: very slow reaction
Trifluoroacetyl	TFA	N	Crystalline; optically active; yields variable; alkali labile	(1) Weak alkali: reaction rate depends on amino acid residue (2) Ethanolic HCl
Alkyl ester	OMe OEt	C	Crystalline hydrochlorides, easily prepared	Mild aqueous alkali
Benzyl	OBZL	C	Crystalline, easily prepared	(1) Hydrogenolysis (2) Sodium in ammonia (3) HBr in acetic acid: slow release
p-Nitrobenzyl	ONB	C	Crystalline, easily prepared	
t-Butyl	OBut	C	Prepared with difficulty	(1) HCl or toluene-*p*-sulphonic acid in toluene (2) HBr in acetic, or trifluoroacetic acid
Triphenyl methyl	OTRI	C	Known in a few cases	Methanolysis[d]

[a] Symbols recommended by the Committee on Nomenclature at the Fifth European Symposium, Oxford, 1962.
[b] Berse *et al.*[161].
[c] Gish and Vigneaud [162]; Shields and Carpenter[163].
[d] Gazis *et al.*[164].

References p. 389

$$PhCH_2OCO.NHCHR_1CONHR_2 \xrightarrow{H_2/Pd} PhCH_3 + CO_2 + H_2NCHR_1CONHR_2$$

Methanol + acetic acid, methanol alone, dimethylformamide and phenol[8] have been used as solvents for the reaction. During hydrogenolysis O-benzyl and S-benzyl groups will also undergo reductive fission as will the nitrogroup of nitro-arginine. Sulphur compounds tend to inactivate the catalyst and this method of cleavage is of limited value for peptides containing cystine, cysteine or methionine

A simpler procedure is to treat the benzyloxycarbonyl derivative with 1 N HBr in anhydrous acetic acid at 20° or a slightly higher temperature. Benzyl bromide and carbon dioxide are split out leaving the hydrobromide of the peptide[9,10].

$$PhCH_2OCO.NHCHR_1CONHR_2 \xrightarrow{HBr} PhCH_2Br + CO_2 + Br^{-+}NH_3CHR_1CONHR_2$$

Other solvents such as nitromethane[11] and carbon tetrachloride[12] may offer advantages in particular cases. Benzyl esters are not split by hydrobromic acid but some hydrolysis may occur.

The *p*-nitro derivatives offer an attractive alternative to the unsubstituted series in that they are all crystalline and undergo hydrogenolysis more readily[13]. Reduction of the nitro group, however, leads to *p*-toluidine as a by-product which has to be separated from the residual peptide. Nevertheless the method has been employed with success in many recent extended syntheses. Melting point data of N-benzyloxycarbonyl derivatives of amino acids are given in Table 53.

Benzyl chloroformate

If no commercial sample is available the reagent may be prepared by the original procedure of Bergmann and Zervas[6]. Dry toluene (500 g) is cooled in an ice-bath and phosgene is bubbled in until 109 g has been absorbed. The process, which takes about 1 h, must be carried out in a fume-cupboard and the exit tube must be protected from atmospheric moisture by fitting a calcium chloride tube. Redistilled benzyl alcohol (108 g) is added rapidly and the reaction allowed to proceed for 30 min at 0°, and at room temperature for a further 2 h. The solution is finally concentrated under reduced pressure at a temperature not exceeding 50°. Yield: about 160 g (90%).

TABLE 53

MELTING POINT DATA OF N-BENZYLOXCARBONYL (Z) DERIVATIVES OF AMINO ACIDS

Amino acid (L-form)	*Melting point (°C)*	*(α), solvent*	*Reference*
Alanine	84		165
Arginine hydrobromide	177–179.5	− 6.9, water	147
Arginine (nitro)	145–146	− 8.0, acetone	166
	134–136		167,168
Asparagine	167–169		169
Aspartic acid	116		6
Aspartic acid (OBZL)	108	+ 12.1, acetic acid	137
Cysteine (S-benzyl)	99		57,48,170,171
Cystine	133–135		172
Glutamic acid	200–201		173,174
Glutamic acid (OEt)	87–88		175
Glutamine	175		176
Glycine	120		6
Histidine	209		6
Isoleucine (ZNO_2)	77.5–80	− 12.6, 1*N* NaOH	13
Leucine (ZNO_2)	60–61	− 15.8, 1*N* NaOH	13
ε-Lysine (OMe) HCl	117		130
Di-lysine	150		130
Di-lysine (ZNO_2)	56–72	− 5.4, pyridine	177
Methionine (OMe)	121–122		56
Phenylalanine	87	+ 5.3, acetic acid	178
Proline	76–77	− 43.5, ethanol	179
		− 61.7, acetic acid	180
Serine	119–120	+ 5.8, acetic acid	181
Threonine (OMe)	89–91		182
Tryptophan	126		183
Tryptophan (ZNO_2)	151–152.5		177
Tyrosine	101	+ 11.1, acetic acid	6
Tyrosine (OZ)	117	− 5.0, acetic acid	184
Valine	59–61	+ 4.0, ethanol	185

N-Benzyloxycarbonylglycine

Glycine (7.5 g, 0.1 mole) is dissolved in 2 *N* NaOH (50 ml) and the solution cooled to 0°. Benzyl chloroformate (24 g) and 4 *N* NaOH (25 ml) are added simultaneously with vigorous stirring during 30 min. After a further 10 min the toluene layer is separated and the aqueous layer extracted once with ether, cooled, and acidified to Congo Red with concentrated HCl. The precipitate is filtered washed with cold water and dried in the air. Yield: 18.19 g (90%). Recrystallized from chloroform the product melts at 120°.

Cleavage of the benzyloxycarbonyl group occurs in 2 *N* HBr in glacial acetic acid at room temperature. The optimum time required is determined by a technique developed by Young[14]. An aliquot of the mixture is removed and the reaction arrested by the addition of Dowex-3 acetate suspended in *n*-butanol–water–acetic acid. A drop of this solution is spotted on paper, chromatographed in the same solvent system and the paper dried. Exposure of the paper to ultraviolet light[15] removes the benzyloxycarbonyl group, if present, and the peptide can then be revealed with ninhydrin as usual.

Toluene-p-sulphonyl (tosyl) derivatives

Beautifully crystalline derivatives are obtained by reaction of amino acids with toluene-*p*-sulphonyl chloride under alkaline conditions.

$$p\text{-}CH_3C_6H_4SO_2Cl + H_2NCHRCO_2H \longrightarrow p\text{-}CH_3C_6H_4SO_2.NHCHRCO_2H$$

Originally employed by Fischer[16] and Schönheimer[17], their general use in peptide synthesis was delayed until a suitable cleavage procedure became available. In 1937 Du Vigneaud and Behrens[18] discovered the reductive action of sodium in liquid ammonia and this has been used extensively in recent work. By this means several protecting groups of different types may be removed simultaneously at the end of a synthesis.

$$p\text{-}CH_3C_6H_4SO_2.NHCHRCONHR_2 \xrightarrow{Na/NH_3} p\text{-}CH_3C_6H_4SO_2H + H_2NCHRCONHR_2$$

Some instances of peptide bond rupture have been observed[19,20] and an alternative procedure using HBr/phenol/acetic acid was proposed by Poduska *et al.*[21] and Weisblat *et al.*[22].

N$^{\alpha}$-Benzyloxycarbonyl-N$^{\omega}$-tosyl-L-arginine[23]

N$^{\alpha}$-Benzyloxycarbonyl-L-arginine (12.5 g) is dissolved in a mixture of acetone (320 ml) and 4 *N* NaOH (40 ml), cooled to 0°, and treated with toluene-*p*-sulphonyl chloride (19 g) in acetone (60 ml) with vigorous stirring. After 2 h the mixture is brought to pH 5, the acetone evaporated *in vacuo*, and the residue extracted twice with ether (30 ml) to remove excess tosyl chloride. The product is extracted with ethyl acetate (150 ml) washed with water and extracted into 5% aqueous bicarbonate. This is then brought to pH 3, extracted with ethyl acetate (150 ml) and washed with 2 *N* HCl and water. Precipitated with petroleum ether the N$^{\omega}$-tosyl derivative melts over the range 75–85°. Yield: 13.5 g (73%).

Cleavage with Na/NH$_3$

The following procedure is from the work of Du Vigneaud and Behrens[18]. The N-tosyl derivative is dissolved in 50 ml of dry liquid ammonia and small pieces of sodium added with shaking until a permanent blue colour remains. Excess sodium is discharged by the addition of ammonium chloride to the reaction mixture, leaving a clear solution with a white precipitate. The ammonia is allowed to evaporate, the residue is taken up in water, and the resulting solution aerated, neutralized to litmus and extracted with ether. The aqueous layer contains the free amino acid or peptide.

Cleavage with HBr/phenol/acetic acid[22]

A solution of toluene-*p*-sulphonanilide (2.5 g, 0.01 mole) and of phenol (2.0 g, 0.02 mole) in 23 g of 30% HBr in acetic acid is allowed to stand at 25° for 16 h. The reaction mixture is poured into anhydrous ether (150 ml) and the precipitated hydrobromide filtered, washed with ether and dried. Yield: 1.48 g (86%). In the absence of phenol bromination products are obtained.

t-Butyloxycarbonyl derivatives

Stability to hydrogenolysis and ease of fission by acids are properties of the *t*-butyloxycarbonyl group which have interested peptide chemists in recent years. It was the protecting group of choice in the development of solid state peptide synthesis (see p. 387). The

References p. 389

derivatives are best prepared by reaction of the amino acid with *t*-butyl *p*-nitrophenyl carbonate[24] or with *t*-butyloxycarbonyl azide[25,26].

$$(CH_3)_3COCON_3 + H_2NCHRCO_2H \longrightarrow (CH_3)_3COCO.NHCHRCO_2H$$

A few amino acids yield derivatives that are difficult to crystallize. Removal of the group is effected by treatment with dilute HCl, HBr in acetic acid, or with anhydrous trifluoroacetic acid[20].

t-Butyloxycarbonyl azide[25]

t-Butyl carbazate is first prepared[27,28] by warming a mixture of *t*-butyl phenyl carbonate (476 g) and 64% hydrazine (252 g). After the strongly exothermic reaction has taken place the mixture is allowed to stand for 12 h, diluted to 1 litre with ether and shaken with aqueous sodium hydroxide (150 g). The carbazate is extracted into ether during 24 h in a continuous extractor, the solvent is removed and the residue distilled. Yield: 221 g (68.3%); b.p. 60–61° (0.5 mm).

Diazotization is then carried out by dissolving the carbazate (100 g) and glacial acetic acid (88 g) in water (125 ml) and treatment with solid sodium nitrite (57.4 g)—or a solution of sodium nitrite in 80 ml of water—added over a period of 50–60 min at 9–13°. After a further period of 30 min at 0° water is added and the golden-yellow azide layer separated. Ether extraction of the aqueous layer provides a further quantity of the azide and the combined organic layers are washed thoroughly with water and 1 *M* sodium bicarbonate. The solution is dried over anhydrous $MgSO_4$ and the solvent removed at 40–45° and at 140 mm. Distillation of the azide is carried out at 90° and at 70 mm. Yield: 80%; b.p. 73–74° (70 mm).

Warning. In view of the recent report[1] of a violent explosion which occurred during the distillation of this reagent this purification step could be omitted and the crude azide used for condensation with the amino acid.

*t-Butyloxycarbonyl-*L*-valine*[26]

L-Valine (20 g), magnesium oxide (8.7 g), water (200 ml) and dioxan (60 ml) are stirred with a solution of *t*-butyloxycarbonyl azide (36 ml) in dioxan (200 ml) and kept at 45° for 22 h. The dioxan is removed *in vacuo* and the residue treated with water and extracted with ethyl acetate. The aqueous phase is acidified with citric acid at 0° and the *t*-butyloxycarbonyl derivative extracted with ethyl acetate. Yield: 25.2 (68%).

The protecting group is removed again by adding the derivative (1 g, as the ethyl ester) to 5 ml of HBr-acetic acid at room temperature. Gas evolution subsides after 30 sec and ether (100 ml) is added to precipitate the ester hydrobromide. The acid is prepared by bubbling HBr into glacial acetic acid at 0°.

Phthaloyl derivatives

This N-protecting group was discovered independently by Kidd and Young[29] and Sheehan and Frank[30]. It has been particularly useful for the synthesis of small peptides. The following methods for the preparation of N-phthaloyl amino acids have been described:

(1) A dry mixture of the amino acid and phthalic anhydride is heated at a temperature of 145–150° [31,32].
(2) A suspension of the amino acid and phthalic anhydride in dioxan[33] is heated at 95–105°.
(3) The amino acid is treated with N-ethoxycarbonylphthalimide in aqueous sodium bicarbonate at room temperature[34].
(4) The amino acid is heated with diphenyl phthalate and triethylamine in phenol[35].

Method (1) requires careful temperature control to avoid appreciable racemization and the reaction is a straightforward condensation:

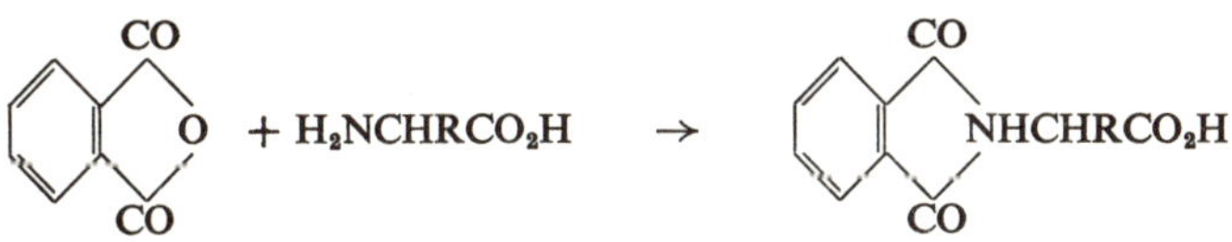

No significant racemization occurs in method (3) but the reagent used to introduce the protecting group has to be prepared from potassium phthalimide and ethyl chloroformate. Melting point data of phthaloyl amino acids prepared by this route are given in Table 54. Reaction with an amino acid proceeds according to the following scheme:

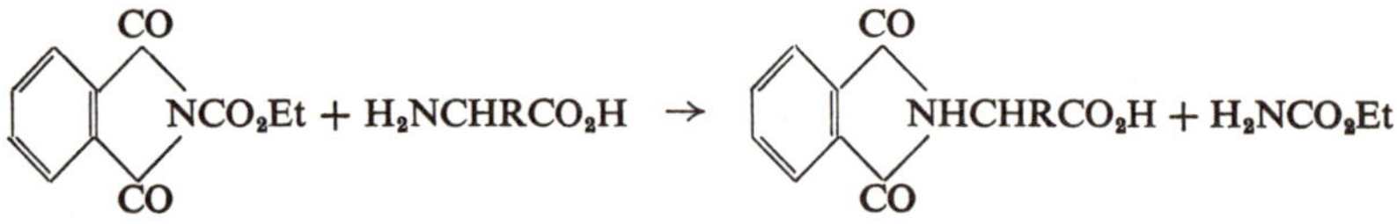

To remove the group the phthaloyl peptide is refluxed for one to two hours with an alcoholic solution of hydrazine; the by-product, phthalazine-1,4-dione, is removed from the mixture by precipitation with mineral acid[36,37].

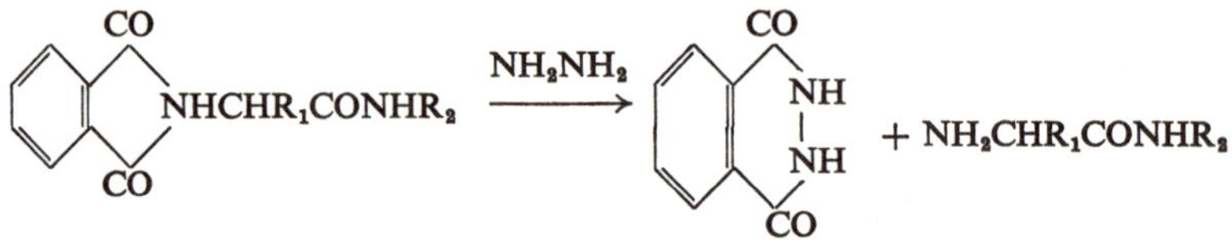

Alternatively, hydrazine acetate[38] or phenylhydrazine in the presence of tri-*n*-butylamine[37] may be used. The derivatives are sensitive to alkali themselves forming phthalamic acids ($HO_2CC_6H_4CONHR$)

TABLE 54

PHTHALOYL AMINO ACIDS PREPARED WITH THE AID OF N-ETHOXYCARBONYL PHTHALIMIDE[34]

Amino acid	*Yield (%)**	*(m.p. °C)*
β-Alanine	91	152
L-Asparagine	85	199
Cystine	92	120
L-Glutamic acid	65	160
Glycine	90	191
DL-Histidine	85	296
L-Leucine	92	110
Nε-L-Lysine	85	232
DL-Methionine	96	102
DL-Phenylalanine	90	178
DL-Serine	95	152

* After 2 days drying over P_2O_5 *in vacuo*.

and ester groups, if present in a peptide, may be converted to hydrazides. *t*-Butyl esters are least likely to undergo reaction with hydrazine[38].

N-Phthaloyl-L-phenylalanine[32]

L-Phenylalanine (9.90 g, 0.06 mole) and finely ground phthalic anhydride (8.95 g, 0.06 mole) are mixed and heated for 30 min with stirring at 145–150°. After cooling, the solid is dissolved in hot methanol (40 ml) and the solution filtered. On addition of water (40 ml) the phthaloyl derivative crystallizes slowly in fine needles, m.p. 183–185°. Yield: 15.8 g (88%).

N-Carboethoxy phthalimide[34]

Potassium phthalimide (92.5 g) is suspended in anhydrous dimethylformamide (250 ml) and freshly distilled ethyl chloroformate (50 ml) is added at 5°. The mixture is left for 1 h at room temperature and then poured into water (3 l). The crystalline N-carboethoxy phthalimide is filtered, washed with water and dried. Recrystallized from hot 90% ethanol it melts at 80°. Yield: 86%.

N-Phthaloyl-L-glutamic acid[34]

L-Glutamic acid (2.4 g) is suspended in water (20 ml) and sodium carbonate (10 H_2O; 11.5 g) added. The mixture is cooled to 0° and N-carboethoxyphthalimide (5 g) added with stirring. After standing for 30 min at room temperature the mixture is filtered and acidified to pH 2.5 with 6 *N* HCl. The oil which separates crystallizes at 0°. Yield: 3.6 g (65%); m.p. 160°; $[\alpha] = 48.3$ (C = 3, in dioxan).

Method of Weygand and Kaeliche[35]

The amino acid (0.01 mole), diphenyl phthalate (0.011 mole), triethylamine (0.01 mole) and phenol (about 5 g) are heated for 5–6 min at 120–130° (internal temperature). The mixture is cooled dissolved in ethyl acetate, and the solution shaken three times with saturated aqueous bicarbonate. The phthaloyl amino acid is then precipitated from the aqueous phase with HCl and crystallized from a suitable solvent, *e.g.* ethyl acetate-petroleum ether.

Cleavage with hydrazine[32]

Phthaloyl-L-phenylalanine (1.41 g, 0.0048 mole) is refluxed in ethanol (15 ml) containing 1 *M* alcoholic hydrazine hydrate (0.70 ml, 0.012 mole) for 2 h. After removal of the solvent *in vacuo*, water (10 ml) is added and acetic acid to pH 6. The suspension is heated on a steam-bath for 1 h, diluted with 30 ml of water, cooled and filtered. The filtrate yields free phenylalanine on concentration. Yield: 0.66 g (84.4%).

Cleavage with phenylhydrazine[37]

Phthaloyl-L-leucine (0.535 g), tri-*n*-butylamine (0.5 ml) and phenylhydrazine (0.4 ml) are refluxed with 3 ml of 96% ethanol for 2 h. Methylethylketone (10 ml) is added and heating continued for 15 min. After cooling, glacial acetic acid (0.2 ml) is added and the crystals of L-leucine filtered and washed. Yield: 0.325 g (82.7%).

Triphenylmethyl (Trityl) derivatives

The triphenylmethyl group has been exploited for both N- and C-protection in peptide synthesis. N-Trityl derivatives were studied independently by three groups of workers[40–42] and many examples of their use are to be found in the recent literature. They may be prepared by direct action of triphenylmethyl chloride on the amino acid in aqueous isopropanol containing diethylamine[42], or indirectly by similar tritylation of an amino acid ester followed by saponification. Some difficulty has been experienced with the saponification of trityl amino acid esters, particularly with phenylalanine and leucine, owing to the bulky nature of the trityl group. Dilute mineral acid or 50% aqueous acetic acid at 50° readily cleaves N-trityl derivatives; a much slower release is achieved by hydrogenolysis.

$$Ph_3CCl + H_2NCHR_1CO_2H \longrightarrow Ph_3CNHCHR_1CO_2H$$

$$Ph_3C.NHCHR_1CONHR_2 \xrightarrow{\text{acid}} Ph_3COH + NH_2CHR_1CONHR_2$$

$$Ph_3C.NHCHR_1CONHR_2 \xrightarrow{H_2/Pd} Ph_3CH + NH_2CHR_1CONHR_2$$

N-Tritylglycine methyl ester[42]

Triphenylchloromethane (2.8 g, 0.01 mole) and triethylamine (2.2 g, 0.022 mole) are added to a suspension of glycine methyl ester hydrochloride (1.25 g, 0.01 mole) in chloroform (15 ml). Tritylation is complete after 6 h at room temperature. The solution is washed twice with water, dried over Na_2SO_4, and the solvent removed *in vacuo*. The residue is recrystallized from methanol.

Cleavage with acid[42]

A solution of N-tritylglycine ethyl ester (1.72 g, 0.005 mole) in 1 *N* ethanolic-HCl (5 ml) is heated for 1–2 min on a water-bath. Glycine ethyl ester hydrochloride crystallizes after evaporation of the ethanol *in vacuo*. Yield: 0.7 g (95%). Trityl ethyl ether (m.p. 80–82), is obtained from the filtrate.

Trifluoroacetyl derivatives

Until quite recently the trifluoroacetyl group was the only N-protecting group of importance which could be removed by alkali. Application in peptide synthesis was demonstrated by Weygand and collaborators[43–45] who obtained the derivatives by the action of trifluoroacetic anhydride on amino acids and peptides. Another useful and milder reagent for trifluoroacetylation is ethyl thioltrifluoroacetate[46,47] the reaction proceeding according to the scheme:

$$F_3CCOSEt + H_2NCHRCO_2Na \xrightarrow{\text{pH } 8\text{–}9} F_3CCO.NHCHRCO_2Na + HSEt$$

Most of the trifluoroacetyl amino acids are crystalline[48] and exposure to weak alkali[43,44,47,49] or ethanolic HCl[49] will remove the protecting group.

N-Trifluoroacetyl-glycine[47]

A solution of glycine (378 mg, 5 mmoles) is brought to pH 10 with 1 N NaOH. Ethyl thioltrifluoroacetate (1.5 ml, 11.5 mmoles) is added and the mixture agitated for 18 h. Addition of 2 ml of 1 *N* HCl precipitates the N-trifluoro-acetyl derivative which is then extracted with three 10-ml portions of ethyl ether. The crystalline product obtained after removal of solvent melts at 110.5–116.5°. Yield: 477 mg (54.8%).

Cleavage by alkali[47]

A solution of N-trifluoroacetyl-L-phenylalanine (262 mg, 1.0

References p. 389

mmole) in 5 ml of 95% ethanol is titrated with 1 *N* NaOH. Excess alkali (5.0 ml) is added and the solution left for 24 h at room temperature. After back titration with 1 *N* HCl the solution is evaporated. Yield: 73 mg (44%).

Other methods for N-protection

There are a variety of methods which for one reason or another cannot be recommended for general use. Some are still under development, others have been found useful in special circumstances. The formyl group, for example, just fails to compete effectively with the N-benzyloxycarbonyl group because of a tendency to undergo undesirable reactions during certain coupling procedures. Formylation of amino acids is usually conducted in a mixture of 98% formic acid and acetic anhydride[50]; mild acid hydrolysis will remove the formyl group.

Schiff's bases formed from amino acids and aromatic aldehydes[51] have limited application; derivatives have been prepared from 5-chlorosalicylaldehyde, 2-hydroxy-1-naphthaldehyde, benzoylacetone and similar ketones[52,53]. They are acid labile. A promising new group which is sensitive to alkali is the *p*-tolylsulphonylethoxycarbonyl group described by Kader and Stirling[54].

PROTECTION OF THE C-TERMINAL GROUP

Most coupling operations are carried out in an organic solvent and for this reason the choice of a C-protecting group is severely restricted. Esterification is in fact the only satisfactory means of rendering the carboxyl component soluble under anhydrous conditions. Although the general problems relevant to this step in peptide synthesis have benefited considerably from our knowledge of the organic chemistry of esters, further research has often been necessary for a true appraisal of the value of a particular type of ester in peptide work. At the removal stage, usually an alkaline hydrolysis, the process is not always straightforward and careful control is needed to avoid undesirable side-reactions. Apart from the danger of racemization of the peptide, hydantoins may be formed from N-benzyloxycarbonyl derivatives, and N-phthaloyl, N-trifluoroacetyl and S-benzyloxycarbonyl groups are cleaved.

Methyl and ethyl esters

Many laboratories still use the classical method[55] of esterifying amino acids or peptides which involves suspension of the compound in anhydrous methanol or ethanol and exposure to anhydrous hydrogen chloride. The ester hydrochlorides usually crystallize from the medium after addition of dry ether. More convenient procedures[56–59] are, however, available and the best of these is treatment of the amino acid with the alcohol containing thionyl chloride[56,57]. The free esters may be liberated by treatment of their hydrochlorides with an equivalent of sodium ethoxide[60] in alcohol, or with a slight excess of ammonia in chloroform[61].

L-*Phenylalanine methyl ester*[57]

Thionyl chloride (9.8 ml) is added to anhydrous methanol (87 ml) at −10°. L-Phenylalanine (18 g) is then added and the mixture refluxed for 2 h. After concentration of the solution *in vacuo* the ester hydrochloride is precipitated by addition of ether. Recrystallized from methanol-ether it melts at 160°. Yield: 21 g (90%).

Benzyl and p-nitrobenzyl esters

The difficulties associated with alkaline hydrolysis of peptide alkyl esters can be avoided by employing benzyl esters which can be removed by hydrogenolysis[62,63] or by treatment with sodium in liquid ammonia[7,64]. Direct esterification of amino acids or peptides with benzyl alcohol is possible in the presence of acid catalysts such as polyphosphoric acid[65], benzenesulphonic acid[66,67] and toluene-*p*-sulphonic acid[68,69]. The reaction is assisted by azeotropic distillation. An indirect route via N-acyl derivatives has been described[10].

Benzyl esters are not completely stable to HBr in acetic acid and their application in syntheses involving N-benzyloxycarbonyl derivatives is consequently limited. On the other hand, the *p*-nitrobenzyl esters[70,71] are considerably more stable to acid and they have played an important role in some recent syntheses[71–74]; like the unsubstituted compounds they may be prepared by both direct[67] and indirect routes.

L-Glutamic acid dibenzylester benzene sulphonate and hydrochloride[67,78]

L-Glutamic acid (2.94 g, 20 mmoles) and benzenesulphonic acid monohydrate (3.7 g, 21 mmoles) are refluxed with benzyl alcohol (14 ml) and carbon tetrachloride (50 ml) for 15 h. During this period the reflux condensate is passed through a continuous extraction apparatus containing silica gel to remove water before the solvent returns to the boiling flask. Ether is added to the warm solution and the crude benzenesulphonate salt recrystallized from methanol-ether; m.p. 115–117°; yield 90%. The hydrochloride is formed by stirring the benzenesulphonate (2 g) in chloroform (9 ml) containing 0.58 ml of triethylamine. The solution is filtered and treated with dry hydrogen chloride. Recrystallized from methanol-ether the product melts at 100–102°. Yield: 90%. The *p*-nitrobenzyl ester is prepared in similar fashion.

t-Butyl esters

t-Butyl esters[39,75] are resistant to hydrogenolysis, to alkali and to hydrazinolysis; cleaved by hydrochloric acid in benzene or by toluene-*p*-sulphonic acid in toluene (unlike the N-benzyloxycarbonyl group); rapidly removed by hydrobromic acid in acetic acid or by trifluoroacetic acid.

When first introduced these esters were prepared by reaction of the amino acid with isobutene under pressure in the presence of a catalytic amount of sulphuric acid[75]:

$$H_2NCHRCO_2H + Me_2C{:}CH_2 \longrightarrow H_2NCHRCOOC(CH_3)_3$$

A similar method[39] starting with N-benzyloxycarbonyl derivatives proved superior but other methods including direct esterification[76] and ester exchange reactions[77], have not been very successful.

N-Benzyloxycarbonyl-L-proline t-butyl ester[39]

N-Benzyloxycarbonyl-L-proline (74.3 g, 0.30 mole) is dissolved in methylene chloride (600 ml) and 3 ml of conc. H_2SO_4 added. The solution is then saturated with isobutene causing an increase in volume of 300 ml. Reaction is complete at the end of 65 h at room temperature, after which water (500 ml) is added and sufficient sodium carbonate to neutralize the acid. The solvent layer is separated, washed with water, and concentrated *in vacuo* at 60°. Recrystallized from petroleum ether-ethyl ether the product melts at 44–45°. Yield: 86.5 g (95%).

Trimethylsilyl esters

A different approach is offered by the use of N,O-disilylated amino acids which have been prepared by reaction of trialkylsilylamines with amino acids[79]:

$$2(CH_3)_3SiNEt_2 + H_2NCHRCO_2H \longrightarrow (CH_3)_3SiNHCHRCO_2Si(CH_3)_3 + 2NHEt_2$$

The derivatives are coupled with an N-protected amino acid and the product is hydrolyzed with water[80]:

$$Z.HNCHR_1CO_2H + Me_3Si.NHCHR_2COOSiMe_3$$

$$\downarrow$$

$$Z.HNCHR_1CO\overset{\overset{\displaystyle SiMe_3}{|}}{N}CHR_2COOSiMe_3$$

$$\downarrow H_2O$$

$$Z.HNCHR_1CONHCHR_2CO_2H$$

β-Methylthioethyl esters

A promising new development[81,82] is based on the discovery of Crane and Rydon[83] that methiodides of β-methylthioethyl esters break down at about pH 10 at room temperature according to the equation,

$$RCOOCH_2CH_2{}^+SMe_2\}\ I^- + OH^- \rightarrow RCOO^- + CH_2{:}CH\overset{+}{S}Me_2\}\ I^- + H_2O$$

The esters are prepared in good yield by heating the N-protected amino acid with β-chloroethylmethyl sulphide, triethylamine and ethyl acetate under reflux.

COUPLING METHODS

A peptide chain may be built up from either the C-terminus or the N-terminus, but the literature of the subject indicates a strong preference for the latter approach. Both methods are, however, suitable for stepwise synthesis of small peptides (up to 5–6 amino acid residues). Synthesis of a complex peptide beyond the octapeptide range is usually accomplished in stages; small units are built up first and, at a later stage, these fragments are coupled together to give the desired peptide. A scheme devised by Kappeler and Schwyzer[156] for the synthesis of β^{1-24}-corticotropin (Fig. 119) illustrates this approach and it will be noted that glycine or proline is the favoured C-terminus for the condensations. Racemization is precluded by this choice since glycine has no asymmetric centre and proline does not undergo structural changes likely to bring about racemization. The coupling methods discussed in this section are based on carboxyl activation (see p. 354) and are suitable for stepwise synthesis of almost any small peptide.

Azide method

This classic method has played a major role in the development of peptide chemistry since Curtius[84] described the use of the azide of benzoylglycine for the synthesis of triglycine in 1902. Acid azides are formed by the action of nitrous acid (from sodium nitrite or butyl nitrite[85]) on an acid hydrazide according to the following scheme:

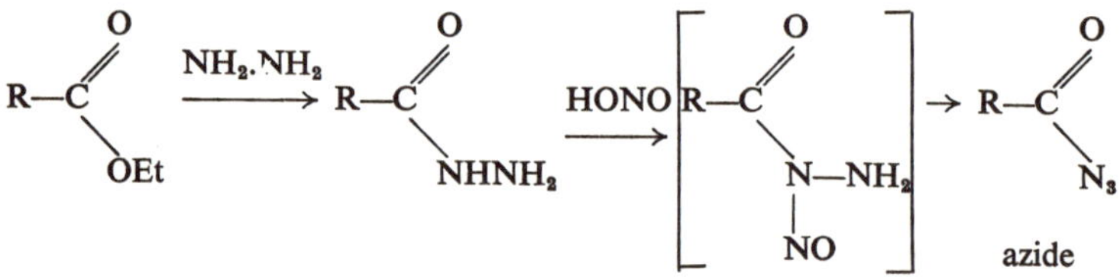

Acid azides of protected amino acids or peptides condense with the free amino group of another amino acid or peptide splitting out hydrazoic acid:

$$\mathrm{Z.HN\overset{\overset{R_1}{|}}{C}HCON_3 + H_2N\overset{\overset{R_2}{|}}{C}HCOOY \longrightarrow Z.HN\overset{\overset{R_1}{|}}{C}HCONH\overset{\overset{R_2}{|}}{C}HCOOY + N_3}$$

The instability of the azides leads to a number of side reactions, the most significant being due to the Curtius degradation. Nitrogen

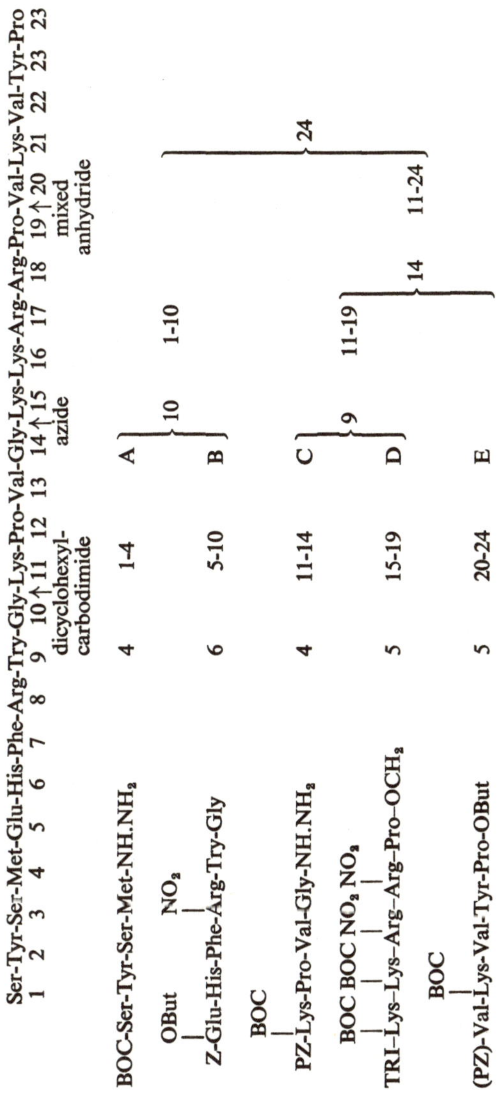

Fig. 119. A scheme for the synthesis of β^{1-24}-corticotropin (Kappeler and Schwyzer[156]).

is evolved and the residual isocyanate then combines with a free amino group to form a urea:

$$R_1CON_3 \longrightarrow R_1NCO + N_2 \xrightarrow{H_2NR} R_1NHCONHR_2$$

The problem was investigated by Harris and Fruton[86] who recommended the use of low temperatures to minimise this tendency. Amide formation has been observed during manipulation of some hydrazides for example, with peptides possessing C-terminal cysteine[78,89–91], with N$^{\alpha}$, N$^{\varepsilon}$-dibenzyloxycarbonyl-lysine[87] and with N-benzyloxycarbonyl-S-benzylcysteinyl tyrosine[88].

Most of the common groups for N-protection have been utilized in the preparation of azides. For those groups which are sensitive to hydrazine a slightly different procedure is necessary and this is illustrated in the scheme below for the synthesis of N-fluoroacetyl amino acid hydrazides[92].

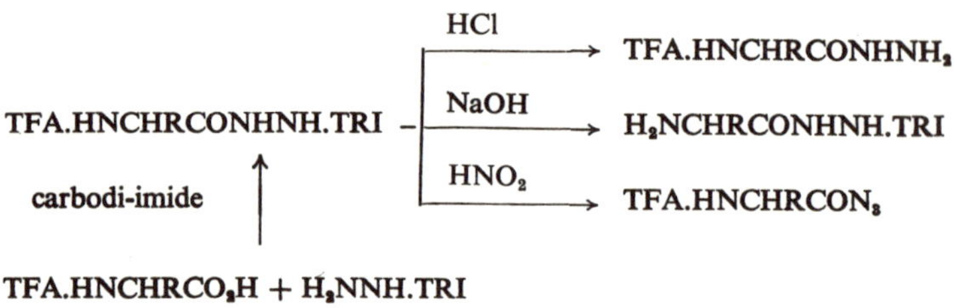

N-Benzyloxycarbonyl[93], N-phthaloyl and N-*t*-butyloxycarbonyl-hydrazides[94–96] have also been prepared and triethylamine salts have been used in the condensation. It will be obvious that substituted hydrazides can themselves play a useful role as C-protecting groups.

By a reaction which remains obscure acid hydrazides can be coupled with C-protected amino acids by treatment with oxidizing agents such as N-bromo-succinimide, hydrogen peroxide or iodine[97]. The yields are good and the method should prove useful with amino acids which are stable to oxidation.

The most satisfactory feature of the azide method of coupling is that no racemization is caused. Despite the trend toward more modern methods for coupling a small number of amino acids, the azide method is invariably used at some stage for the synthesis of longer chains.

Synthesis of the tripeptide L-*serylglycyl*-L-*glutamic acid*

N-Benzyloxycarbonyl-L-*serine hydrazide*[98]

L-Serine (3.2 g) is esterified with methanolic-HCl and the N-benzyloxycarbonyl derivative prepared. The derivative is not isolated but treated directly in ethanolic solution with hydrazine hydrate (1.5 ml). Crystallization of the hydrazide is complete after 24 h at room temperature. Yield: 3.4 g; m.p. 181°.

N-Benzyloxycarbonyl-L-*serylglycine ethyl ester*[98]

Glacial acetic acid (3 ml) and concentrated HCl (1 ml) are added to a suspension of N-benzyloxycarbonyl-L-serine hydrazide (3.5 g) in water (40 ml). The solution is cooled to +4° and sodium nitrite (1.5 g) in water (10 ml) is added slowly. The resulting syrupy azide is extracted into ethyl acetate and the solution washed quickly with cold water, bicarbonate, and water, and finally dried over anhydrous sodium sulphate. Azide equivalent to 1 g of hydrazide is dissolved in ethyl acetate and added to a solution of glycine ethyl ester (from 5 g of the hydrochloride) in ether. After 24 h at room temperature the mixture is washed with dilute HCl, water, bicarbonate, and water. On addition of petroleum ether to the concentrated solution crystals, m.p. 105–107° are obtained. Yield: 0.8 g.

N-Benzyloxycarbonyl-L-*seryl glycine hydrazide*[86]

N-Benzyloxycarbonyl-L-serylglycine ethyl ester (8.0 g) is treated with hydrazine hydrate (2.5 ml) in anhydrous methanol (100 ml). The hydrazide crystallizes after 1 h. Recrystallized from dilute acetic acid it melts at 188°. Yield: 7.6 g.

N-Benzyloxycarbonyl-L-*serylglycyl*-L-*glutamic acid diethyl ester*

The dipeptide ester (2.3 g) prepared above is dissolved in a mixture of water (35 ml), acetic acid (3.3 ml) and concentrated HCl (1.7 ml) and converted to the azide using 0.6 g of sodium nitrite in water. Coupling with L-glutamic acid diethyl ester (from 4 g of the hydrochloride) is carried out in ethyl acetate solution at 0° during

References p. 389

12 h, and at room temperature for a further 24 h. After working up as before the syrup is brought to crystallization with ethyl acetate-petroleum ether. Yield: 1.7 g; m.p. 107°.

L-*Serylglycyl*-L-*glutamic acid*

The protected tripeptide diethyl ester is hydrolyzed as follows. 1 *N* NaOH (6.4 ml) is added in three portions at 10 minute intervals to a solution of the diethyl ester (1.3 g) in methanol (50 ml). The solution is adjusted to pH 4 and concentrated. Hydrogenolysis of the N-benzyloxycarbonyl tripeptide gives the free peptide which can be crystallized from aqueous ethanol. Yield: 0.14 g.

Mixed anhydrides with carboxylic acids

Mixed anhydrides are formed when N-protected amino acids or peptides are allowed to react under anhydrous conditions with acid chlorides in the presence of an acid binder, usually a tertiary amine:

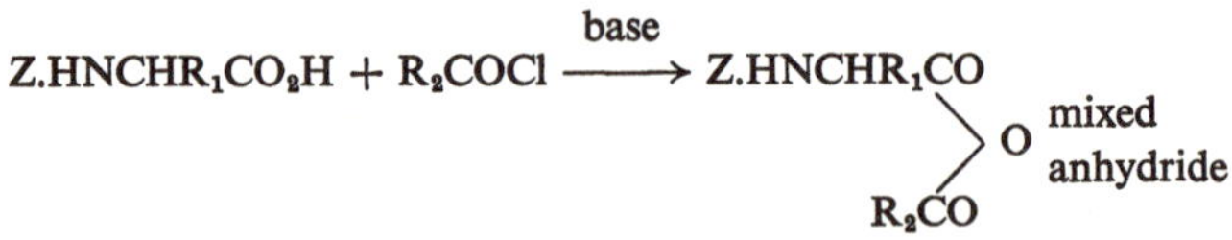

Aminolysis of a mixed anhydride can theoretically proceed in two directions according to the mode of cleavage of the oxygen bridge. This is illustrated by the following scheme:

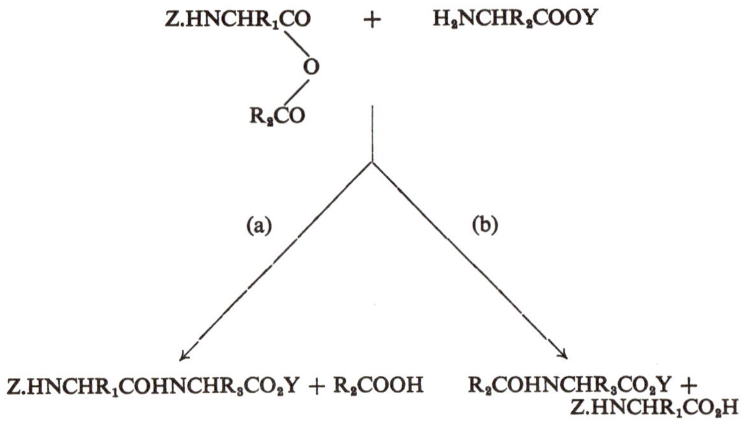

In peptide synthesis the obvious aim is to secure preferential acylation of the amino group by route (a). The course of reaction is influenced by such factors as the nature of R_2 and the solvent employed (discussed by Emery and Gold[99]). Favourable cleavage has been obtained using benzoic[100] and isovaleric anhydrides[101], but mixed carbonic anhydrides exhibit more desirable properties; cleavage occurs almost entirely in the desired direction and the liberated monalkyl carbonate decarboxylates spontaneously:

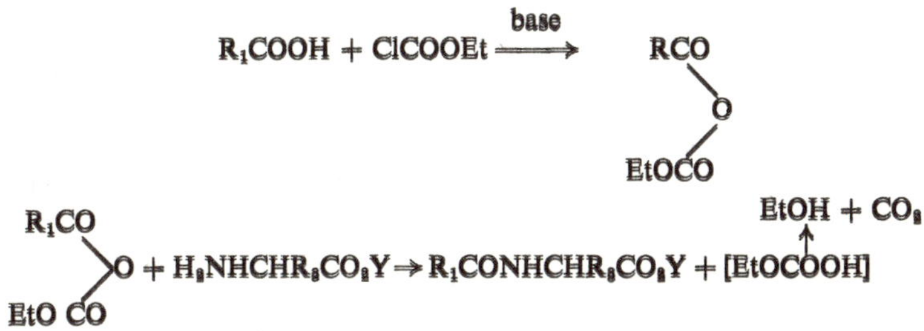

Wieland *et al.*[102] were responsible for the pioneering work with benzoic acid and the carbonic anhydride procedure was developed soon afterwards by Boissonas[103] and other workers[104,105]. The most popular reagents are ethyl, isobutyl and *n*-butyl chloroformates (also termed chlorocarbonates), the last two giving slightly higher yields on coupling. Although coupling can be carried out in aqueous media, organic solvents are preferable; the mixed anhydrides are not isolated as they rapidly disproportionate[103]. Racemization during coupling presents a problem and this has been investigated thoroughly by Smart *et al.*[106], and by Vaughan and Eichler[107]. Correct choice of solvent has an important bearing on retention of optical activity[108], and coupling should be conducted at a low temperature (−15°) with the minimum of delay.

*N-Phthaloyl-glycyl-*DL*-phenylalanyl-glycyl-*L*-leucine*[108]

*N-Phthaloyl-glycyl-*DL*-phenylalanine*

Ethyl chloroformate (0.79 ml) is added to a solution of phthaloylglycine (1.68 g) and tri-*n*-butylamine (1.95 ml) in dioxan (5 ml) at 0°. After 10 min the mixture is added to a solution of DL-phenylalanine (1.36 g) in 4 *N* NaOH (2.1 ml). Carbon dioxide evolution

References p. 389

ceases after about 1 min and the mixture is then rendered acid to Congo Red by the addition of 2.8 ml of 3 *N* HCl. The precipitated oil is extracted with ether, the solution is concentrated and the residue recrystallized from isoamyl alcohol. Yield: 2.06 g (71.5%); m.p. 197–199°.

Coupling with glycyl-L-leucine

A solution of N-phthaloyl-glycyl-DL-phenylalanine (3.52 mg) and tri-*n*-butylamine (0.24 ml) in dioxan (2.5 ml) is treated with ethyl chloroformate (0.10 ml) and coupled with glycyl-L-leucine (188 mg) in 1 ml of 1 *N* NaOH. A further 1 ml of 0.1 *N* NaOH is added, the dioxan extracted with ether, and the solution finally rendered acid to Congo Red. The phthaloyl-tetrapeptide is recrystallized from methanol. Yield: 342 mg (65.5%); m.p. 195–196°.

Activated esters

Certain alkyl and acyl esters are unusually reactive towards amines and in this respect simulate mixed anhydrides. 'Activation' is attributable to the high electronegativity of the substituent in the alcohol moiety. Interest in this type of coupling quickened soon after Wieland *et al.*[109] demonstrated that phenyl thiolesters (Z.HNCHRCOSPh) could be coupled with the sodium salts of amino acids in aqueous solution. Cyanomethyl esters and *p*-nitrophenyl esters were introduced in 1955 by Schwyzer *et al.*[110] and Bodansky[111].

Cyanomethyl esters are prepared in good yield by reaction of a carboxyl component with a slight excess of acetonitrile in the presence of triethylamine[112].

$$\text{RCOOH} + \text{ClCH}_2\text{CN} \xrightarrow{\text{NEt}_3} \text{RCO—OCH}_2\text{CN}$$

Asymmetry is preserved and these reagents have found some application in peptide synthesis.

Outstanding success has been achieved with the *p*-nitrophenyl esters which can be prepared by treatment of N-protected amino acids with *p*-nitrophenyl sulphite or phosphite[113], or with *p*-nitrophenol in the presence of N,N^1-dicyclohexylcarbodi-imide[114–116] (see p. 377). The latter method has been the most rewarding and the products are usually crystalline solids of a high degree of purity. Coupling is a simple operation involving interaction of equimolar quantities of the amino component with the *p*-nitrophenyl ester at room temperature.

$$\text{Z.HNCH(R}_1\text{)COOC}_6\text{H}_4\text{NO}_2 + \text{H}_2\text{NCH(R}_2\text{)CO}_2\text{Et} \rightarrow \text{Z.HNCH(R}_1\text{)CONHCH(R}_2\text{)CO}_2\text{Et} + \text{HOC}_6\text{H}_4\text{NO}_2$$

Results of a comparative study of the reactivity of a variety of phenols undertaken by Pless were recently reported[117]. For practical details of the method see p. 380.

Carbodi-imides

In this method, and in the one that follows, carboxyl activation is effected *in situ*, *i.e.* in the presence of both amino acid components. The third component, a highly reactive compound, is added to the system and coupling ensues. A very popular method of peptide synthesis due to Sheehan and Hess[118] uses dicyclohexylcarbodi-imide as the direct coupling agent; condensation takes place at room temperature in an organic solvent with precipitation of dicyclohexylurea.

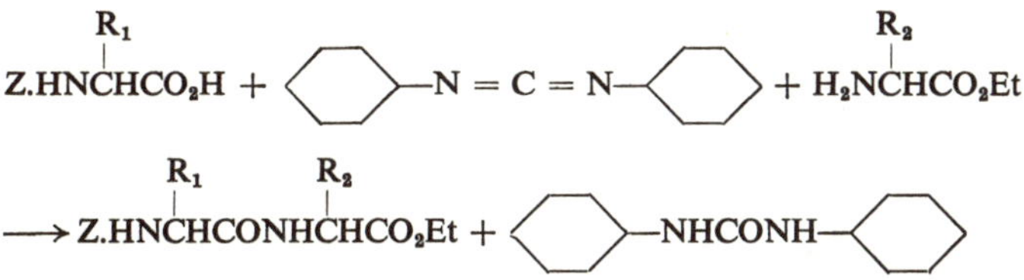

Racemization is slight to negligible[119] and the major drawback is the tendency to side reactions. Depending on the reaction conditions and on the nature of the carboxyl component, a substantial amount of acyl urea may be formed:

$$\text{Z.HNCHR}_1\text{CO}_2\text{H} + \text{RN}=\text{C}=\text{NR} \longrightarrow \text{Z.NHCHR}_1\text{COO{-}C(=NR)NHR}$$

$$\longrightarrow \text{Z.HNCHR}_1\text{CO{-}N(R){-}CONHR}$$

To facilitate removal of the di-substituted urea (the preferred by-product), Sheehan *et al.*[120,121] later introduced reagents which could be extracted by dilute acid or water of the following structures:

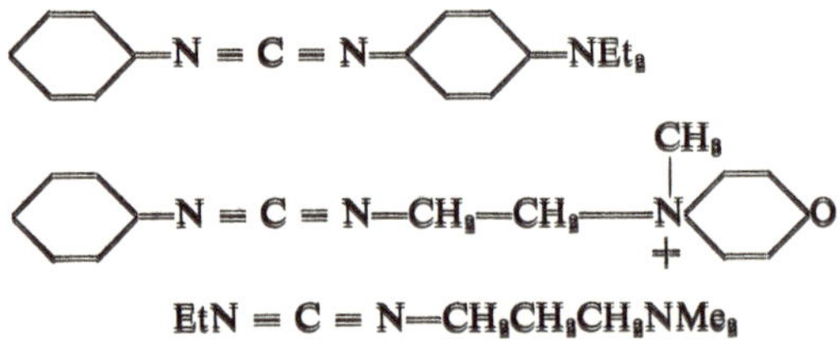

Coupling with 1-ethyl-3-(3-dimethylaminopropyl)carbodi-imide has been highly successful.

The following experimental details taken from the work of Bodansky and Du Vigneaud[114] on oxytocin illustrate the use of both activated esters and carbodi-imides.

S-Benzyl-N-benzyloxycarbonyl-L-cysteine p-nitrophenyl ester

S-Benzyl-N-benzyloxycarbonyl-L-cysteine (7 g) is dissolved in anhydrous tetrahydrofuran (22 ml) and cooled to 0°. *p*-Nitrophenol (4 g) is added and then 4.2 g of dicyclohexylcarbodi-imide; the latter acts as a dehydrating agent and brings about esterification of the cysteine. After a reaction period of 0.5 h at room temperature the precipitated urea derivative is filtered off and washed with solvent. The mother liquor and washings are concentrated and the residue triturated with 50 ml of 0.2 *N* $KHCO_3$. The *p*-nitrophenyl ester is dried *in vacuo* over P_2O_5. Yield: 8.2 g (88%); m.p. 85–88°.

S-Benzyl-N-benzyloxycarbonyl-L-cysteinyl-L-tyrosine methyl ester

(1) The above *p*-nitrophenyl ester (11.7 g) and tyrosine methyl ester (5.5 g) are dissolved in 12.5 ml of dimethylformamide and left for three days. The mixture is diluted with ethyl acetate and the solu tion extracted with 1 *N* HCl, water, 0.25% $KHCO_3$, and water. Concentration of the dried solution, followed by addition of hexane, affords crystals of the dipeptide derivative.

(2) An alternative procedure uses the carbodi-imide method of coupling. S-Benzyl-N-benzyloxycarbonyl-L-cysteine (140 g) and tyrosine methyl ester (80 g) are dissolved in 200 ml of dimethyl-

formamide. Dicyclohexylcarbodi-imide (94 g) is added at 0° in four portions, and after 24 h the reaction mixture is diluted with ethyl acetate (600 ml) and filtered. On working up as above 214 .5g (quantitative yield) of the protected peptide ester m.p. 92–94° are obtained. The ester is saponified by treatment with 1 *N* NaOH in methanol for 1 h at room temperature. The yield of the protected peptide is 78%.

Pyrophosphites

In a method introduced by Anderson *et al.*[9] tetraethyl pyrophosphite is the coupling agent and diethyl phosphorate (diethyl phosphite) acts as the solvent:

$$\text{Z.HN}\overset{\displaystyle R_1}{\underset{|}{\text{C}}}\text{HCO}_2\text{H} + (\text{EtO})_2\text{POP(OEt)}_2 + \text{H}_2\text{N}\overset{\displaystyle R_2}{\text{C}}\text{HCO}_2\text{Et}$$

$$\longrightarrow \text{Z.HN}\overset{\displaystyle R_1}{\text{C}}\text{HCONH}\overset{\displaystyle R_2}{\text{C}}\text{HCO}_2\text{Et} + 2\,(\text{EtO})_2\text{POH}$$

The yields are good but the method has been known to lead to racemization and the best safeguard is to use the benzyloxycarbonyl group or the phthaloyl group for N-protection. Other pyrophosphites have been used as coupling agents, the bis-*o*-phenylene compound being the most stable[122].

Tetraethylpyrophosphite[9]

This reagent is prepared from diethyl phosphite, triethylamine and diethyl chlorophosphite. Equimolar quantities are allowed to react at 0° for 15 min in benzene and the precipitated triethylamine salt removed by filtration under nitrogen. Yield: 41%; b.r. 79–81 (0.15 mm).

Coupling with tetraethylpyrophosphite[9]

In general the N-protected amino acid or peptide (0.010 mole) and the amino acid ester or peptide ester(0.010 mole) are mixed with diethyl phosphite (7 ml; 0.050 mole) and treated with tetraethyl pyrophosphite (0.011 mole) on a steam bath for 30 min. Water is added to precipitate the peptide derivative which is then washed and crystallized from a suitable solvent.

References p. 389

Coupling with bis-O-phenylene pyrophosphite[122]

The reagent is prepared from catechol and phosphorus trichloride in a simple reaction; it distils at 90° (18 mm) and melts at 30°. Equimolar quantities of the carboxyl and amino components are treated with 10% excess reagent in anhydrous pyridine. The mixture is heated on the steam bath for 30 min, cooled, and poured on to ice. The product is washed and crystallized.

N-Carboxyanhydrides

Since the classical studies of Leuchs[123] in 1903 on the N-carboxy-anhydride of glycine, chemists have been concerned mainly with the polymerization potentialities[124] of this series of compounds. A simplified scheme for the polymerization reaction is as follows:

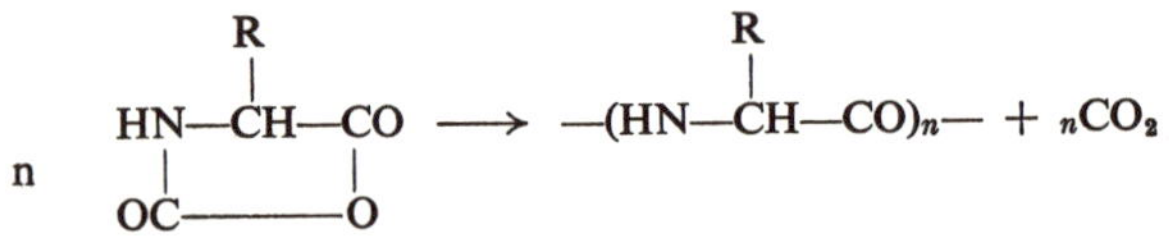

Various attempts[125,126] to use these anhydrides in a stepwise synthesis of peptides have been only partially successful; a high measure of control is possible, however, under the conditions employed by Bailey[127]. In this method one equivalent of an amino acid ester or peptide ester is precooled to—40° and added to a solution of the N-carboxyanhydride and one equivalent of a tertiary base at —40°. Triethylamine or methyldioctylamine are recommended for the reaction which proceeds according to the scheme:

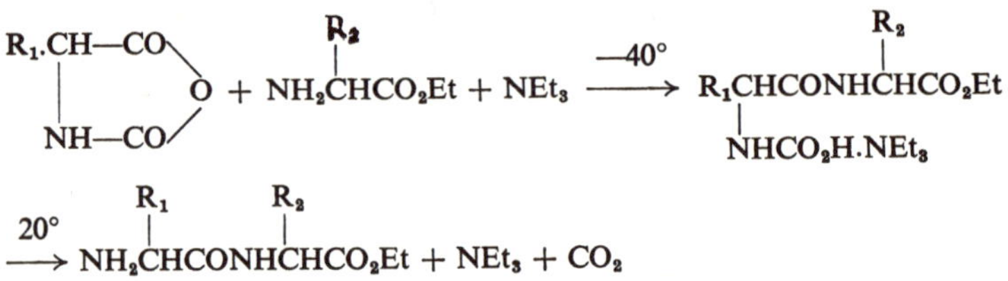

A different course of reaction takes place in the presence of a large excess of amine leading to a ureido acid.

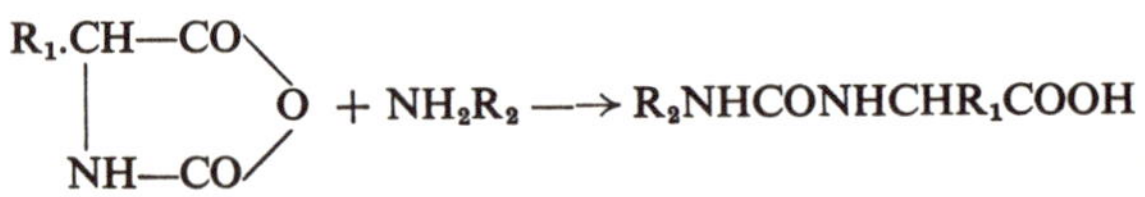

The extent to which this by-product is a contaminant is largely governed by the properties of the amine employed[128].

N-Carboxyanhydrides of amino acids are conveniently prepared by the action of phosgene on the amino acid in dioxan[129], or by cyclization of the acid chloride of the N-benzyloxycarbonyl derivative[130,131]. No racemization has been detected during the coupling procedure; the main disadvantage of the method is the tendency to form polymeric by-products.

L-*Cystyldiglycine*[127]

No tertiary amine is employed at the coupling stage. Di(N-benzyloxycarbonyl)-L-cystine (5 g) in anhydrous dioxan (30 ml) is shaken with phosphorus pentachloride (11 g) with cooling during 15 min. The solution is filtered and warmed to 40° during 1 h when di(anhydro-N-carboxy)-L-cystine separates as fine needles. Yield: 2.8 g (74%).

The anhydride (0.76 g) dissolved in anhydrous tetrahydrofuran (3 ml) is added to glycine ethyl ester (from 1.2 g of hydrochloride) in the same solvent (2 ml) at −10°. After 2 h the mixture is allowed to warm to room temperature and ether (1.5 ml) is added. The precipitate is discarded and L-cystyldiglycine ethyl ester crystallizes from the filtrate on storage at 0°. The free peptide is obtained from the ester by hydrolysis with 0.37 *N* barium hydroxide during 2 min at 10°. The product is recrystallized from aqueous ethanol. Yield: 78%.

L-*Tyrosylglycylglycine*[127]

O-Acetyl-L-tyrosine N-carboxyanhydride (0.25 g) in anhydrous tetrahydrofuran (5 ml) is pre-cooled to −40° and added to glycylglycine ethyl ester (from 0.20 g of hydrochloride) and methyldioctylamine (0.32 ml) in 2 ml of chloroform. Reaction takes place during 6 h at temperatures in the range −40° to −65°. On warming to room temperature the intermediate carbamate decomposes liberating the tripeptide ester and the tertiary base. The solvent is removed *in vacuo*, water is added and the tertiary base extracted into ether. The aqueous layer provides the free peptide after treatment with alkali as above. Yield: 77%.

References p. 389

Other coupling methods

Many more coupling methods fall into the category of carboxyl activation but few have been extensively used and some are still being developed. Brief mention will be made of three of the more ingenious proposals.

Something akin to an activated ester coupling is provided by the method of Woodward *et al.*[182] in which the properties of the costly reagent N-ethyl-5-phenyl-isoxazolium-3′-sulphonate are exploited. Its use in peptide synthesis is illustrated by the scheme:

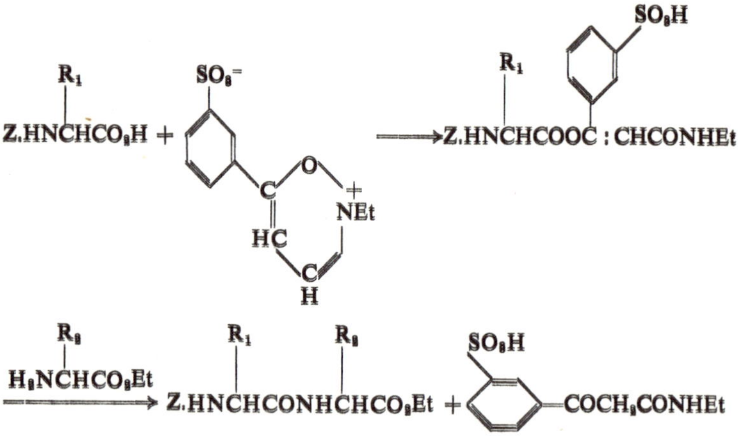

Oxazolidones are produced by the interaction of N-protected amino acids with aldehydes and best yields appear to be obtained using the chloral derivative[183,184]. The oxazolidone condenses with an amino acid ester according to the equation:

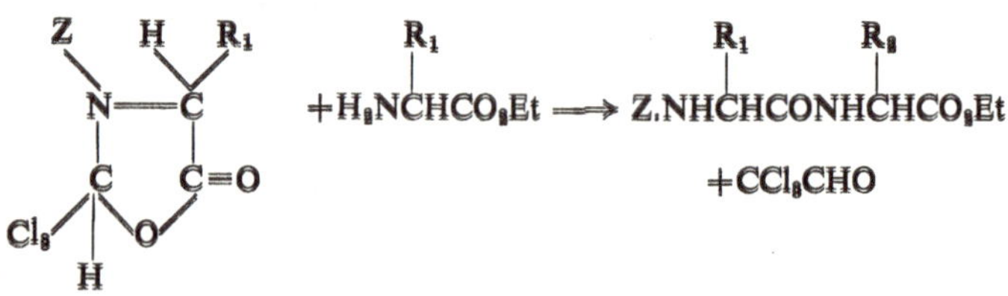

A principle known as 'amino acyl insertion' has been exploited for peptide synthesis[185]. A diacyl hydrazine, prepared from an N-protected amino acid hydrazide and an N-carboxyanhydride, rearranges in the presence of organic acids or bases giving a near quantitative yield of a dipeptide[186]:

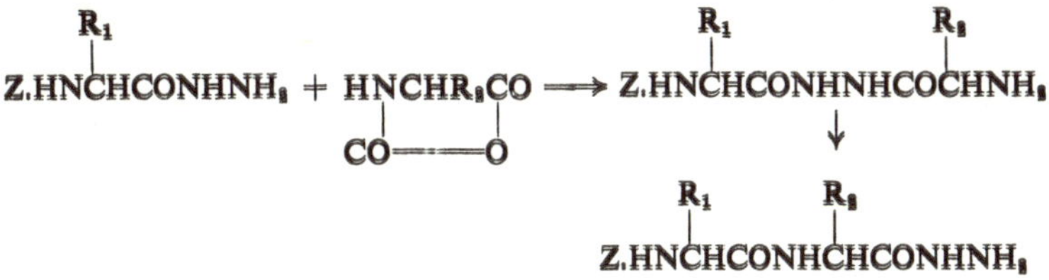

SIDE-CHAIN PROTECTION

The coupling methods which have been discussed will work quite well for syntheses involving simple amino acids. Difficulties arise, however, with amino acids which contain reactive functional groups, and measures must therefore be taken to protect the side chains of aspartic and glutamic acids, lysine and arginine, and cysteine. Occasionally it is necessary to mask the hydroxyl groups of serine, threonine and tyrosine.

Aspartic and glutamic acid residues

The simplest expedient is to work with the benzyl or *t*-butyl esters which may be removed subsequently by hydrogenolysis or acid respectively. β-Benzyl N-benzyloxycarbonylaspartate is prepared by partial saponification of the dibenzyl ester[181,187]; the isomeric α-benzyl ester[68] is available from the anhydride of N-benzyloxycarbonylaspartic acid[6] according to the scheme:

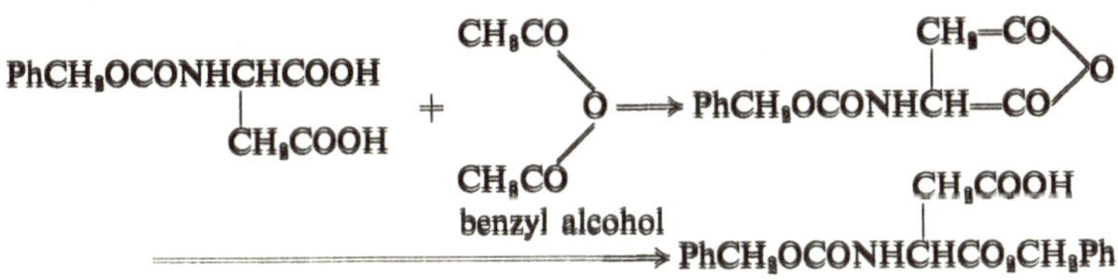

The corresponding derivatives of glutamic acid are not so readily obtainable. The best route to γ-benzyl N-benzyloxycarbonylglutamate is by partial esterification of glutamic acid followed by N-acylation[188]. The N-phthaloyl derivative is available from the anhydride by treatment with benzyl alcohol.

References p. 389

The γ-*t*-butyl ester of N-benzyloxycarbonyl glutamic acid is made from the α-ethyl ester and isobutylene[139] and the corresponding derivative of aspartic acid is prepared in similar fashion[140]. Peptides of glutamine and asparagine can be converted to their glutamic and aspartic analogues by partial acid hydrolysis; for example, warming to 40° in concentrated hydrochloric acid removes the β-amide group of peptides containing asparagine.

Lysine residues

The ε-amino group of lysine can be protected by direct acylation or by selective removal of the α-amino group from a fully substitutes lysine. The copper chelate of lysine (see p. 166) effectively masks the α-amino group and this compound has been used for the preparation of N^{ε}-benzyloxycarbonyl-[141], N^{ε}-tosyl-[90] and N^{ε}-formyl[142] derivatives. The N^{ε}-formyl group is cleaved by very mild acid hydrolysis[50]; $N^{\alpha\varepsilon}$-bistrityl lysine loses the N^{α}-trityl moiety on treatment with methanolic-hydrogen chloride[143]. The merits of different combinations of N-protecting groups for lysine have been discussed fully by Rudinger[144].

Arginine residues

The strongly basic guanidino group of arginine is usually blocked by nitration[145,146] (mixed concentrated nitric and fuming sulphuric acids are used) and the nitro-group removed eventually by catalytic hydrogenolysis. Sometimes it is possible to employ the hydrochloride or hydrobromide of N^{α}-benzyloxycarbonyl arginine; thus, N^{α}-benzyloxycarbonylarginine has been coupled with leucine methyl ester[147] (using the tetraethyl pyrophosphite method) or with tryptophan methyl ester hydrochloride dissolved in dimethylformamide containing diethylphosphite[148–150] (carbodi-imide method).

Cysteine residues

Synthesis of simple peptides containing cystine presents little difficulty and they are easily converted to derivatives of cysteine by reduction of the disulphide bond. Larger peptides are best handled in the reduced form throughout a synthesis but the cysteine side chain must be protected by a benzyl group. The latter may be introduced in two different ways; by reaction of cysteine with benzyl chloride in liquid ammonia in the presence of sodium[151,152], or by treatment of cysteine with benzyl chloride in dilute alcohol in the presence of sodium hydroxide[153]. The thio-ether link is cleaved by sodium in liquid ammonia.

S-Benzyloxycarbonyl and S-trityl are also useful groups for cysteine protection. N-S-bistritylcystine is readily formed by interaction of cysteine and trityl chloride in a water-ether suspension in the presence of diethylamine[154]; the N-trityl moiety is removed with hydrochloric acid in aqueous acetone, whereas the S-trityl linkage undergoes scission in warm aqueous acetic acid[154]. The S-benzylthiomethyl group (—CH_2SCH_2Ph) is particularly stable and requires mercuric chloride for the regeneration of the cysteine thiol group[155].

SOLID-PHASE PEPTIDE SYNTHESIS

Although methods for amino acid protection and coupling have increased in number from year to year the basic art of peptide synthesis has remained substantially unchanged since Bergmann introduced the benzyloxycarbonyl procedure in the early thirties. The discovery of solid-phase peptide synthesis by Merrifield[157] in 1962 therefore constitutes a major event in the development of the subject and should provide the stimulus for further innovations and advances.

Basic to this new process is the concept that it should be possible to anchor an amino acid or peptide chemically to an insoluble supporting particle, elongate the chain at the free end, and eventually release the completed polypeptide from the solid support. This idea was explored experimentally and shown to work initially for a simple tetrapeptide[158] and later for a more complex nonapeptide[159]. Advantages of the process are: higher efficiency at the coupling stages; a reduction in the number of steps and consequent decrease in operation time; ease of removal of by-products leading to high overall yields. The method has been automated and is being used in a study of the bradykinins[160].

An outline of the synthesis of bradykinin is shown below (Fig. 120). Polystyrene cross-linked with 2% of divinyl benzene is chloromethylated and then esterified with the triethylammonium salt of *t*-butyloxycarbonyl-nitro-L-arginine. The protecting group is removed with 1 *N* hydrochloric acid in acetic acid and the amino component released by the addition of excess triethylamine. This is coupled

References p. 389

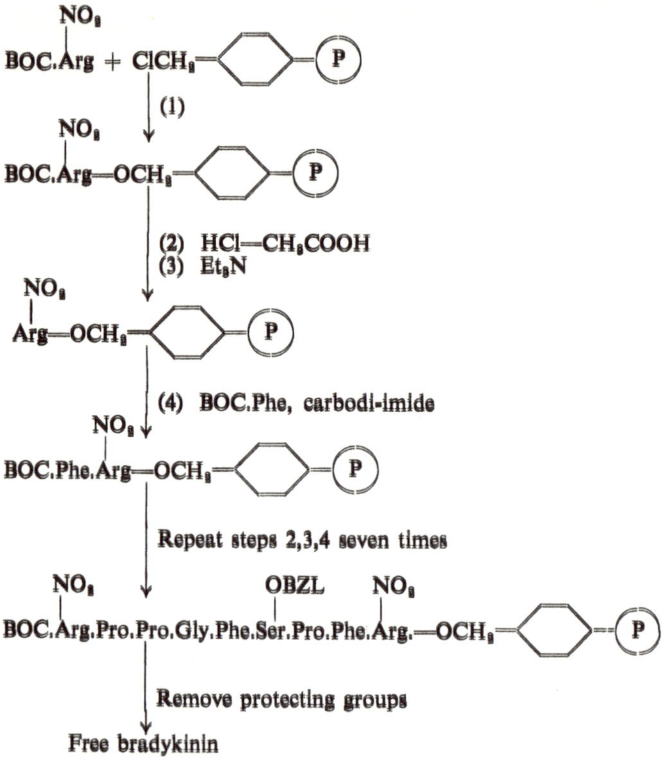

Fig. 120. Solid-phase synthesis of bradykinin.

with the N-butyloxycarbonyl derivative of L-phenylalanine with the aid of dicyclohexylcarbodi-imide. By repeating the cycle seven more times with the appropriate amino acid derivative the nonapeptide is assembled on the polymer in stages. The link with the polymer is finally broken by passing a stream of hydrogen bromide through a suspension of the peptide polymer in trifluoroacetic acid. This step also cleaves the protecting groups with the exception of the nitrogroups of the two arginines which must be removed by hydrogenolysis.

The overall yield of pure bradykinin is 68%. No measurable amount of racemization takes place and nineteen amino acid residies have been introduced into a peptide in this way. The opinion has been expressed that the total synthesis of proteins might be achieved by an automated solid-phase process.

REFERENCES

1 K. Hofmann and P. G. Katsoyannis, in: H. Neurath (Ed.), *The Proteins*, Vol. 1, Chap. 2, Academic Press, New York, 1963.
2 J. S. Fruton, *Advan. Protein Chem.*, 5 (1949) 1.
3 M. Goodman and G. W. Kenner, *Advan. Protein Chem.*, 12 (1957) 466.
4 J. P. Greenstein and M. Winitz, in: *Chemistry of the Amino Acids*, Chap. 10, John Wiley, New York and London, 1961.
5 H. N. Rydon, *Roy. Inst. Chem. (London), Lecture Series*, 1962, No. 5.
6 M. Bergmann and L. Zervas, *Ber.*, 65 (1932) 1192.
7 P. G. Katsoyannis, D. T. Gish, G. P. Hess and V. Du Vigneaud, *J. Am. Chem. Soc.*, 80 (1958) 2558.
8 E. Brand, B. F. Erlanger and H. Sachs, *J. Am. Chem. Soc.*, 74 (1952) 1851.
9 G. W. Anderson, J. Blodinger and A. D. Welcher, *J. Am. Chem. Soc.*, 74 (1952) 5309.
10 D. Ben-Ishui and A. Berger, *J. Org. Chem.*, 17 (1952) 1564.
11 N. F. Albertson and F. C. McKay, *J. Am. Chem. Soc.*, 75 (1953) 5323.
12 A. Katchalsky and M. Paecht, *J. Am. Chem. Soc.*, 76 (1954) 6042.
13 F. H. Carpenter and D. T. Gish, *J. Am. Chem. Soc.*, 74 (1952) 3818; 75 (1953) 950, 5872.
14 G. T. Young, in: Discussion to *Peptides, Proceedings of the Fifth European Symposium, Oxford*, 1962, Pergamon Press, Oxford, 1963, p. 56.
15 J. A. Barltrop and P. Schofield, *Tetrahedron Letters*, (1962) 697.
16 E. Fischer, *Ber.*, 48 (1915) 93.
17 R. Schönheimer, *Z. Physiol. Chem.*, 154 (1926) 203.
18 V. Du Vigneaud and O. K. Behrens, *J. Biol. Chem.*, 117 (1937) 27.
19 K. Hofmann and H. Yajima, *J. Am. Chem. Soc.*, 83 (1961) 2289.
20 R. Schwyzer, W. Rittel, H. Kappeler and B. Iselin, *Angew. Chem.*, 72 (1960) 915.
21 K. Poduska, J. Rudinger and F. Šorm, *Collection Czech. Chem. Commun.*, 20 (1955) 1174.
22 D. I. Weisblat, B. J. Magerlain and D. R. Myers, *J. Am. Chem. Soc.*, 75 (1953) 3630.
23 E. Schnabel and C. H. Li, *J. Am. Chem. Soc.*, 82 (1960) 4576.
24 G. W. Anderson and A. C. McGregor, *J. Am. Chem. Soc.*, 79 (1957) 6180.
25 L. A. Carpino, *J. Am. Chem. Soc.*, 79 (1957) 98.
26 R. Schwyzer, P. Seiber and H. Kappeler, *Helv. Chim. Acta*, 42 (1959) 2622.
27 L. A. Carpino, C. A. Giza and B. A. Carpino, *J. Am. Chem. Soc.*, 81 (1959) 955.
28 L. A. Carpino, *J. Am. Chem. Soc.*, 79 (1957) 4427.
29 D. A. Kidd and F. E. King, *Nature*, 162 (1948) 776.
30 J. C. Sheehan and V. S. Frank, *J. Am. Chem. Soc.*, 71 (1949) 1856.
31 M. Fling, F. N. Minard and S. W. Fox, *J. Am. Chem. Soc.*, 69 (1947) 2466.
32 J. C. Sheehan, D. W. Chapman and R. W. Roth, *J. Am. Chem. Soc.*, 74 (1952) 3822.
33 J. C. Sheehan, M. Goodman and G. P. Hess, *J. Am. Chem. Soc.*, 78 (1956) 1367.

[34] G. H. L. Nefkens, *Nature*, 185 (1960) 309; G. H. L. Nefkens, G. I. Tesser and R. J. F. Nivard, *Rec. Trav. Chim.*, 79 (1960) 688.
[35] F. Weygand and J. Kaelicke, *Chem, Ber.*, 95 (1961) 1031.
[36] W. Grassman and E. Schulte-Uebbing, *Ber.*, 83 (1950) 244.
[37] I. Schumann and R. A. Boissonas, *Helv. Chim. Acta*, 35 (1952) 2235.
[38] R. Schwyzer, A. Costopanagiotis and P. Sieber, *Helv. Chim. Acta*, 46 (1963) 870.
[39] G. W. Anderson and F. M. Callahan, *J. Am. Chem. Soc.*, 82 (1960) 3359.
[40] A. Hillman-Elies, G. H. Hillman and H. Jatzkewitz, *Z. Naturforsch.*, 8b (1953) 445.
[41] G. Amiard, R. Heymes and L. Velluz, *Bull. Soc. Chim. France*, 1955, p. 1.
[42] L. Zervas and D. M. Theodoropoulos, *J. Am. Chem. Soc.*, 78 (1956) 1359.
[43] F. Weygand and E. Csendes, *Angew. Chem.*, 64 (1952) 136.
[44] F. Weygand and E. Leising, *Chem. Ber.*, 87 (1954) 248.
[45] F. Weygand and R. Geiger, *Chem. Ber.*, 89 (1956) 647.
[46] M. Hauptschein, C. S. Stokes and E. A. Nodiff, *J. Am. Chem. Soc.*, 74 (1952) 4005.
[47] E. E. Schallenberg and M. Calvin, *J. Am. Chem. Soc.*, 77 (1955) 2779.
[48] W. S. Fones and M. Lee, *J. Biol. Chem.*, 210 (1954) 227.
[49] F. Weygand and M. Reicher, *Chem. Ber.*, 88 (1955) 26.
[50] J. C. Sheehan and D. D. Yang, *J. Am. Chem. Soc.*, 80 (1958) 1154.
[51] F. C. McIntire, *J. Am. Chem. Soc.*, 69 (1947) 1377.
[52] J. C. Sheehan and V. J. Grenda, *J. Am. Chem. Soc.*, 84 (1962) 2417.
[53] E. Dane, *Angew. Chem.*, 21 (1962) 837.
[54] A. J. Kader and C. J. M. Stirling, *Proc. Chem. Soc.*, 1962, p. 363.
[55] E. Fischer, *Hoppe-Seylers, Z. Physiol. Chem.*, 33 (1901) 151, 412.
[56] M. Brenner and W. Huber, *Helv. Chim. Acta*, 36 (1953) 1109.
[57] R. A. Boissonas, S. Guttman, P.-A. Jaquenoud and J.-P. Waller, *Helv. Chim. Acta*, 39 (1956) 1421.
[58] E. Taschner and C. Wasielewski, *Ann.*, 640 (1961) 136, 142.
[59] A. F. Olechnowitz and G. Zimmerman, *Angew. Chem.*, 67 (1955) 209.
[60] E. Fischer and U. Suzuki, *Ber.*, 38 (1905) 4173.
[61] G. Hillmann, *Z. Naturforsch.*, 1b (1946) 682.
[62] M. Bergmann and J. S. Fruton, *J. Biol. Chem.*, 117 (1937) 189.
[63] M. Bergmann, L. Zervas and L. Salzmann, *Ber.*, 66 (1933) 1288.
[64] R. H. Sifferd and V. Du Vigneaud, *J. Biol. Chem.*, 108 (1935) 753.
[65] B. F. Erlanger and R. M. Hall, *J. Am. Chem. Soc.*, 76 (1954) 5781.
[66] H. K. Miller and H. Waelsch, *J. Am. Chem. Soc.*, 74 (1952) 1092.
[67] J. E. Shields, W. H. McGregor and F. H. Carpenter, *J. Org. Chem.*, 26 (1961) 1491.
[68] J. D. Cipera and R. V. V. Nicholls, *Chem. Ind. (London)*, (1955), 16.
[69] L. Zervas, M. Winitz and J. P. Greenstein, *J. Org. Chem.*, 22 (1957) 1515.
[70] H. Schwarz and K. Arakawa, *J. Am. Chem. Soc.*, 81 (1959) 5691.
[71] R. Schwyzer and P. Sieber, *Helv. Chim. Acta*, 42 (1959) 972.
[72] K. Arakawa and F. M. Bumpus, *J. Am. Chem. Soc.*, 83 (1961) 728.
[73] St. Guttmann, *Helv. Chim. Acta*, 44 (1961) 721.
[74] P. G. Katsoyannis and K. Suzuki, *J. Am. Chem. Soc.*, 83 (1961) 4057.

[75] R. ROESKE, *Chem. Ind. (London)*, (1959) 1121.
[76] E. TASCHNER, A. CHIMIAK, B. BATOR AND T. SOKOTOWSKA, *Ann.*, 646 (1961) 134.
[77] E. TASCHNER, C. WASIELEWSKI AND J. F. BIERNAT, *Ann.*, 646 (1961) 119.
[78] J. A. MACLAREN, W. E. SAVIGE AND J. M. SWAN, *Australian J. Chem.*, 11 (1958) 345.
[79] K. RUHLMANN, *Chem. Ber.*, 94 (1961) 1876.
[80] L. BIRHOFER AND A. RITTER, *Chem. Ber.*, 93 (1960) 424; L. BIRKOFER, W. KONKEL AND A. RITTER, *Chem. Ber.*, 94 (1961) 1263.
[81] P. MAMALIS AND H. N. RYDON, *J. Chem. Soc.*, 1955, 1049.
[82] H. N. RYDON AND J. E. WILLETT, in: G. T. YOUNG (Ed.), *Peptides, Proceedings of the Fifth European Symposium, Oxford*, 1962, Pergamon Press, Oxford, 1963, p. 23.
[83] C. W. CRANE AND H. N. RYDON, *J. Chem. Soc.*, 1947, 766.
[84] T. CURTIUS, *Ber.*, 35 (1902) 3226.
[85] J. HONZL AND J. RUDINGER, *Collection Czech. Chem. Commun.*, 26 (1961) 2333.
[86] J. I. HARRIS AND J. S. FRUTON, *J. Biol. Chem.*, 191 (1951) 143.
[87] V. PRELOG AND P. WIELAND, *Helv. Chim. Acta*, 29 (1946) 1128.
[88] C. W. ROBERTS, *J. Am. Chem. Soc.*, 76 (1954) 6203.
[89] B. HEGEDÜS, *Helv. Chim. Acta*, 31 (1948) 737.
[90] R. ROESKE, F. H. C. STEWART, R. J. STEDMAN AND V. DU VIGNEAUD, *J. Am. Chem. Soc.*, 78 (1956) 5883.
[91] G. F. HOLLAND AND L. A. COHEN, *J. Am. Chem. Soc.*, 80 (1958) 3765.
[92] F. WEYGAND AND W. STEGLICH, *Ber.*, 92 (1959) 313.
[93] K. HOFMANN, M. Z. MAGEE AND A. LINDENMANN, *J. Am. Chem. Soc.*, 72 (1950) 2814; K. HOFMANN, A. LINDENMANN, M. Z. MAGEE AND N. H. KHAN, *J. Am. Chem. Soc.*, 74 (1952) 470.
[94] R. SCHWYZER, E. SURBECK–WEGMANN AND H. DIETRICH, *Chimia (Aarau)*, 14 (1960) 360.
[95] R. A. BOISSONAS, ST. GUTTMANN AND P.-A. JAQUENOUD, *Helv. Chim. Acta*, 43 (1960) 1349.
[96] W. KLEE AND M. BRENNER, *Helv. Chim. Acta*, 44 (1961) 2151.
[97] Y. WOLMAN, P. M. GALLOP, A. PATCHORNIK AND A. BERGER, *J. Am. Chem. Soc.*, 84 (1962) 1889.
[98] J. S. FRUTON, *J. Biol. Chem.*, 146 (1942) 465.
[99] A. R. EMERY AND V. GOLD, *J. Chem. Soc.*, (1950) 1443.
[100] K. C. HOOPER, H. N. RYDON, J. A. SCHOLFIELD AND G. S. HEATON, *J. Chem. Soc.*, (1956) 3148.
[101] C. RESSLER AND V. DU VIGNEAUD, *J. Am. Chem. Soc.*, 76 (1954) 3107.
[102] T. WIELAND AND R. SEHRING, *Ann.*, 569 (1950) 122; T. WIELAND, W. KERN AND R. SEHRING, *Ann.*, 569 (1950) 117.
[103] R. A. BOISSONAS, *Helv. Chim. Acta*, 34 (1951) 874.
[104] J. R. VAUGHAN, JR., *J. Am. Chem. Soc.*, 73 (1951) 9547.
[105] T. WIELAND AND H. BERNHARD, *Ann.*, 572 (1951) 190.
[106] N. A. SMART, G. T. YOUNG AND M. W. WILLIAMS, *J. Chem. Soc.*, (1960) 3902.
[107] J. R. VAUGHAN AND J. A. EICHLER, *J. Am. Chem. Soc.*, 76 (1954) 2474.

108 J. R. VAUGHAN, *J. Am. Chem. Soc.*, 74 (1952) 6137.
109 T. WIELAND, W. SCHÄFER AND E. BOKELMANN, *Ann.*, 573 (1951) 99.
110 R. SCHWYZER, B. ISELIN AND M. FEURER, *Helv. Chim. Acta*, 38 (1955) 69.
111 M. BODANSKY, *Nature*, 175 (1955) 685.
112 R. SCHWYZER, M. FEURER, B. ISELIN AND H. KÄGI, *Helv. Chim. Acta*, 38 (1955) 80.
113 B. ISELIN, W. RITTEL, P. SIEBER AND R. SCHWYZER, *Helv. Chim. Acta*, 40 (1957) 373.
114 M. BODANSKY AND V. DU VIGNEAUD, *J. Am. Chem. Soc.*, 81 (1959) 2504.
115 D. F. ELLIOTT AND D. W. RUSSELL, *Biochem. J.*, 66 (1957) 49P.
116 M. ROTHE AND F. M. KUNITZ, *Ann.*, 609 (1957) 88.
117 J. PLESS, in: G. T. YOUNG (Ed.), *Peptides, Proceedings of Fifth European Symposium, Oxford*, 1962, Pergamon Press, Oxford, 1963, p. 69.
118 J. C. SHEEHAN AND G. P. HESS, *J. Am. Soc.*, 77 (1955) 1067; see also H. G. KHARANA, *Chem. Ind. (London)*, (1955) 1087.
119 G. W. ANDERSON AND F. M. CALLAHAN, *J. Am. Chem. Soc.*, 80 (1958) 2902.
120 J. C. SHEEHAN AND J. J. HVLAVKA, *J. Org. Chem.*, 21 (1956) 439.
121 J. C. SHEEHAN, P. A. CRUICKSHANK AND G. L. BOSHART, *J. Org. Chem.*, 26 (1961) 2525.
122 P. C. CROFTS, J. H. H. MARKES AND H. N. RYDON, *J. Chem. Soc.*, (1958) 4250; (1959) 3610.
123 H. LEUCHS, *Ber.*, 36 (1903) 857.
124 For reviews see: C. H. BAMFORD, A. ELLIOTT AND W. E. HANBY, *Synthetic Polypeptides*, Academic Press, New York, 1956; E. KATCHALSKI AND M. SELA, *Advan. Protein Chem.*, 13 (1958) 243; M. A. STAHMANN (Ed.), *Polyaminoacids, Polypeptides and Proteins*, Univ. Wisconsin Press, Madison, 1962.
125 F. WESSELY AND F. SIGMUND, *Z. Physiol. Chem.*, 157 (1926) 91.
126 M. HUNTER AND V. DU VIGNEAUD, *J. Biol. Chem.*, 124 (1938) 699.
127 J. L. BAILEY, *J. Chem. Soc.*, (1950) 3461.
128 K. D. KOPPLE, *J. Am. Chem. Soc.*, 79 (1957) 662.
129 A. C. FARTHING, *J. Chem. Soc.*, (1950) 3213.
130 M. BERGMANN, L. ZERVAS AND W. F. ROSS, *J. Biol. Chem.*, 111 (1935) 245.
131 A. BERGER AND E. KATCHALSKI, *J. Am. Chem. Soc.*, 73 (1951) 4084.
132 R. B. WOODWARD AND R. A. OLOFSON, *J. Am. Chem. Soc.*, 83 (1961) 1007; R. B. WOODWARD, R. A. OLOFSON AND H. MAYER, *J. Am. Chem. Soc.*, 83 (1961) 1010.
133 F. MICHEEL AND W. MECKSTROTH, *Chem. Ber.*, 92 (1959) 1075.
134 F. MICHEEL AND H. HANEKE, *Chem. Ber.*, 92 (1959) 309; 95 (1962) 1009.
135 M. BRENNER AND SIX COLLABORATORS, *Helv. Chim. Acta*, 40 (1957) 1497.
136 M. BRENNER AND W. HOFER, *Helv. Chim. Acta*, 44 (1961) 1794, 1798; R. WEBER, W. HOFER, W. HEER AND M. BRENNER, *Helv. Chim. Acta*, 44 (1961) 2154.
137 M. FRANKEL AND A. BERGER, *J. Org. Chem.*, 10 (1951) 1513.
138 W. E. HANBY, S. G. WALEY AND J. WATSON, *J. Chem. Soc.*, 1950, p. 3239.
139 R. SCHWYZER AND H. KAPPELER, *Helv. Chim. Acta*, 44 (1961) 1991.
140 R. SCHWYZER AND A. DIETRICH, *Helv. Chim. Acta*, 44 (1961) 2003.

[141] A. Neuberger and F. Sanger, *Biochem. J.*, 37 (1943) 515.
[142] K. Hofmann, E. Stutz, G. Spühler, H. Yajima and E. T. Schwartz, *J. Am. Chem. Soc.*, 82 (1960) 3727.
[143] G. Amiard and B. Goffinet, *Bull. Soc. Chim. France*, 1957, 1133; B. Bezas and L. Zervas, *J. Am. Chem. Soc.*, 33 (1961) 719.
[144] J. Rudinger, *Collection Czech. Chem. Commun.*, 24 (1959), Special Issue, p. 95.
[145] M. Bergmann, L. Zervas and H. Rinke, *Z. Physiol. Chem.*, 224 (1934) 40.
[146] K. Hofmann, A. Rheiner and W. D. Peckham, *J. Am. Chem. Soc.*, 75 (1953) 6083.
[147] G. W. Anderson, *J. Am. Chem. Soc.*, 75 (1953) 6081.
[148] R. A. Boissonas, St. Guttmann, R. L. Huguenin, P.-A. Jaquenoud and Ed. Sandrin, *Helv. Chim. Acta*, 41 (1958) 1867.
[149] R. A. Boissonas, St. Guttmann, P.-A. Jaquenoud, Ed. Sandrin and J.-P. Waller, *Helv. Chim. Acta*, 44 (1961) 123.
[150] St. Guttmann and R. A. Boissonas, *Helv. Chim. Acta*, 42 (1959) 1257.
[151] V. Du Vigneaud, L. F. Audrieth and H. S. Loring, *J. Am. Chem. Soc.*, 52 (1930) 4500.
[152] J. L. Wood and V. Du Vigneaud, *J. Biol. Chem.*, 130 (1939) 109.
[153] M. Frankel, D. Gertner, H. Jacobson and A. Zilkha, *J. Chem. Soc.*, (1960) 1390.
[154] G. Amiard, R. Heymes and L. Velluz, *Bull. Soc. Chim. France*, (1956) 698.
[155] P. J. E. Pimlott and G. T. Young, *Proc. Chem. Soc.*, (1958) 257.
[156] H. Kappeler and R. Schwyzer, *Helv. Chim. Acta*, 41 (1961) 491.
[157] R. B. Merrifield, *Federation Proc.*, 21 (1962) 412.
[158] R. B. Merrifield, *J. Am. Chem. Soc.*, 85 (1963) 2149.
[159] R. B. Merrifield, *J. Am. Chem. Soc.*, 86 (1964) 304; *Biochemistry*, 3 (1964) 1385.
[160] R. B. Merrifield, *J. Org. Chem.*, 29 (1964) 3100.
[161] C. Berse, R. Boucher and L. Piche, *J. Org. Chem.*, 22 (1957) 805.
[162] D. T. Gish and V. Du Vigneaud, *J. Am. Chem. Soc.*, 79 (1957) 3579.
[163] J. E. Shields and F. H. Carpenter, *J. Am. Chem. Soc.*, 83 (1961) 3066.
[164] E. Gazis, B. Bezas, G. C. Stelakatos and L. Zervas, in: G. T. Young (Ed.), *Peptides, Proceedings of Fifth European Symposium, Oxford*, 1962, Pergamon Press, Oxford, 1963, p. 17.
[165] H. Werbin and A. D. McLaren, *J. Am. Chem. Soc.*, 73 (1951) 501.
[166] C. Berse and L. Piche, *J. Org. Chem.*, 21 (1956) 808.
[167] K. Hofmann, W. D. Peckam and A. Rheiner, *J. Am. Chem. Soc.*, 78, (1956) 238.
[168] H. O. Van Orden and E. L. Smith, *J. Biol. Chem.*, 208 (1954) 751.
[169] W. J. LeQuesne and G. T. Young, *J. Chem. Soc.*, 1952, p. 24.
[170] E. Sandheimer and R. W. Holley, *J. Am. Chem. Soc.*, 76 (1954) 2467.
[171] R. A. Boissonas, S. Guttmann, P.-A. Jaquenoud and J. P. Waller, *Helv. Chim. Acta*, 38 (1955) 1491.
[172] V. Du Vigneaud and G. L. Miller, *J. Biol. Chem.*, 116 (1952) 649.
[173] K. Schlogl, J. Derkosch and E. Wawersich, *Monatsh. Chem.*, 85 (1954) 607.

[174] T. WIELAND AND H. WEIDENMULLER, *Ann.*, 597 (1955) 111.
[175] Y. SAKATO, T. HASHIZUME AND Y. KISHIMOTO, *J. Agr. Chem. Soc. Japan*, 18 (1942) 524.
[176] T. MATSUKAWA AND K. MASUDA, *J. Pharm. Soc. Japan*, 73 (1954) 397; *Chem. Abst.* 48, 2597.
[177] D. T. GISH AND F. H. CARPENTER, *J. Am. Chem. Soc.*, 75 (1953) 950.
[178] D. W. CLAYTON, J. A. FARRINGTON, G. W. KENNER AND J. M. TURNER, *J. Chem. Soc.*, (1957) 1398.
[179] M. ZAORAL, *Chem. Listy*, 48 (1954) 1583.
[180] A. BERGER, J. KURTZ AND E. KATCHALSKI, *J. Am. Chem. Soc.*, 76 (1954) 5552.
[181] E. BAER AND J. MAURUKAS, *J. Biol. Chem.*, 212 (1955) 25.
[182] C. A. DEKKER, S. P. TAYLOR AND J. S. FRUTON, *J. Biol. Chem.*, 180 (1949) 155.
[183] E. L. SMITH, *J. Biol. Chem.*, 175 (1948) 39.
[184] E. KATCHALSKI AND M. SELA, *J. Am. Chem. Soc.*, 75 (1953) 5284.
[185] J. R. VAUGHAN, JR., AND J. A. EICHLER, *J. Am. Chem. Soc.*, 75 (1953) 5556.

SUBJECT INDEX